PLATE I

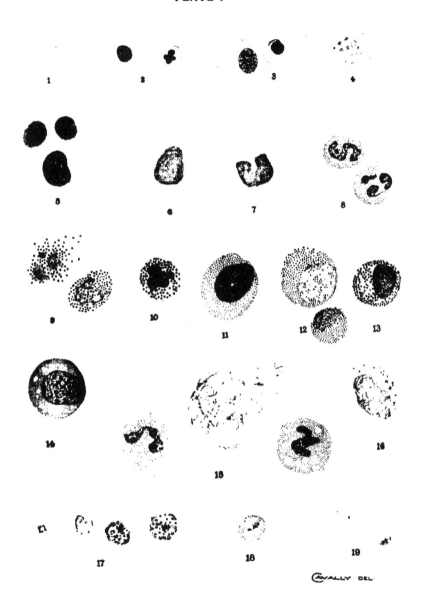

CAVALLY DEL

EXPLANATION OF PLATE I

Stained with Wright's stain. All drawn to same scale.

1, Normal red corpuscle for comparison; 2, normoblasts, one with lobulated nucleus; 3, megaloblast and microblast. The megaloblast shows a considerable degree of polychromatophilia; 4, blood-plaques, one lying upon a red corpuscle; 5, lymphocytes, large and small; 6, large mononuclear leukocyte; 7, transitional leukocyte; 8, polymorphonuclear neutrophilic leukocytes; 9, eosinophilic leukocytes, one ruptured; 10, basophilic leukocyte; 11, neutrophilic myelocyte. The granules are sometimes less numerous and less distinct than here shown; 12, eosinophilic myelocytes; 13, basophilic myelocyte; 14, "irritation" or "stimulation" form, with small vacuoles; 15, degenerated leukocytes: two polymorphonuclear neutrophiles, one ruptured, one swollen and vacuolated; and a "basket cell" composed of an irregular meshwork of nuclear material; 16, large mononuclear leukocyte containing pigment-granules; from a case of tertian malaria; 17, four stages in the asexual cycle of the tertian malarial parasite: the second and fourth were drawn from the same slide taken from a case of double tertian; 18, red corpuscle containing tertian parasite and showing malarial stippling; 19, estivo-autumnal malarial parasites: two small ring forms within the same red cell, and a crescent with remains of the red corpuscle in its concavity.

Clinical Diagnosis

A MANUAL OF LABORATORY METHODS

BY

JAMES CAMPBELL TODD, Ph. B., M. D.

PROFESSOR OF CLINICAL PATHOLOGY, UNIVERSITY OF COLORADO

Illustrated

Fourth Edition, Revised and Reset

PHILADELPHIA AND LONDON

W. B. SAUNDERS COMPANY

1918

PRINTED IN AMERICA

PRESS OF
W. B. SAUNDERS COMPANY
PHILADELPHIA

TO

MY FATHER

Joe H. Todd, M. D.

THESE PAGES ARE

AFFECTIONATELY DEDICATED

PREFACE TO THE FOURTH EDITION

In the present edition, as in the preceding one, the scope of this book has been somewhat extended and its size increased. It is hoped that its value has thereby been enhanced without sacrifice of the simplicity and conciseness which were its original aim. As before, chief emphasis has been laid upon methods and microscopic morphology.

Much of the new material is the outgrowth of questions which have arisen in class-room and laboratory. To one who sees a great deal of the work of students in the clinical laboratory, it soon becomes evident that errors in microscopic diagnosis spring much less frequently from ignorance of the typical appearance of microscopic structures than from imperfect preparation of the material, faulty manipulation of the microscope, or failure to recognize extraneous structures, artifacts, and other misleading appearances. Such sources of error have been given especial attention.

In order to keep the size of the volume within bounds, room has been made for the new matter by omissions and condensations so far as these have seemed wise, but great care has been exercised to avoid omission of essential details. It has, in fact, been found necessary to elaborate many subjects which seemed to have been too briefly stated in past editions for clear understanding. Very brief descriptions of methods and microscopic structures are attractive, but only too often they are worse than useless. They necessarily omit details which seem unimportant in themselves, but which in reality are essential to guard the reader against errors. They give him an unfounded confidence which is doomed to disillusionment when he actually attempts the work in the laboratory.

The changes and additions are widely scattered throughout the book, hence most of them evade special mention. The use of colorimeters and of the pocket spectroscope and methods of matching blood for transfusion have been given at some length. There have also been included sections dealing with the new Bass and Johns concentration method for malarial parasites; the fractional method of gastric analysis; vital staining of blood-corpuscles; resistance of red corpuscles; the mastic reaction in the spinal fluid; the Wilber and Addis. method for urobilin as an aid in diagnosis of pernicious anemia; and estimation of amylase in urine and feces in diagnosis of pancreatic disease. The chapter upon sero-diagnostic methods has been revised by Professor Whitman from whose pen it originated.

In a book which deals largely with clinical microscopy, accurate pictures of microscopic structures should play a large, if not predominant, part. They give information which cannot be conveyed in any other way. For this reason the illustrations have been carefully revised. The poorer illustrations of previous editions have been omitted and 90 new black and white pictures have been added, making an increase of 57 in the total. The majority of the new pictures are photomicrographs by the author. Inadequate as is the photomicrograph in some fields, its superiority in clinical microscopy cannot be questioned. Of the colored plates, four are new in this edition. Figure 1 of Plate II was made under the direction of Dr. Stella M. Gardner of Chicago. The remainder of the new colored pictures were painted under the author's supervision. All were drawn with painstaking accuracy from actual specimens, in most cases with the aid of photomicrographs.

To all those of his present and former pupils whose suggestive questions have been an influence in shaping the book, the author is duly grateful; and in particular he takes pleasure in acknowledging his indebtedness to Robert C. Lewis, Professor of Physiology and Biochemistry in the University of Colorado, for helpful suggestions concerning the chemical examination of the urine and of the gastric contents.

J. C. T.

HENRY S. DENISON RESEARCH LABORA-
TORIES, UNIVERSITY OF COLORADO.
BOULDER, COLORADO,
June, 1918.

PREFACE

This book aims to present a clear and concise statement of the more important laboratory methods which have clinical value, and a brief guide to interpretation of results. It is designed for the student and practitioner, not for the trained laboratory worker. It had its origin some years ago in a short set of notes which the author dictated to his classes, and has gradually grown by the addition each year of such matter as the year's teaching suggested. The eagerness and care with which the students and some practitioners took these notes and used them convinced the writer of the need of a volume of this scope.

The methods offered are practical; and as far as possible are those which require the least complicated apparatus and the least expenditure of time. Simplicity has been considered to be more essential than absolute accuracy. Although in many places the reader is given the choice of several methods to the same end, the author believes it better to learn one method well than to learn several only partially.

More can be learned from a good picture than from any description, hence especial attention has been given to the illustrations, and it is hoped that they will serve truly to *illustrate*. Practically all the microscopic structures mentioned, all apparatus not in general use, and many of the color reactions are shown in the pictures.

Although no credit is given in the text, the recent medical periodicals and the various standard works have been freely consulted. Among authors whose writings have been especially helpful may be mentioned v. Jaksch, Boston, Simon, Wood, Emerson, Purdy, Ogden, Ewald, Ehrlich and Lazarus, Da Costa, Cabot, Osler, Stengel, and McFarland.

The author wishes hereby to express his indebtedness to Dr. J. A. Wilder, Professor of Pathology in the Denver and Gross College of Medicine, for aid in the final revision of the manuscript; and to W. D. Engel, Ph.D., Professor of Chemistry, for suggestions in regard to detection of drugs in the urine. He desires to acknowledge the care with which Mr. Ira D. Cassidy has made the original drawings, and also the uniform courtesy of W. B. Saunders Company during the preparation of the book.

J. C. T.

Denver, Colorado.

CONTENTS

INTRODUCTION

CHAPTER I

CHAPTER II

CHAPTER III

CHAPTER IV

CHAPTER V

CHAPTER IX

CHAPTER X

CLINICAL DIAGNOSIS

INTRODUCTION

USE OF THE MICROSCOPE

THERE is probably no laboratory instrument whose usefulness depends so much upon proper manipulation as the microscope, and none is so frequently misused by beginners. Some suggestions as to its proper use are, therefore, given at this place. It is presumed that the reader is already familiar with its general construction (Fig. 1).

For those who wish to understand the principles of the microscope and its manipulation—and best results are impossible without such an understanding—a careful study of some standard work upon microscopy, such as those of

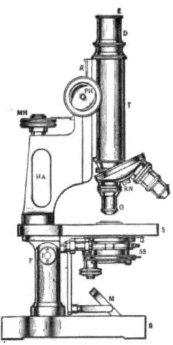

FIG. 1.—Handle-arm microscope: E, Eye-piece; D, draw-tube; T, body-tube; RN, revolving nose-piece; O, objective; PH, pinion head for coarse focusing; MH, micrometer head for fine focusing; HA, handle-arm; SS, substage; S, stage; M, mirror; B, base; R, rack; P, pillar; I, inclination joint.

2

Carpenter, Spitta, and Sir A. E. Wright, is earnestly recommended. It is also recommended that the beginner provide himself with some slides of diatoms, for example, *Pleurosigma angulatum, Surirella gemma*, and *Amphipleura pellucida*, costing fifty cents each, and with some good preparations of stained and unstained blood. The blood slides can easily be made from one's own blood, as described in Chapter III. Faithful practice upon such test-objects, in the light of the principles of microscopy, will enable the student to reach, intelligently, an accuracy in manipulation to which the ordinary laboratory worker attains only slowly and by rule of thumb. He will soon find that the bringing of an object into accurate focus is ·by no means all of microscopy.

Source of Light.—Good work cannot be done without proper illumination and this is therefore the first and most important consideration for one who wishes to use the microscope effectively.

The light which is generally recommended as best is that from a white cloud, the microscope being placed by preference at a north window, to avoid direct sunlight. At any other window a white window-shade is desirable. Such light is satisfactory for all ordinary work. Artificial light is, however, imperative for those who must work at night, and is a great convenience at all times. Properly regulated artificial light, moreover, offers decided advantages over daylight for critical work. Almost any strong light which is diffused through a frosted globe will give fair results. The inverted Welsbach light with such a globe is excellent as is also the Mazda incandescent lamp with frosted bulb. Such a bulb may conveniently be inclosed within a tin or paste-board box with small openings in the back for ventilation and a circular window in the front to transmit the light. At the University of Colorado, where the students do most of their microscopic work by artificial light, the lamp shown in Fig. 2 is very popular. It has the advantage that the

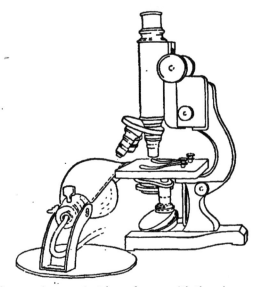

Fig. 2.—A convenient lamp for use with the microscope.

eyes are shaded from the glare, while at the same time there is abundant light for drawing or writing upon the table beside the microscope. Its cost, with Mazda bulb, is $1.25. All such lights have a yellow tinge, to counteract which a blue glass disk, usually supplied with the microscope, is placed in a supporting ring beneath the condenser. Recently a blue glass "daylight" bulb has been put upon the market. The following plan is much used abroad, and gives results equal to the best daylight: A Welsbach lamp or strong electric light is used, and a spheric glass globe—a 6-inch round-bottom flask answers admirably—is placed between it and the microscope, to act as a condenser (Fig. 3). The flask should be at a distance equal to its diameter from both the light and the mirror of the microscope. In order

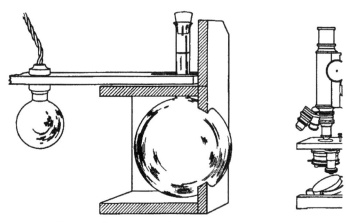

Fig. 3.—Illumination with water-bottle condenser.

to filter out the yellow rays the flask is filled with water to which have been added a few crystals of copper sulphate and a little ammonia.

Within the past few years manufacturers have paid more attention than formerly to means of artificial illumination and most of them now offer several types of lamp. Two good types are shown in Figs. 4 and 5. Both can be fitted with light filters made of the newly-invented "daylight glass," which, when used with the nitrogen-filled tungsten lamp, transmits a light practically indistinguishable from daylight either visually or spectrophotometrically.

The microscope lamp should not stand at so great

FIG. 4.—Small microscope lamp with daylight-glass filter.

FIG. 5.—An excellent type of microscope lamp suitable both for ordinary work and for dark-ground illumination.

a distance from the microscope that its image fails to fill the aperture of the condenser—a condition which one can readily detect by removing the ocular and looking down the tube.

Forms of Illumination.—After one has arranged the microscope in proper relation to the source of light, whether this be daylight or any of the artificial sources mentioned above, the next problem is to secure an evenly illuminated field of view without mottling or any trace of shadows. This is accomplished by manipulating the mirror and the condenser. Following this the direction and the amount of light must be considered in relation to the character of the object under examination.

Illumination may be either *central* or *oblique*, depending upon the direction in which the light enters the microscope. To obtain **central illumination,** the mirror should be so adjusted that the light from the source selected is reflected directly up the tube of the microscope. This is easily done by removing the eye-piece and looking down the tube while adjusting the mirror. The eye-piece is then replaced, and the light reduced as much as desired by means of the diaphragm.

Oblique illumination is obtained in the more simple instruments by swinging the mirror to one side, so that the light enters the microscope obliquely. The more complicated instruments obtain it by means of a rack and pinion, which moves the diaphragm laterally. Beginners frequently use oblique illumination without recognizing it, and are thereby much confused. If the light be oblique, an object in the center of the field will appear to sway from side to side when the fine adjustment is turned back and forth.

The **amount of light** admitted is also important. It is regulated by the diaphragm.

The bulk of routine work is done with central illumi-

nation, and, therefore, every examination should begin
with it. Each of the forms of illumination, however—
central and oblique, subdued and strong—has its special
uses and demands some consideration here. The well-
known rule, "Use the least light which will show the
object well," is good, but it does not go far enough.

In studying any microscopic structure one considers:
(1) its color, (2) its outline, and (3) its surface contour.
No one form of illumination shows all of these to the

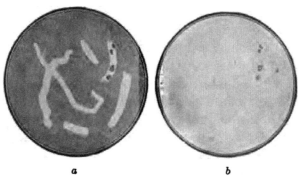

a   b

FIG. 6.—a, Hyaline casts, one containing renal cells; properly subdued illumination; b, same as a; strong illumination. The casts are
lost in the glare, and only the renal cells are seen. (From Greene's
"Medical Diagnosis.")

best advantage. It may, therefore, be necessary to
change the illumination many times during a microscopic examination. *To see color best, use central illumination with strong light.* The principle is that by
which a stained glass window shows the purest color
when the light is streaming through it. Strong central
light is, therefore, to be used for structures such as
stained bacteria, whose recognition depends chiefly upon
their color, and, alternating with other forms, for stained

structures in general. *To study the outline of an object use very subdued central illumination.* The diaphragm is closed to the point which trial shows to be best in each case. This illumination is required by delicate colorless objects, such as hyaline tube-casts and cholesterin crystals, which are recognized chiefly by their outline. The usual mistake of beginners is to work with the diaphragm too wide open. Strong light will often render semitransparent structures entirely invisible (Fig. 6). *To study surface contour use oblique light of a strength suited to the color or opacity of the object.* In routine work oblique illumination is resorted to only to study more fully some object which has been found with central illumination, as, for instance, to demonstrate the cylindric shape of a hyaline tube-cast.

Dark-ground illumination consists in blocking out the central rays of light and directing the peripheral rays against the microscopic object from the side. Only those rays which strike the object and are reflected upward pass into the objective. The object thus appears bright upon a black background. By means of this form of illumination very minute structures can be seen, just as particles of dust in the atmosphere become visible when a ray of sunlight enters a darkened room.

Dark-ground illumination for low-power work can be obtained by means of the ring stops with central disks which accompany most microscopes when purchased. The stop is placed in a special ring beneath the condenser. When the regular stop is not at hand, one can use the glass disk which is generally supplied with the microscope or an extra-large round cover-glass, in the center of which is pasted a circular disk of black paper.

The size of the black disk depends upon the aperture of the objective with which it is to be used, and can be ascertained by trial.

For oil-immersion work a special condenser is necessary. This is sold under the name of reflecting condenser, "dunkelfeld," dark-field illuminator, etc. With some makes it is placed upon the stage of the microscope; with others it is substituted for the regular condenser. It requires an intense light, like that given by a nitrogen-filled tungsten lamp or a small arc-light. Direct sunlight may be used. The condenser must be accurately centered. The space between it and the slide is usually filled in with immersion oil but water answers almost as well and makes cleaning easier. For this work the aperture of the oil-immersion objective must be reduced by placing in it a "funnel stop" obtainable from the maker of the objective.

The chief use of dark-ground illumination in clinical work is for demonstration of *Treponema pallidum* in fresh material (see Fig. 221).

In the "ultramicroscope" dark-ground illumination by means of ultra-violet light is utilized. The image is invisible to the eye and must be obtained by photography.

The Condenser.—For the work of the clinical laboratory a substage condenser is a necessity. Its purpose is to condense the light upon the object to be examined. For critical work the light must be focused on the object by raising or lowering the condenser by means of the screw provided for the purpose. The image of the light source will then appear in the plane of the object. This is best seen by using a low-power objective and ocular.

Should the image of the window-frame or other nearby object appear in the field and prove annoying, the condenser may be raised or lowered a little. It is often advised to remove the condenser for certain kinds of work, but this is not necessary and is seldom desirable in the clinical laboratory.

The condenser is constructed for parallel rays of light. With daylight, therefore, the plane mirror should be used; while for the divergent rays of ordinary artificial light the concave mirror, which tends to bring the rays together, is best.

It is very important that the condenser be accurately centered in the optical axis of the instrument, and most high-grade instruments have centering screws by which it can be adjusted at any time. The simplest way to recognize whether the condenser is centered is to close the diaphragm beneath it to as small an opening as possible, then remove the eye-piece and look down the tube. If the diaphragm opening does not appear in the center of the field, the condenser is out of center.

The use of the condenser is further discussed in the following sections.

Objectives and Eye-pieces.—Unfortunately, different makers use different systems of designating their lenses. The best system, and the one chiefly used in this country, is to designate objectives by their focal lengths in millimeters, and eye-pieces by their magnifying power, indicated by an " X." Most foreign makers use this system for their high-grade lenses, but still cling to arbitrary letters or numbers for their ordinary output.

Objectives are of two classes—achromatic and apochromatic. Those in general use are of the achromatic type, and they fulfil all requirements for ordinary work. Apochromatic objectives are more highly corrected for chromatic and spheric aberration, and represent the highest type of microscope lenses produced. They are very desirable for photomicrography and research, but for routine laboratory work do not offer advantages commensurate with their great cost. They require the use of special "compensating" eye-pieces.

Objectives are "corrected" for use under certain fixed conditions, and *they will give the best results only when used under the conditions for which corrected.* The most **important corrections** are: (*a*) For tube length; (*b*) for thickness of cover-glass; and (*c*) for the medium between objective and cover-glass.

(*a*) The tube length with which an objective is to be used is usually engraved upon it—in most cases it is 160 mm. The draw-tube of the microscope should be pulled out until the proper length is obtained, as indicated by the graduations on its side. When a nose-piece is used, it adds about 15 mm. to the tube length, and the draw-tube must be pushed in for that distance, unless, as is the case with the newer American instruments, the graduations upon the draw tube are correct with the nose-piece in place.

(*b*) The average No. 2 cover-glass is about the thickness for which most objectives are corrected—usually 0.17 or 0.18 mm. One can get about the right thickness by buying No. 2 covers and discarding the thick ones; or by buying No. 1 covers and discarding the thinner ones. Slight differences in cover-glass thickness can be compensated by increasing the length of tube when the cover is too thin, and decreasing it when the cover is too thick. This should be done with a spiral motion while supporting the body-tube with the other hand. The amount of correction necessary will depend upon the focal length and numeric aperture of the objective. With a 4-mm. objective of 0.85 numeric aperture a difference of 0.03 mm. in cover-glass thickness requires a change of 30 mm. in the tube length. Many high-grade objectives are supplied with a "correction collar," which accomplishes the same end. While for critical work, especially with apochromatics, cover-glass thickness is very important, one pays little attention to it in the clinical laboratory. A high-power dry lens always requires a cover, but its exact thickness is unimportant in routine work. Very low-power and oil-immersion objectives may be used without any cover-glass.

(c) The correction for the medium between objective and cover-glass is very important. This medium may be either air or some fluid, and the objective is hence either a "dry" or an "immersion" objective. The immersion fluid generally used is an especially prepared cedar oil, which gives great optical advantages because its index of refraction is the same as that of crown glass. It is obvious that only objectives with very short working distance, as the 2 mm., can be used with an immersion fluid.

To use an oil-immersion objective a suitable field for study should first be found with the low power. A drop of immersion oil is then placed upon the cover, and the objective lowered into it. A slight flash of light will be seen when the front lens touches the oil. The objective is then brought to a focus in the usual way.

In order to avoid bubbles the oil must be placed upon the cover carefully and without stirring it about. Bubbles are a frequent source of trouble, and should always be looked for when an immersion objective does poor work. They are readily seen by removing the eye-piece and looking down the tube. If they are present, the oil must be removed and a new drop applied. Immediately after use both objective and slide should be wiped clean with lens-paper or a soft linen handkerchief. In an emergency glycerin may be used instead of cedar oil, but, of course, with inferior results.

Curvature of field, through which it is impossible to focus both center and periphery sharply at the same time, is a very noticeable defect; but it is less serious than appears at first sight, particularly for visual work. It is easily compensated by frequent use of the fine focusing adjustment. Complete flatness of field cannot be attained without sacrifice of other and more desirable properties. Some of the finest objectives made, notably the apochromatics, show decided curvature.

The **working distance** of an objective should not be confused with its focal distance. The former term refers to the distance between the front lens of the objective, when it is in focus, and the cover-glass. It is always less than the focal distance, since the "focal point" lies somewhere within the objective; and it varies considerably with different makes. Long working distance is a very desirable feature. Some oil-immersion objectives have such short working distance that only very thin cover-glasses can be used.

A useful **pointer** can be made by placing a straight piece of a hair across the opening of the diaphragm of the eye-piece, cementing one end with a tiny drop of balsam, and cutting the hair in two in the middle with small scissors. When the eye-piece is in place, the hair appears as a black line extending from the periphery to the center of the microscopic field. If the pointer does not appear sharply defined it is out of focus and the diaphragm must be raised or lowered a little within the ocular.

The **formation of the microscopic image** demands brief consideration (Fig. 7). The rays of light which are reflected upward from the mirror and which pass through the object are brought to a focus in a magnified, inverted real image. This can be focused to appear at different levels, but when the microscope is used in the ordinary way it is formed at about the level of the diaphragm in the ocular. It can be seen by removing the ocular, placing a piece of ground glass on the top of the tube, and focusing upon it. When viewing this image a roll of paper or a cylindric mailing tube should be used to exclude extraneous light. This image, in turn, is magnified by the eye-lens of the ocular, producing a second real image, which is again inverted, and, therefore, shows the object right side up. This can be seen upon a ground glass held a few inches above the ocular, provided strong artificial light be used and the room darkened. The eye, when it looks into the microscope, sees, not this real image, but rather an inverted *virtual image* which appears about 250 mm. (10 inches) in front of the eye.

Numeric Aperture.—This expression, usually written N.A., indicates the amount of light which enters an objective from a point in the microscopic field. In optical language, N.A. is the sine of one-half the angle of aperture multiplied by the index of refraction of the medium between the cover and the front lens. Nu-

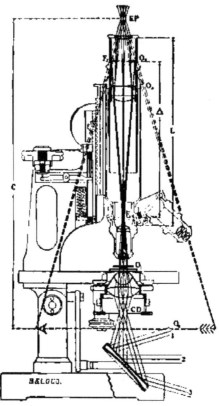

FIG. 7.—Diagram Showing Path of Light Rays; F_1, Upper focal plane of objective; F_2, lower focal plane of eye-piece; Δ, optical tube length = distance between F_1 and F_2; O_1, object; O_2, real image in F_2, transposed by the collective lens, to O_3, real image in eye-piece diaphragm; O_4, virtual image formed at the projection distance C, 250 mm. from EP, eyepoint; CD, condenser diaphragm; L, mechanical tube length (160 mm.); 1, 2, 3, three pencils of parallel light coming from different points of a distant illuminant, for instance, a white cloud, which illuminate three different points of the object.

meric aperture is extremely important, because upon it depends *resolving power*, which is the most important property of an objective.[1]

Resolving power is the ability to separate minute details of structure. For example, the dark portions of a good half-tone picture appear gray or black to the un-aided eye, but a lens easily resolves this apparently uniform surface into a series of separate dots. Resolving power does not depend upon magnification. The fine lines and dots upon certain diatoms may be brought out clearly and crisply (*i.e.*, they are resolved) by an objective of high numeric aperture, whereas with an objective of lower numeric aperture, but greater magnifying power, the same diatom may appear to have a smooth surface, with no markings at all, no matter how greatly it is magnified. Knowing the N.A., it is possible to calculate how closely lines and dots may lie and still be resolved by a given objective. To state the numeric aperture, therefore, is to tell what the objective can accomplish, provided, of course, that spheric and chromatic aberrations are satisfactorily corrected. An objective's N.A. is usually engraved upon the mounting.

It is an important fact, and one almost universally overlooked by practical microscopists, that the proportion of the numeric aperture of an objective which is *utilized* depends upon the aperture of the cone of light delivered by the condenser. In practice, the numeric aperture of an objective is reduced nearly to that of

---

[1] Resolving power really depends upon two factors, the N. A. and the wave length of light, but the latter can be ignored in practice. The great resolving power of the ultramicroscope depends upon its use of light of short wave length.

the condenser (which is indicated by lower-case letters, n.a.).[1] The condenser should, therefore, have a numeric aperture at least equal to that of the objective with which it is to be used. Lowering the condenser below its focal distance and closing the diaphragm beneath it have the effect of reducing its working aperture. A condenser, whatever its numeric aperture, cannot deliver through the air a cone of light of greater N.A. than 1. From these considerations it follows that the proper adjustment of the substage condenser is a matter of great importance when using objectives of high N.A., and that, to gain the full benefit of the resolving power of such objectives, the condenser must be focused on the object under examination, it must be oiled to the under surface of the slide in the same way as the immersion objective is oiled to the cover-glass, and the substage diaphragm must be wide open. The last condition introduces a difficulty in that colorless structures will appear "fogged" in a glare of the light, making a satisfactory image impossible when the diaphragm is more than three-quarters open (see Fig. 6). Wright suggests that the size of the light source be so regulated by a diaphragm that its image, thrown on the slide by the condenser, coincides with the real field of the objective, and maintains that in this way it is possible to reduce the glare of light and to dispel the fog without closing the diaphragm.

One can easily determine how much of the aperture of an objective is in use by removing the eye-piece, look-

---

[1] The N.A. of the objective is not reduced wholly to that of the condenser, because, owing to diffraction phenomena, a small part of the unilluminated portion of the back lens is utilized.

3

ing down the tube, and observing what proportion of the
back lens of the objective is illuminated. The relation
of the illuminated central portion to the unilluminated
peripheral zone indicates the proportion of the numeric
aperture in use. The effect of raising and lowering the
condenser and of oiling it to the slide can thus be easily
seen.

Another property of an objective which depends
largely upon N.A. is **depth of focus,** the ability to render
details in different planes clearly at the same time. The
higher the N.A. and the greater the magnification, the
less the depth of focus. Any two objectives of the same
focal length and same N.A. will have exactly the same
depth of focus. Depth of focus can be increased by
closing down the diaphragm, and thus reducing the
N.A. Great depth is desirable for certain low-power
work, but for high powers it does not offer advantages
to balance the loss of N.A. by which it is attained. In
some cases, indeed, it is a real disadvantage.

Magnification.—The degree of magnification should
always be expressed in *diameters*, not *times*, which is a
misleading term. The former refers to increase of
diameter; the latter, to increase of *area*. The comparatively low magnification of 100 diameters is the same as
the apparently enormous magnification of 10,000 times.

According to the system of rating magnification in
use in this country the magnifying power of an objective is ascertained by dividing the *optical tube-length* by
the focal length of the objective. The optical tube-
length is usually somewhere near 165 mm., but it varies
with the different objectives and the makers' catalogs
must be consulted for an accurate statement of magni-

fying power. One maker, at least, follows the commend able plan of engraving both the focal length of the objective and its initial magnification upon its barrel.

This system of rating magnification measures the enlarged image at the level of the diaphragm in the ocular, and this image is in turn magnified by the ocular so that when an objective and ocular are used together the total magnification is the product of the two. In the case, for example, of the 1.9 mm. oil immersion objective, whose initial magnification is 95 diameters, the total magnification with the 5✕ ocular is 475 diameters. These figures hold good, however, only when the ocular is rated upon the same system as the objective; thus, the 4✕ ocular of the Zeiss firm, which uses a different system, is equivalent to a 6✕ ocular of American make.

It is easy to find the magnifying power of any combination of objective and ocular by actual trial. Place the counting slide of the hemacytometer upon the microscope and focus the ruled lines. Now adjust a sheet of paper upon the table close to the microscope in such a position that when the left eye is in its proper place at the ocular the paper will lie in front of the right eye at the normal visual distance, i.e., 250 mm. (10 inches). (The paper may be supported upon a book, if necessary.) If both eyes are kept open, the ruled lines will appear to be projected on the paper. With a pencil, mark on the paper the apparent location of the lines which bound the small squares used in counting red blood corpuscles and measure the distance between the marks. Divide this distance by 0.05 mm., which is the actual distance between the lines on the slide. The quotient gives the magnification. If, to take an example, the lines in the image on the paper are 5 mm. apart, the magnification is 100 diameters. The figures obtained in this way will vary somewhat as one is near or far sighted, unless the defect of vision is corrected with glasses.

In practice, magnification can be increased in one of three ways:

(*a*) *Drawing Out the Tube.*—Since the increased tube length interferes with spheric correction, it should be used only with the knowledge that an imperfect image will result.

(*b*) *Using a Higher Power Objective.*—As a rule, this is the best way, because resolving power is also increased; but it is often undesirable because of the shorter working distance, and because the higher objective often gives greater magnification than is desired, or cuts down the size of the real field to too great an extent.

(*c*) *Using a Shorter Eye-piece.*—This is the simplest method. It has, however, certain limitations. When too high an eye-piece is used, there results a hazy image in which no structural detail is seen clearly. This is called "empty magnification," and depends upon the fact that the objective has not sufficient resolving power to support the high magnification. It has been aptly compared to the enlargement, by stretching in all directions, of a picture drawn upon a sheet of rubber. No new detail is added, no matter how great the enlargement. The extent to which magnification can be satisfactorily increased by eye-piecing depends wholly upon the resolving power of the objective, and consequently upon the N.A. The greatest total or combined magnification which will give an *absolutely* crisp picture is found by multiplying the N.A. of an objective by 400. The greatest magnification which can be used at all

satisfactorily is 1000 times the N.A. For example:
The ordinary 1.9-mm. objective has a N.A. of 1.30; the
greatest magnification which will give an absolutely
sharp picture is 520 diameters, which is obtained approximately by using a 5.5× eye-piece. Higher eye-
pieces can be used, up to a total magnification of 1300
diameters (12.5× eye-piece), beyond which the image
becomes wholly unsatisfactory.

The Microscope in Use.—Optically, it is a matter of
indifference whether the instrument be used in the
vertical position or inclined. Examination of fluids requires the horizontal stage, and since much of the work
of the clinical laboratory is of this nature it is well to
accustom one's self to the use of the vertical microscope.
While working one should sit as nearly upright as is
possible compatible with comfort, and the height of
the seat should be adjusted with this in view.

It is always best to "focus up," which saves annoy-
ance and probable damage to slides and objectives.
This is accomplished by bringing the objective nearer
the slide than the proper focus, and then, with the eye
at the eye-piece, turning the tube up until the object
is clearly seen. *The fine adjustment should be used only
to get an exact focus with the higher power objectives after
the instrument is in approximate focus.* It should not be
turned more than one revolution.

There will be less fatigue to the eyes if both are kept
open while using the microscope, and if no effort is made
to see objects which are out of distinct focus. Fine
focusing should be done with the fine adjustment, not
with the eye. An experienced microscopist keeps his

fingers almost constantly upon one or other of the focusing adjustments.

Although the ability to use the eyes interchangeably
is sometimes very desirable, greater skill in recognizing
objects will be acquired if the same eye be always used.
The left eye is the more convenient, because the right
eye is thus left free to observe the drawing one may wish
to do with the right hand. After a little practice one
can cause the microscopic image to appear as if projected upon a sheet of paper placed close to the microscope under the free eye. This gives the effect of a
camera lucida, and it becomes very easy to trace outlines. When one is accustomed to spectacles, they
should not be removed.

It is very desirable that one train himself to work
with the low-power objective as much as possible,
reserving the higher powers for detailed study of the
objects which the low power has found. This makes
both for speed and for accuracy. A search for tube-
casts, for example, with the 4-mm. objective is both
time-consuming and liable to failure. Even such
minute structures as nucleated red corpuscles in a
stained blood-film are more quickly found with an
8-mm. or even a 16-mm. objective combined with a
high ocular than with the oil-immersion lens.

To be seen most clearly, an object should be brought
to the center of the field. Acuity of vision will be
greatly enhanced and fatigue lessened if all light except
that which enters through the microscope be excluded
from both eyes. To this end various eye-shades have
been devised and some workers go so far as to work
inside a small tent constructed of strips of wood covered

with black cloth, the source of illumination being placed outside the tent.

One often wishes to mark a particular field upon a permanent preparation so as to refer to it again. The vernier of the mechanical stage cannot be relied upon, because it is impossible to replace the stage in exactly the same position after it has been removed and because its position is frequently changed by the slight knocks which it receives. There are on the market several "object markers" by which a desired field can be marked with ink, or by a circle scratched on the cover-glass by a minute diamond, while the slide is in place on the microscope. The circle is easily located with a low power. In the absence of these, one can, while using the low power, place minute spots with a fine pen at the edge of the field on opposite sides.

A good marking material is a cement which the author has long used for making cells, ringing cover-glasses, etc. To a few ounces of white shellac in wood alcohol add an equal volume of gasoline, shake thoroughly, and let stand for twenty-four hours, or until well separated into two layers. Pipet off the clear lower portion, add 5 to 10 drops of castor oil to each ounce, and color with any analin dye dissolved in absolute alcohol. When too thick, thin with alcohol. This makes a beautiful, transparent, easy-flowing cement which does not crack and which is not readily attacked by xylol. Glycerin mounts which the writer ringed with it twenty years ago are still in perfect condition.

Many good workers advise against the use of spring clips to hold the slide against the stage of the microscope. Manipulation of the slide with the fingers alone certainly gives good training in delicacy of touch, and is desirable when examining infectious material which might contaminate the clips, or when one must detect slight pressure of the objective upon the cover-glass as in studying a hanging-drop preparation. For the majority of examinations, however, it is more satisfactory to use a clip at one end of the slide, with just sufficient pressure to hold the slide without interfering with its freedom of movement.

Occasionally when one wishes a very low-power objective for some special work it may be desirable to unscrew the front lens of the 16-mm. objective and use the back lens only. This procedure is not recommended for critical work, and it should not be tried with high-power objectives, *which must never be taken apart*.

To attach an objective it should be supported in position against the nose-piece by means of the index-finger and middle finger, which grasp it as one would a cigar. It is then screwed into place with the fingers of the other hand.

Care of the Microscope.—The microscope is a delicate instrument and should be handled accordingly. Even slight disturbance of its adjustments may cause serious trouble. It is so heavy that one is apt to forget that parts of it are fragile. It seems unnecessary to say that when there is unusual resistance to any manipulation, force should never be used to overcome it until its cause has first been sought; and yet it is no uncommon thing to see students, and even graduates, push a high-power objective against a microscopic preparation with such force as to break not only the cover-glass, but even a heavy slide.

It is most convenient to carry a microscope with the fingers grasping the pillar and the arm which holds the tube; but since this throws a strain upon the fine adjustment, it is safer to carry it by the base. In the more recent instruments a convenient handle-arm is provided. To bend the instrument at the joint, the force should be applied to the pillar and never to the tube or the stage.

The microscope should be kept scrupulously clean, and dust must not be allowed to settle upon it. When not in use the instrument should be kept in its case or under a cover. An expensive glass bell-jar is not needed, and, in fact, is undesirable, except for display. It is heavy and awkward to handle, and when lifted is almost certain (unless great care is exercised) to strike the microscope. It is particularly liable to strike the mechanical stage and disturb its adjustment. The simplest, cheapest, lightest, and probably the best cover for the microscope is a truncated cone or pyramid of pasteboard, covered with creton or similar material. This is easily made at home. In the absence of a special cover a square of lintless cloth may be draped over the microscope.

Lens surfaces which have been exposed to dust only should be cleaned with a camel's-hair brush. A small brush and a booklet of lens-paper should always be at hand in the microscope case. Those surfaces which are exposed to finger-marks should be cleaned with lens-paper, or a soft linen handkerchief, moistened with water if necessary. The rubbing should be done very gently and with a circular motion. Particles of dirt which are seen in the field are upon the slide, the eye-piece, or the condenser. Their location can be determined by moving the slide, rotating the eye-piece, and lowering the condenser. Dirt upon the objective cannot be seen as such; it causes a diffuse cloudiness. When the image is hazy, the objective probably needs cleaning; or in case of an oil-immersion lens, there may be bubbles in the oil.

Oil and balsam which have dried upon the lenses —an insult from which even dry objectives are not immune—may be removed with alcohol or xylol; but these solvents must be used sparingly and carefully, as there is danger of softening the cement between the components of the lens. Some manufacturers now claim to use a cement which resists xylol. Care must be taken not to get any alcohol upon the brass parts, as it will remove the lacquer. Balsam and dried oil are best removed from the brass parts with xylol.

When the vulcanite stage becomes brown and discolored the black color can be restored by rubbing well with petrolatum.

Measurement of Microscopic Objects.—Of the several methods, the most convenient and accurate is the use of a micrometer eye-piece. In its simplest form this is similar to an ordinary eye-piece, but it has within it a glass disk upon which is ruled a graduated scale. When this eye-piece is placed in the tube of the microscope, the ruled lines appear in the microscopic field, and the size of an object is readily determined in *terms of the divisions of this scale*. The value of these divisions in millimeters manifestly varies with different magnifications. Their value must, therefore, be determined separately for each objective. This is accomplished through use of a stage micrometer—a glass slide with carefully ruled scale

divided into subdivisions, usually hundredths of a milli
meter. The stage micrometer is placed upon the stage
of the microscope and brought into focus. The tube
of the microscope is then pushed in or pulled out until
two lines of the one scale exactly coin-
cide with two lines of the other. From
the number of divisions of the eye-piece
scale which then correspond to each
division of the stage micrometer the
value of the former in micra or in frac-
tions of a millimeter is easily calculated.
*This value, of course, holds good only for
the objective and the tube-length with which
it was found.* The counting slide of the
hemacytometer will answer in place of a
stage micrometer, the lines which form
the sides of the small squares used in
counting red blood corpuscles being 0.05
mm. apart. When using the counting
chamber with an oil-immersion lens a
cover must be used; otherwise the oil
will fill the ruled lines and cause them to
disappear. Any eye-piece can be converted into a micrometer eye-piece by
placing a micrometer disk—a small circular glass plate with ruled scale—ruled
side down upon its diaphragm. If the

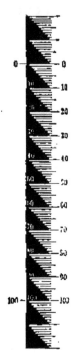

FIG. 8.—Scale
of the step mi-
crometer eye-
piece.

lines upon this are at all hazy the disk has probably
been inserted upside down or else the diaphragm is
out of its proper position. Usually it can be pushed
up or down as required. The new "step" micrometer
eye-piece is very satisfactory. The step-like arrange-

ment of the scale (Fig. 8) makes it easy to read and the divisions are such that they read in micra or easy multiples of micra with little or no change from the regular tube-length.

The following method of micrometry is less accurate, but is fairly satisfactory for comparatively coarse objects, such as the ova of parasites. A ruled scale corresponding to the magnified image of the hemacytometer ruling is drawn upon cardboard in the manner described for ascertaining magnifications (see p. 35) except that the card is placed upon the table beside the microscope and not necessarily at a distance of ten inches from the eye. This card may then be used as a micrometer, and should be inscribed with the value of its graduations, and the objective, ocular, and tube length with which it is to be used. In the example cited upon p. 35 the lines on the card are 5 mm. apart, corresponding to an actual distance of 0.05 mm. To measure an object, the cardboard is placed in the position which it occupied when made (upon the table at the right of the microscope). The lines and the objects on the slide can then be seen together, and the space covered by any object indicates its size. The graduations made as above indicated are too coarse for most work, and they should be subdivided. If five subdivisions are made, each will have a value of 10 μ.

Tuttle has suggested that in feces and other examinations a little lycopodium powder be mixed with the material. The granules are of uniform size—30 μ in diameter—and are easily recognized (Fig. 9). They furnish a useful standard with which the size of other structures can be compared. Care must be exercised not to use too much powder. The lycopodium is conveniently kept in a gelatin capsule, and a faint cloud can be dusted over the slide by gently scraping the edge of the lid upon the rim of the capsule.

The principal microscopic objects which are measured clinically are animal parasites and their ova and abnormal blood-corpuscles. The metric system is used almost exclusively. For very small objects o.oor mm. has been adopted as the unit of measurement, under the name

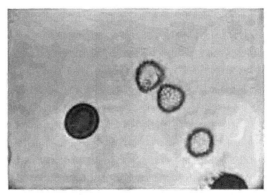

Fig. 9.—Egg of *Tænia saginata.* Lycopodium granules used as micrometer (× 250).

micron. It is represented by the Greek letter μ. For larger objects, where exact measurement is not essential, the diameter of a red blood-corpuscle (7 to 8 μ) is sometimes taken as a unit.

Photomicrography.—Although high-grade photomicrography requires expensive apparatus and considerable skill in its use, fairly good pictures of microscopic structures can be made by any one with simple instruments.

Any camera with focusing screen or a Kodak with plate attachment may be used. It is best, but not necessary, to remove the photographic lens. The camera is placed with the lens (or lens-opening, if the lens has been removed) looking into the eye-piece of the microscope, which may be in either the vertical or the horizontal position. One can easily rig up a standard to which the camera can be attached in the proper position by means of a tripod screw. A light-tight connection can be made of a cylinder of paper or a cloth sleeve with draw-strings. The image will be thrown upon the ground-glass focusing screen, and is focused by means of the fine adjustment of the microscope. The degree of magnification is ascertained by placing the ruled slide of the blood-counting instrument upon the microscope and measuring the image on the screen. The desired magnification is obtained by changing objectives or eye-pieces or lengthening the camera-draw.

Focusing is comparatively easy with low powers, but when using an oil-immersion objective it is a difficult problem unless the source of light be very brilliant. If one always uses the same length of camera and microscope tube, a good plan is as follows: Ascertain by trial with a strong light how far the fine adjustment screw must be turned from the correct eye focus to bring the image into sharp focus upon the ground-glass screen. At any future time one has only to focus accurately with the eye, bring the camera into position, and turn the fine adjustment the required distance to right or left. When the camera-draw is 10 inches little or no change in the focusing adjustment will be necessary.

The light should be as intense as possible in order to

shorten exposure, but any light that is satisfactory for ordinary microscopic work will answer. The light must be carefully centered. It is nearly always necessary to insert a colored filter between the light and the microscope. Pieces of colored window-glass are useful for this purpose but much better filters can be purchased at trifling cost. The writer has had best results with the Wratten "micro" filters. These may be purchased in the form of gelatin sheets which can be cemented between glass plates with balsam. The screen should have a color complementary to that which it is desired to bring out strongly in the photograph: for blue structures, a yellow screen; for red structures, a green screen. For the average stained preparation, a picric-acid yellow or a yellow green will be found satisfactory.

Very fair pictures can be made on Kodak film, but orthochromatic plates (of which Cramer's "Iso" and Seed's "Ortho" are examples) give much better results. Panchromatic plates like the Wratten "M" are still better but are more difficult to handle because more sensitive to red light. In order to avoid halation all plates should if possible be "backed." The length of exposure depends upon so many factors that it can be determined only by trial. It will probably vary from a few seconds to fifteen minutes. Plates are developed in the usual way. Either the tray or tank method may be used, but in order to secure good contrast it is often desirable to overdevelop somewhat. Metol-hydrochinon is an excellent developer, as it gives good contrast with full detail.

The photograph from which Fig. 10 was made was taken with a Kodak and plate attachment on an "Iso" plate, the source of light being the electric lamp and condensing lens illustrated in Fig. 3. It was focused by the method described above. The screen was a picric-acid stained photographic plate. Exposure, three and a half-minutes. The picture loses considerable detail in reproduction.

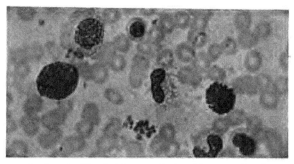

FIG. 10.—Leukemic blood (about × 650). Photograph taken with a Kodak, as described in the text.

Choice of a Microscope.—It is poor economy to buy a cheap instrument.

For the work of a clinical laboratory the microscope should preferably be of the handle-arm type, and should have a large stage. It should be provided with a substage condenser (preferably of 1.40 n.a.), three or more objectives on a revolving nose-piece, and two or more eye-pieces. After one has learned to use them, the new mon-objective binocular microscopes are extremely satisfactory, giving an impression of stereoscopic vision.

The most generally useful objectives are: 16 mm., 4 mm., and 2 mm. oil immersion. The 4-mm. objective may be obtained with N.A. of 0.65 to 0.85. If it is to be used for blood-counting, the former is preferable, since its working distance is sufficient to take the thick cover of the blood-counting instrument. For coarse objects a 32-mm. objective is very desirable. The eyepieces most frequently used are 5✕ and 10✕. A very low power (2✕) and a very high (15✕) will sometimes be found useful. The microm-eter eye-piece is almost a necessity. A mechanical stage, preferably of the attachable type, is almost indispensable for blood and certain other work. A new, simple and comparatively inexpensive stage of this type is shown in Fig. 11.

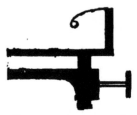

FIG. 11.—An inexpensive mechanical stage with rack and pinion movement in one direction.

A first-class monocular microscope, of either American or foreign make, equipped as just described, will cost in the neighborhood of $70 to $80, exclusive of the mechanical stage.

Practical Exercises.—The following is a brief outline of certain exercises which the author has found useful in teaching microscopy. The student must learn as early as possible what can be expected of his microscope with proper manipulation. When he sits down to work his first glance should tell him whether the instrument is giving its best results. If the microscopic picture falls short of the best, he must locate the difficulty and correct it before proceeding.

1. Clean the microscope and study its parts, familiarizing yourself with the names, purposes, and movements of each (see Fig. 1).

2. Practice the manipulations necessary to locate particles

4

of dust or dirt which appear in the microscopic field (see p. 41).

3. Place the microscope before a window, focus upon a dusty slide and adjust condenser and mirror so that the image of the window-frame or, better, of trees just outside the window appear in the microscopic field. Try the effect of raising and lowering the condenser, and of changing from plane to concave mirror, upon these images. Note that they cause an unevenly illuminated or mottled field when a little out of focus.

4. Insert a "pointer" in one of the oculars (see p. 29).

5. Study illumination. Use a slide of some colorless structures such as cholesterol crystals and two preparations of blood, one a dried film stained with eosin or any blood stain (see p.307) and mounted in balsam, the other an unstained wet preparation made is described for the malarial parasite (see p. 355). Study only the areas in which the corpuscles are well separated.

(1) Place one of these on the microscope, bring to a focus, and practice the manipulations necessary to secure (see p. 22)—

(a) Central illumination.

(b) Oblique illumination.

(c) Strong and subdued illumination.

The field in each case must be evenly lighted throughout, without mottling. Continue until you can adjust any desired form of illumination quickly and surely, and can recognize each by a glance into the microscope.

(2) Using the three slides mentioned above, ascertain the best form of illumination to study (see p. 23)—

(a) Outlines.

(b) Color.

(c) Surface contour. The unstained normal and crenated red corpuscles are excellent objects for study of surface contour.

(3) Try dark-ground illumination by means of the sub-stage disk (see p. 24). Study the unstained blood-smear and draw a few corpuscles. Also examine the cholesterol crystals, a drop of diluted milk and a bit of lens paper. Use the 16-mm. objective for this.

6. With central illumination, focus upon a slide and observe how much of the numeric aperture is in use (see p. 33). Try the effect upon numeric aperture of—

(1) Opening and closing the diaphragm.

(2) Raising and lowering the condenser.

(3) Using the oil-immersion objective—

(a) Without oil.

(b) With oil between objective and cover-glass.

(c) With oil between slide and condenser.

7. Upon the same species of diatom compare two objectives of 3-mm. focus (therefore of same magnifying power), one of N.A. 1.4 and the other of N.A. 0.85. They will be adjusted by the instructor. Note the superior resolving power of the lens of high N.A. (see p. 32).

8. Practice using the oil-immersion objective (see p. 28) upon an unstained film preparation of blood or a slide strewn with diatoms. These are nearly colorless and hence difficult to see. If there is difficulty in finding the specimen, move the slide about while lowering the objective to a focus. Moving objects will catch the eye as the objective approaches the correct focus. If a cover-glass is used, its edge can be easily found, but it must be borne in mind that when the upper surface of the cover is in focus, objects beneath it are so far out of focus as to be invisible.

Produce some bubbles in the oil by stirring it about on the slide; observe their effect on the image of the blood cells or diatoms, and learn to detect their presence (see p. 29).

9. Image formation (see p. 30). Mount a bit of paper printed with very small type using oil or balsam to render it transparent. Focus upon this with a low power objective. Remove the ocular and lay a piece of ground glass across the top of the tube. This forms a screen upon which an image can be focused by means of the coarse adjustment. Note whether it is right side up or reversed. Repeat this with the ocular in place, holding the ground glass some inches above the ocular. These exercises, especially the last, are best done in a darkened room with strong artificial illumination, but extraneous light can usually be sufficiently excluded by viewing the image through a pasteboard mailing cylinder.

10. Find by trial the magnification produced by your 16-mm. objective with the 4✕ or 6✕ ocular (see p. 35). Compare your result with that listed by the maker of the microscope.

11. Micrometry.

(1) Evaluate the scale of your micrometer eye-piece with a high-power objective, and measure accurately 10 red blood corpuscles and 10 leukocytes (see pp. 42, 43).

(2) Prepare a cardboard micrometer and measure 10 lycopodium granules (see p. 44).

12. Focus upon a stage micrometer or hemacytometer slide and measure the diameter of the real field of each of your objectives with each of the oculars. Note the effect of increasing the tube-length.

13. Study the following structures, chiefly with a view to best illumination. Examine separately the color, outline and surface contour of each. Many of these are met as accidental contaminations in microscopic preparations and one must learn to recognize them. Make drawings of each.

Fluids are examined by placing a drop in the center of a clean slide and applying a cover-glass. The drop should be

large enough to fill the space between the slide and cover,
but not large enough to float the cover about. Fibers or
insoluble powder may be placed in a drop of water and
covered.

(1) With 16 mm. and 4 mm. objectives examine the
upper surface of a new cover-glass without cleaning.   Usually it will show dirt and often crystals;
if not, make finger prints upon it and produce
faint scratches by rubbing two covers together.

(2) Air bubbles produced by shaking a little diluted
mucilage.

(3) Fresh milk diluted with three or four volumes of
water.   Prepare three slides.

(a) Examine one untreated.

(b) Treat one with solution of Sudan III.   (For
method see p. 201.)   Note color assumed by
the fat globules.   This is one of the most
useful tests for microscopic fat.

(c) Treat one with dilute acetic acid.   Note
clumping of globules similar to that of typhoid bacilli in the Widal test.

(4) A drop of diluted India-ink.   Note the dancing ·
motion of the smaller particles ("Brownian
motion").

(5) Starch granules.   Gently scrape the freshly cut
surface of a potato with a knife, place a drop
of the cloudy fluid upon a slide with a drop of
water, and apply a cover-glass.   Make two
preparations.

(a) Examine one untreated.   Note the variously
sized starch granules, oval, colorless, concen-
trically striated.   Make sure that you find
the best form of illumination to bring out
the striations clearly.   The starch granules
themselves are easy to see because of their

broad dark outlines. This means that they are "highly refractive"—a term much used in describing microscopic structures—or, more correctly, that their index of refraction differs greatly from that of the medium in which they are mounted.

(b) Treat one with dilute Lugol's or Gram's iodin solution.

Note the change in color of the granules. This is the standard test for starch.

(6) Yeast which has been growing in a dextrose solution. Make two preparations.

(a) Examine one unstained. Note "budding."

(b) Treat one with iodin solution. Compare color of yeast with that taken by starch.

(7) Mold from moldy food. Note hyphæ and spores. Try the effect of iodin.

(8) Various fibers and other structures mounted in a drop of water.

(a) Cotton.

(b) Wool.

(c) Linen.

(d) Silk.

(e) Feather tip.

(f) Some dust from a carpeted room. Colored fibers from the carpet are frequently found in urine.

(g) A hair.

(9) A drop of decomposing urine. Note bacteria of various kinds, some motile, some non-motile. Make an effort to distinguish true motility from that due to currents in the fluid and to "Brownian motion."

(10) Some of the scum from the bottom of a stagnant pool. Note the abundance of microscopic life.

Look especially for diatoms, amebæ and ciliated organisms.

(11) Test your proficiency in using the microscope by trying to resolve diatoms. For the 4-mm. objective use *Pleurosigma angulatum*. The dots should be clearly seen. For the oil-immersion lens use *Surirella gemma*. The fine lines between the ribs should be seen as rows of dots. As a most critical test, both of the oil-immersion lens and of your skill in manipulation use *Amphipleura pellucida*. Select a diatom of large size. Use oblique illumination and endeavor to bring out the cross striations. Try the same with central light, although you are not likely to succeed. These striations consist of rows of extremely minute dots which can be seen only under the most favorable conditions such as are not attained in clinical work.

CHAPTER I

THE SPUTUM

Preliminary Considerations.—Before beginning the study of the sputum, the student will do well to familiarize himself with the structures which may be present in the normal mouth, and which frequently appear in the sputum as contaminations. Nasal mucus and material obtained by scraping the tongue and about the teeth should be studied as described for unstained sputum. A drop of Lugol's solution should then be placed at the edge of the cover, and, as it runs under, the effect upon different structures noted. Another portion should be spread upon slides or covers and stained by some simple stain and by Gram's method. The structures likely to be encountered are epithelial cells of columnar and squamous types; leukocytes, chiefly mononuclear, the so-called salivary corpuscles; food-particles; *Leptothrix buccalis*; great numbers of saprophytic bacteria; and frequently spirochetes and endamebæ. These structures are described later.

When **collecting the sample** for examination, the morning sputum, or the whole amount for twenty-four hours should be saved. In beginning tuberculosis tubercle bacilli can often be found in that first coughed up in the morning when they cannot be detected at any other time of day. Sometimes, in these early cases, there are only a few mucopurulent flakes which contain the bacilli, or only a small purulent mass every few days, and these may easily be overlooked by the patient.

Patients should be instructed to rinse the mouth well in order to avoid contamination with food-particles which may prove confusing in the examination, and to make sure that the sputum comes from the lungs or bronchi and not from the nose and nasopharynx. Many persons find it difficult to distinguish between the two. It is desirable that the material be raised with a distinct expulsive cough, but this is not always possible. Material from the upper air-passages can usually be identified by the large proportion of mucus and the character of the epithelial cells.

The sputum of infants and young children is usually swallowed and therefore cannot be collected. In such cases examination of the feces for tubercle bacilli will sometimes establish a diagnosis of tuberculosis.

As a receptacle for the sputum, a clean, wide-mouthed bottle with tightly fitting cork may be used. The patient must be particularly cautioned against smearing any of it upon the outside of the bottle. This is probably the chief source of danger to those who examine sputum. Disinfectants should not be added. Although some of them (phenol, for example) do not interfere with detection of tubercle bacilli, they generally so alter the character of the sputum as to render it unfit for other examinations.

The following outline is suggested for the **routine examination**:

1. Spread the material in a thin layer in a large Petri dish or between two plates of glass. The use of glass plates is messy, but is to be recommended for careful work.

The top plate should be much smaller than the lower one, or have some sort of handle.

2. Examine all parts carefully with the naked eye (best over a black background) or with a hand lens. The portions most suitable for further examination may thus be easily selected. *This macroscopic examination should never be omitted.*

3. Transfer various portions, including all suspicious particles, to clean slides, cover, and examine unstained with the microscope (see p. 63).

4. Slip the covers from some or all of the above unstained preparations, leaving a thin smear on both slide and cover.

5. Dry and fix the smears and stain one or more by each of the following methods:

    (*a*) For tubercle bacilli (see p. 76).

    (*b*) Gram's method (see p. 572).

6. When indicated, make special examinations for—

    (*a*) Capsules of bacteria (see p. 87).

    (*b*) Eosinophilic cells (see p. 91).

    (*c*) Much's granules (see p. 83).

    (*d*) Presence of albumin (see p. 94).

After the examination the sputum must be destroyed by heat or chemicals, and everything which has come in contact with it must be sterilized. The utmost care must be taken not to allow any of it to dry and become disseminated through the air. If flies are about, it must be kept covered. It is a good plan to conduct the examination upon a large newspaper, which can then be burned. Contamination of the work table is thus avoided. If this is not feasible, the table should be washed off with 10 per cent. lysol or other disinfectant solution, and allowed to dry slowly, as soon as the sputum work is finished.

Examination of the sputum is most conveniently considered under four heads: I. Physical examination. II. Microscopic examination. III. Chemic examination. IV. Characteristics of the sputum in various diseases.

I. PHYSICAL EXAMINATION

1. Quantity.—The quantity expectorated in twenty-four hours varies greatly. It may be so slight as to be overlooked entirely in beginning tuberculosis. It is usually small in acute bronchitis and lobar pneumonia. It may be very large—sometimes as much as 1000 c.c.—in advanced tuberculosis with large cavities, edema of the lung, bronchiectasis, and following rupture of an abscess or empyema. It is desirable to obtain a general idea of the quantity, but accurate measurement is unnecessary.

2. Color.—Since the sputum ordinarily consists of varying proportions of mucus and pus, it may vary from a colorless, translucent mucus to an opaque, whitish or yellow, purulent mass. A yellowish green is frequently seen in advanced phthisis and chronic bronchitis. In jaundice, in caseous pneumonia, and in slowly resolving lobar pneumonia it may assume a bright green color, due to bile or altered blood-pigment.

A red or reddish-brown color usually indicates the presence of blood. Bright red blood, most commonly in streaks, is strongly suggestive of phthisis. It may be noted early in the disease and generally denotes an extension of the tuberculous process. One must, however, be on his guard against blood-streaked mucus or muco-pus originating in nasopharyngeal catarrh. Tuberculous patients not infrequently mistake this

for true sputum. Blood-stained sputum is also sometimes seen in bronchiectasis. A rusty red sputum is the
rule in croupous pneumonia, and was at one time considered pathognomonic of the disease. Exactly similar
material may be raised in pulmonary infarction.
"Prune-juice" sputum is said to be characteristic of
"drunkard's pneumonia." It at least indicates a
dangerous type of the disease, as it is apparently
referable to coincident edema of the lung. A brown
color, due to altered blood-pigment, follows hemorrhages from the lungs, and is present, to greater or less
degree, in chronic passive congestion of the lungs,
which is most frequently due to a heart lesion.

Gray or black sputum is observed among those who
work much in coal-dust, and is occasionally seen in
smokers who are accustomed to "inhale."

3. Consistence.—According to their consistence,
sputa are usually classified as serous, mucoid, purulent,
seropurulent, mucopurulent, etc., which names explain
themselves. As a rule, the more mucus and the less pus
and serum a sputum contains, the more tenacious it is.

The rusty sputum of croupous pneumonia is extremely tenacious, so that the vessel in which it is contained may be inverted without spilling it. The same
is true of the almost purely mucoid sputum ("sputum
crudum") of beginning acute bronchitis, and of that
which follows an attack of asthma. A purely serous
sputum, usually slightly blood tinged, is fairly characteristic of edema of the lungs.

Formerly much attention was paid to the so-called
"nummular sputum." This consists of definite mucopurulent masses which flatten out into coin-like disks

and sink in water It is fairly characteristic of advanced tuberculosis.

4. Dittrich's Plugs.—While these bodies sometimes appear in the sputum, they are more frequently expectorated alone. They are caseous masses, usually about the size of a pin-head, but sometimes reaching that of a bean. The smaller ones are yellow, the larger ones gray. When crushed, they emit a foul odor. Microscopically, they consist of granular débris, fat-globules, fatty acid crystals, and bacteria. They are formed in the bronchi, and are sometimes expectorated by healthy persons, but are more frequent in putrid bronchitis and bronchiectasis. The laity commonly regard them as evidence of tuberculosis. The similar caseous masses which are formed in the crypts of the tonsils are sometimes also included under this name.

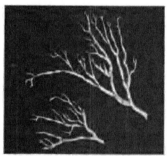

Fig. 12.—Bronchial casts as seen when carefully spread out and viewed over a black background. Natural size.

5. Bronchial Casts.—These are branching, tree-like casts of the bronchi, frequently, but not always, composed of fibrin (Fig. 12). In color they are usually white or grayish, but may be reddish or brown, from the presence of blood-pigment. Their size varies with that of the bronchi in which they are formed. Casts 15 or more centimeters in length have been observed, but they are usually very much smaller. Ordinarily they are coiled into a ball or tangled mass and can be recognized only by floating out in water—best over a black background—when their tree-like structure becomes evident. The naked-eye examination will usually suffice; occasionally a hand lens may be required.

Bronchial casts appear in the sputum in croupous pneumonia, in fibrinous bronchitis, and in diphtheria when the process extends into the bronchi. In diphtheria they are usually large. In fibrinous or chronic plastic bronchitis they are of medium size and usually of characteristic structure. Their demonstration is essential for the diagnosis of this disease. In some cases they may be found every day for considerable periods; in others, only occasionally. In almost every case of croupous pneumonia the casts are present in the sputum in variable numbers during the stage of hepatization and beginning resolution. Here they are usually small (0.5 to 1 cm. in length) and are often not branched.

II. MICROSCOPIC EXAMINATION

The portions most likely to contain structures of interest should be very carefully selected, as already described. *The few minutes spent in this preliminary examination will sometimes save hours of work later.* Opaque, white or yellow particles are most frequently bits of food, but may be cheesy masses from the tonsils; small cheesy nodules, derived from tuberculous cavities and containing many tubercle bacilli and elastic fibers;

Curschmann's spirals, or small fibrinous casts, coiled into little balls; or shreds of mucus with great numbers of entangled pus-corpuscles. The food-particles most apt to cause confusion are bits of bread, which can be recognized by the blue color which they assume when touched with iodin solution.

Some structures are best identified without staining; others require that the sputum be stained.

A. UNSTAINED SPUTUM

A careful study of the unstained sputum should be included in every routine examination. Unfortunately it is almost universally neglected. It best reveals certain structures which are seen imperfectly or not at all in stained preparations. It gives a general idea of the other structures which are present, such as pus-corpuscles, eosinophiles, epithelial cells, and blood and thus suggests appropriate stains to be used later. It enables one to select more intelligently the portions to be examined for tubercle bacilli.

The particle selected for examination should be transferred to a clean slide, covered with a clean cover-glass, and examined with the 16-mm. objective, followed by the 4 mm. The oil-immersion lens should not be used for this purpose. It is convenient to handle the bits of sputum with a wooden tooth-pick or with a wooden cotton-applicator, which may be burned when done with. The platinum wire used in bacteriologic work is unsatisfactory because not usually stiff enough. A little practice is necessary before one can handle particles of sputum readily. The bit desired should be separated from the bulk of the sputum by cutting it free with the toothpick and drawing it out upon the dry portion of the glass dish. It can then be picked up by rotating the end of a fresh tooth-pick against it. *The slide must never be dipped into the sputum, nor must any of the sputum be allowed to reach its edges in spreading.*

The more important structures to be seen in unstained sputum are: elastic fibers, Curschmann's spirals, Charcot-Leyden crystals, pigmented cells, myelin globules, the ray fungus of actinomycosis, and molds. Forming the background for these are usually pus-corpuscles, granular detritus, and mucus in the form of translucent, finely fibrillar or jelly-like masses. The pus cells appear as finely granular grayish or yellowish balls about 12 μ in diameter and without visible nuclei (see Figs. 20, 21). They are best studied in stained preparations.

1. Elastic Fibers.—These are the elastic fibers of the pulmonary substance, where they are distributed in the walls of the alveoli, the bronchioles, and the blood-vessels. When found in the sputum they always indicate destructive disease of the lung, provided they do not come from the food, which is a not infrequent source. They are found most commonly in phthisis; rarely in other diseases. Advanced cases of tuberculosis often show great numbers, and, rarely, they may be found in early tuberculosis when the bacilli cannot be detected. After the diagnosis is established they furnish a valuable clue as to the existence and rate of lung destruction. In gangrene of the lung, contrary to the older teaching, elastic tissue is probably always present in the sputum, usually in large fragments.

The fibers should be searched for with a 16-mm. ob-

jective, although a higher power is needed to identify
them with certainty.   They may usually be more clearly
seen if a drop of 10 to 20 per cent. caustic soda solu-

FIG. 13.—Elastic fibers in tuberculous sputum, unstained, as seen with
a low power objective (× 100).

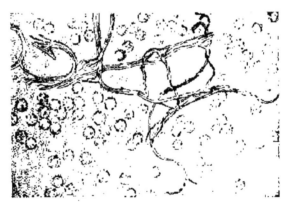

FIG. 14.—Elastic fibers and pus-corpuscles in tuberculous sputum,
unstained (× 300).

tion be mixed with the sputum on the slide before the
cover is applied.   Under the 4 mm. they appear as
slender, highly refractive, wavy fibers with double contour, and often curled or split ends.   Frequently they

5

are found in alveolar arrangement, retaining the original outline of the alveoli of the lung (Figs. 13 and 14). This arrangement is positive proof of their origin in the lung.

Leptothrix buccalis, which is a normal inhabitant of the mouth, may easily be mistaken for elastic tissue. It can be distinguished by running a little Lugol's solution under the cover-glass (see p. 535). Fatty-acid crystals, which are often present in Dittrich's plugs and in sputum which has lain in the body for some time, also simulate elastic tissue when very long, but they are more like stiff, straight or curved needles than wavy threads. They show varicosities when the cover-glass is pressed upon. The structures which most frequently confuse the student are the cotton fibrils which are present as a contamination in most sputa. These are usually coarser than elastic fibers, and flat, with one or two twists, and often have longitudinal striations and frayed-out ends. In stained preparations students frequently report the fibrils of precipitated mucus as elastic tissue.

Elastic fibers from the food are coarser, less frequently wavy and not arranged in alveolar order.

To find elastic fibers when not abundant, boil the sputum with a 10 per cent. solution of caustic soda until it becomes fluid; add several times its bulk of water, and centrifugalize, or allow to stand for twenty-four hours in a conical glass. Examine the sediment microscopically. The fibers will be pale and swollen and, therefore, somewhat difficult to recognize. Too long boiling will destroy them entirely.

The above procedure, although widely recommended,

will rarely or never be necessary if the sputum is carefully examined in a thin layer against a black background macroscopically and with a hand-lens, and if
all suspicious portions are further studied with the
microscope.

2. Curschmann's Spirals.—These peculiar structures are found most frequently in bronchial asthma, of
which they are fairly characteristic. Although not pres-

FIG. 15.—Curschmann's spirals in asthmatic sputum as seen when
pressed out between two plates of glass and viewed over a black background. Each is embedded in a mass of grayish mucus. Natural size.

ent in every attack, they probably occur at some time
in every case. Sometimes they can be found only near
the end of the attack. They may occasionally be met
with in chronic bronchitis and other conditions. Their
nature has not been definitely determined.

Macroscopically, they are whitish or yellow, wavy
threads, frequently coiled into little balls (Fig. 15).
Their length is rarely over 1.5 cm., though it sometimes exceeds 5 cm. They can sometimes be definitely
recognized with the naked eye. Under a 16-mm.
objective they appear as mucous threads with a bright,
colorless central line—the so-called central fiber—

about which are wound many fine fibrils (Figs. 16 and 17). The spiral fibrils are sometimes loosely, sometimes tightly wound. Eosinophiles are usually present

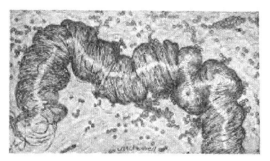

Fig. 16.—End of a large, tightly wound Curschmann's spiral in sputum from a case of bronchial asthma. Unstained (× 70).

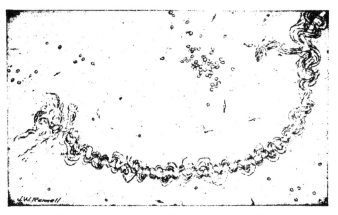

Fig. 17.—Slender, loosely wound Curschmann's spiral in sputum from a case of bronchial asthma. A few Charcot-Leyden crystals are also shown. Unstained (× 70).

within them, and sometimes Charcot-Leyden crystals, also. Not infrequently the spirals are imperfectly formed, consisting merely of twisted strands of mucus enclosing leukocytes.   The central fiber is absent from these.

3. Charcot-Leyden   Crystals. —Of   the   crystals which may be found in the sputum, the most interesting are the Charcot-Leyden crystals.   They may be absent when the sputum is expectorated, and appear in large numbers after it has stood for some time.   They are rarely found except in cases of bronchial asthma, and

Fig. 18.—Charcot-Leyden crystals and eosinophilic leukocytes in sputum from a case of bronchial asthma.   Unstained (× 475).   The magnification is greater than is usually used in studying these structures.

were at one time thought to be the cause of the disease. They frequently adhere to Curschmann spirals.   Their exact nature is unknown.   Their formation seems to be in some way connected with the presence of eosinophilic cells.   Outside of the sputum they are found in the feces in association with animal parasites, and in the coagulated blood in leukemia.

They are colorless, pointed, often needle-like crystals (Fig. 18).   Formerly they were described as octahedral, but are now known to be hexagonal in cross-section.

Their size varies greatly, the average length being about three or four times the diameter of a red blood-corpuscle.

Other crystals—hematoidin, cholesterin, and, most frequently, fatty-acid needles (see Fig. 49)—are common in sputum which has remained in the body for a considerable time, as in abscess of the lung and bronchiectasis. The fatty-acid crystals are regularly found in Dittrich's plugs. They might be mistaken for elastic fibers (see p. 66). Sometimes they form rounded masses with the individual crystals radially arranged and they then bear considerable resemblance to the clumps of *Actinomyces bovis*.

4. Pigmented Cells.—Granules of pigment are sometimes seen in ordinary pus-corpuscles but the more common pigment-containing cells are large mononuclear cells whose origin is in some doubt. They were formerly thought to be the flattened epithelial cells which line the pulmonary alveoli. The present tendency is to identify them with the large mononuclear leukocytes which are known to take up pigment granules readily. Two kinds of pigmented cells deserve mention: those which contain blood-pigment chiefly hemosiderin; and those which contain carbon.

To those which contain blood-pigment the name **heart failure cells** has been given, because they are most frequently found in long-continued passive congestion of the lungs resulting from poorly compensated heart disease. The presence of these cells in considerable numbers, by directing one's attention to the heart, will sometimes clear up the etiology of a chronic bronchitis. They are sometimes so numerous as to give the sputum a brownish tinge. Such cells are also found in the sputum in pulmonaiy infarction and for some time after a pulmonary hemorrhage. In fresh unstained sputum heart failure cells appear as round, grayish or colorless bodies filled with variously sized rounded granules of yellow to brown pigment (see Plate II, Fig. 1). Sometimes the pigmentation takes the form of a diffuse staining. The nucleus is usually obscured by the pigment. The cells are large, averaging about 35 μ in diameter.

To demonstrate the nature of the brown pigment apply a 10 per cent. solution of potassium ferrocyanid for a few minutes and follow with weak hydrochloric acid. Iron-containing pigment assumes a blue color. Many of the granules, will, however, fail to respond. The test may be applied either to wet preparations or to dried smears.

Carbon-laden cells are less important (see Plate II, Fig. 1). They are especially abundant in the sputum of anthracosis where angular black granules, both intracellular and extracellular, may be so numerous as to color the sputum. Similar cells with smaller carbon particles are often abundant in the morning sputum of those who inhale tobacco smoke to excess.

5. Myelin Globules.—These have little or no clinical significance but require mention because of the danger of confusing them with more important structures, notably blastomyces. They are colorless, round or oval globules of various sizes, often resembling fat-droplets but more frequently showing peculiar concentric or irregularly spiral markings (Figs. 19 and 27). Such globules are abundant in the scanty morning sputum of apparently healthy persons, but may be found in any mucoid sputum. They lie both free in the sputum and contained within the large cells which have long been known as alveolar cells but which are possibly large mononuclear leukocytes.

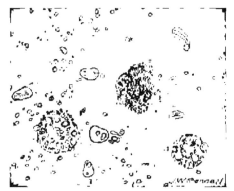

FIG. 19.—Myelin globules, free and contained within cells. From a "normal morning sputum" (× 350).

6. Actinomyces Bovis (Ray-fungus).—In the sputum of pulmonary actinomycosis and in the pus from actinomycotic lesions elsewhere, small, yellowish, "sulphur" granules can be detected with the unaided eye. Without a careful macroscopic examination they are almost certain to be overlooked. The fungus can be seen by crushing one of these granules between slide and cover, and examining with a low power. It consists of a network of threads having a more or less radial arrangement (Figs 20 and 21). In cattle, and to a less extent in man, the filaments at the periphery of the nodule present club-shaped extremities. It can be brought out more clearly by running a little solution of eosin in alcohol and glycerin under the cover. This organism, also called *Streptothrix actinomyces*, apparently stands midway between the bacteria and the molds. It stains by Gram's method.

Actinomycosis of the lung is rare. The clinical picture is that of tuberculosis.

FIG. 20.—A "sulphur granule" crushed beneath the cover glass. From the pus of a case of actinomycosis of submaxillary lymph nodes. Unstained (× 60).

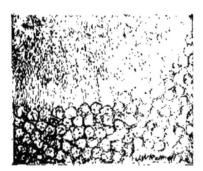

FIG. 21.—A portion of Fig. 20, more highly magnified (× 300.)

7. Molds and Yeasts.—The hyphæ and spores of various molds are occasionally met with in the sputum. They are usually the result of contamination, and have little significance. The hyphæ are rods, usually jointed or branched (see Fig. 79) and often arranged in a meshwork (mycelium); the spores are highly refractive spheres and ovoids. Both stain well with the ordinary stains.

In the extremely rare condition of systemic blastomycosis the specific yeasts have been found in the sputum in large numbers. It is advisable to add a little 10 per cent. caustic soda solution and examine unstained.

8. Animal Parasites.—These are extremely rare in the sputum in this country. A trichomonad, perhaps identical with *Trichomonas intestinalis* has been seen in the sputum of putrid bronchitis and gangrene of the lung, but its causal relationship is doubtful. In Japan, infection with the lung fluke-worm, *Paragonimus westermannii*, is common, and the ova are found in the sputum. The lung is not an uncommon seat for echinococcus cysts, and hooklets and scolices may appear, as may also *Endamœba histolytica*, when a hepatic abscess has ruptured into the lung. Larvæ of *Strongyloides intestinalis* and of the hook-worm have been reported. Ciliated body-cells, with cilia in active motion, are not infrequently seen, and may easily be mistaken for infusoria. All the above-mentioned parasites are described in Chapter VI.

B. STAINED SPUTUM

Structures which are best seen in stained sputum are bacteria and cells.

A number of smears should be made upon slides or covers. These films must, of course, be thin, but it is easily possible to get them too thin. This is a common error of students who have just finished a course in bacteriology and who have there been accustomed to work with scarcely perceptible films of bacteria. It is a good plan to slide off the cover-glass from the preparation used for the unstained microscopic examination. If this is properly done satisfactory smears will be left on both slide and cover. They are then dried in the air, and fixed in the flame, as described on page 571, or better, by immersion for one or two minutes in pure wood alcohol or saturated solution of corrosive sublimate. Fixation will ordinarily kill the bacteria and the smears may be kept indefinitely; but smears on slides when fixed by heat are often not sterile, and should be handled accordingly. One of the smears should be stained with some general stain, like Löffler's methylene blue or pyronin-methyl green (see p. 642), which will give a good idea of the various cells and bacteria present. Special stains may then be applied, as indicated, but a routine examination should, in all cases, include a stain by the method for the tubercle bacillus and by Gram's method.

1. **Bacteria.**—Saprophytic bacteria from mouth contamination are frequently present in large numbers and will prove confusing to the inexperienced. The presence of squamous cells in their neighborhood will suggest their source. Among the pathogenic organisms are: tubercle bacilli; staphylococci and streptococci; pneumococci; bacilli of Friedländer; influenza bacilli; and *Micrococcus catarrhalis*. Of these the tubercle bacillus is the only one whose recognition has great clinical value and the only one which is easily identified in stained smears. Their cultural characteristics are described in Chapter VIII.

(1.) **Tubercle Bacillus.**—The presence of tubercle bacilli may be taken as positive evidence of the existence of tuberculosis somewhere along the respiratory tract, most likely in the lung. In laryngeal tuberculosis they are not easily found in the sputum, but can frequently be detected in swabs made directly from the larynx.

The importance of carefully selecting the portion for examination cannot be too strongly emphasized. It is always best to select the more purulent portions of the sputum, keeping away from the mucoid parts. If bits of necrotic tissue are present they may show immense numbers of tubercle bacilli, when other portions of the specimen contain very few. One must, however, be on his guard against bits of food which resemble these "caseous particles." The specimen should be examined while fresh. It will usually liquefy upon standing, and this, by preventing the selection of particles favorable for examination, will greatly reduce one's chances of finding bacilli.

Recognition of the tubercle bacillus depends upon the fact that it stains with difficulty; but that when once stained, it retains the stain tenaciously, even when treated with a mineral acid, which quickly removes the stain from other bacteria. This "acid-fast" property is due to the presence of a waxy or lipoid substance. A number of the best staining methods are included here. Since Gabbet's method is convenient, inexpensive and widely used in office work, it is given in greater detail than the others. The author, however, would recommend Pappenheim's method for routine work, as least likely to give trouble to the inexperienced. Students rarely fail to get perfect results at the first trial.

Tubercle bacilli can often be found in very poorly
prepared slides, but for dependable results when bacilli
are scarce properly spread, fixed, and stained preparations free from precipitated stain are absolutely essential. The person who is content with an imperfect
preparation.because it is "good enough for diagnosis"
will succeed only in the most obvious cases.

Gabbet's Method.—1. Spread suspicious particles thinly
and evenly upon a slide or a cover-glass held in the grasp of
cover-glass forceps. In general, slides are more satisfactory, but cover-glasses are easier to handle while staining.
Do not grasp a cover too near the edge or the stain will not
stay on it well. Tenacious sputum will spread better if
gently warmed while spreading.

2. Dry the film in the air.

3. Fix the film by immersing in saturated aqueous solution of corrosive sublimate or in methyl alcohol for two or
three minutes, and then rinse well in water. This is much
to be preferred, particularly for beginners, to the usual practice of fixing in the flame (see p. 572). Should the film be
washed off during future manipulations, fixation has been
insufficient.

4. Apply as much carbol-fuchsin (see p. 639) as will stay
on, and hold over a flame so that it will steam for three
minutes or longer, replacing the stain with a dropper as it
evaporates. If the stain is allowed to evaporate completely, the preparation is ruined. If the bacilli are well
stained in this step, there will be little danger of decoloriz-
ing them later. Too great heat will interfere with the
staining of some of the bacilli, probably by destroying the
waxy substance upon which the acidfast property depends.

Recently it has been shown that fifteen to twenty minutes
staining at room temperature will suffice, and this may be

recommended on the score of avoiding precipitates. With
most batches of carbol-fuchsin even five minute's staining
is sufficient.

5. Wash the film in water.

6. Apply Gabbet's stain (see p. 641.) to the under side
of the cover-glass to remove excess of carbol-fuchsin, and
then to the film-side. Allow this to act for one-fourth
to one-half minute.

7. Wash in water.

8. If, now, the thinner portions of the film are blue, proceed to the next step; if they are still red, repeat steps 6
and 7 until the red has disappeared. Too long application
of Gabbet's stain will decolorize the tubercle bacilli.

9. Place the preparation between layers of filter-paper
and dry by rubbing with the fingers, as one would in blotting
ink. Warm over the flame until thoroughly dry.

10. Put a drop of Canada balsam upon a clean slide, place
the cover-glass film side down upon it, and examine with an
immersion objective. Cedar oil or water may be used in
place of balsam for temporary preparations. Smears on
slides may be examined directly with an oil-immersion lens,
no cover being necessary.

Ziehl-Neelson Method.—The objection is often made to
the above method that decolorization is masked by the blue
in Gabbet's stain. Although this will not make trouble if
step 8 is carefully carried out, most experienced workers
prefer the Ziehl-Neelson method. This resembles Gabbet's
method, with the following exceptions: After the staining
with carbol-fuchsin the smear is washed in 5 per cent. nitric
acid (or, better, a mixture of 3 c.c. concentrated hydrochloric
acid and 97 c.c. 70 per cent. alcohol) until decolorized,
washed in water, stained lightly with Löffler's methylene
blue, again washed, and finally dried and mounted.

Pappenheim's Method.—This is the same as Gabbet's

PLATE II.

Fig. 1.—Heart-failure cells and carbon-laden cells in unstained sputum. Two small squamous epithelial cells and four red blood-corpuscles are included for comparison of size. × 200.

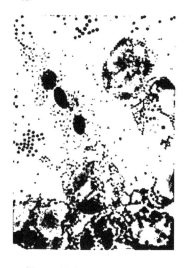

Fig. 2.—Eosinophilic leukocytes and staphylococci in asthmatic sputum. Eosin and methylene-blue. × 1000.

Fig. 3.—Tubercle bacilli, streptococci, pus corpuscles, and mucous threads in tuberculous sputum. Ziehl-Neelson method. × 1000.

Fig. 4.—Much's granules from two fields of a slide stained as described in the text. A group of half-digested staphylococci is also shown. × 1500.

method, except that Pappenheim's methylene blue solution (see p. 641) is substituted for Gabbet's stain.

The method is very satisfactory for routine work. Decolorization of the tubercle bacillus is practically impossible: it retains its red color, even when soaked overnight in Pappenheim's solution. The stain was originally recommended as a means of differentiating the smegma bacillus, which is decolorized by it; but it is not to be absolutely relied upon for this purpose.

In films stained by these methods tubercle bacilli, if present, will be seen as slender red rods upon a blue background of mucus (appearing as delicate threads and strands), granular detritus, and cells (Plate II, Fig. 3). They vary considerably in size, averaging 3 to 4 μ in length—about one-half the diameter of a red blood-corpuscle. Beginners must be warned against mistaking the edges of cells, or particles which have retained the red stain, for bacilli. The appearance of the bacilli is almost always typical, and if there seems room for doubt, the structure in question is probably not a tubercle bacillus. They may lie singly or in groups. They are very frequently bent and often have a beaded appearance. It is possible that the larger, beaded bacilli indicate a less active tuberculous process than do the smaller, uniformly stained ones. Sometimes they are present in great numbers—thousands in a field of the 2-mm. objective. Sometimes, even in advanced cases, several cover-glasses must be examined to find a single bacillus. At times they are so few that none are found in stained smears, and special methods are required to detect them. The number may bear some relation to the severity of the disease, but this relation is by no means constant. The mucoid sputum from an incipient case sometimes contains great numbers, while sputum from large tuberculous cavities at times contains very few. Failure to find them is not conclusive, though their absence is much more significant *when the sputum is purulent than when it is mucoid.*

When it is desired to record the approximate number of bacilli present, the Gaffky table as modified by Brown may be employed, using an oil-immersion lens and 4× ocular:

    I. One to four bacilli to the slide.
    II. Average of one in many fields.
    III. Average of one in a field.
    IV. Average of two to three in a field.
    V. Average of four to six in a field.
    VI. Average of seven to twelve in a field.
    VII. Average of thirteen to twenty-five in a field.
    VIII. Average of about fifty in a field.
    IX. Average of about one hundred in a field.
    X. Enormous numbers in a field.

Since the sputum raised at various times in the day, and even different parts of the same sample, may vary greatly in bacillary content, such a table is of little value unless the twenty-four-hour sputum is collected and uniformly mixed before preparing the slides.

When bacilli are not found in suspected cases, one of the following methods should be tried:

1. **Antiformin Method.**—This has lately come into use, and has superseded the older methods of concentration. The chief difficulty with the older methods, such as boiling with caustic soda, is that the bacilli are so injured in the process that they do not stain characteristically.

Antiformin is a trade name for a preparation consisting essentially of equal parts of a 15 per cent. solution of caustic soda and a 20 per cent. solution of sodium hypochlorite. It keeps fairly well. Substitutes appear to be less satisfactory than the original preparation.

Löffler's method is probably the best for clinical work. It kills the bacilli, so that there is no danger in handling the material. Upon this account, however, it is not applicable to isolation of tubercle bacilli for cultures.

Place 10 to 20 c.c. of the sputum in a small flask, with an equal amount of 50 per cent. antiformin, and heat to the boiling-point. The sputum will be thoroughly liquefied, usually within a few seconds. For each 10 c.c. of the resulting fluid add 1.5 c.c. of a mixture of 1 volume of chloroform and 9 volumes of alcohol. Shake vigorously for several minutes or until emulsification has taken place. The object is to impregnate the lipoid capsule of the bacilli with chloroform, thus increasing their specific gravity. Pour off the emulsion into centrifuge tubes and centrifugalize at high speed for about fifteen minutes. The chloroform will go to the bottom, and the sediment which collects on its surface in a thin firm layer will contain the tubercle bacilli. Pour off the supernatant liquid and transfer the sediment to glass slides, removing the excess of fluid with filter-paper. To facilitate removal of the disk of sediment *in toto* Williamson recommends the use of a centrifuge tube, the lower ½ inch of which is of uniform caliber and the bottom of which is open and plugged with a rubber stopper. Add a little egg-albumen solution (see p. 87) or, better, some of the original sputum, to cause the film to adhere to the slide, mix well, spread into a uniform layer, and finally dry, fix, and stain by the Ziehl-Neelson method. Löffler recommends 0.1 per cent. solution of malachite green for counterstain.

2. **Animal Inoculation.**—Inoculation of guinea-pigs is the court of last appeal in detection of tubercle bacilli, but even

6

this is not infallible for it has been shown that the injected material must contain 10 to 150 bacilli in order to produce tuberculosis in the guinea-pig, the number required depending upon their virulence. The method is described on page 534.

There are a number of bacilli, called **acid-fast bacilli,** which stain in the same way as the tubercle bacillus. They stain with difficulty, and when once stained, retain the color even when treated with a mineral acid; but, unlike the tubercle bacillus, most of them can be decolorized with alcohol. Of these, the smegma bacillus is the only one likely ever to cause confusion. It, or a similar bacillus, is sometimes found in the sputum of gangrene of the lung. It occurs normally about the glans penis and the clitoris, and is often present in the urine and in the wax of the ear. The method of distinguishing it from the tubercle bacillus is given later (see p. 236).

Other bacteria than the acid-fast group are stained blue by Gabbet's and the Ziehl-Neelson method. Those most commonly found are staphylococci, streptococci, and pneumococci. Their presence in company with the tubercle bacillus constitutes *mixed infection*, which is much more serious than single infection by the tubercle bacillus. It is to be remembered, however, that a few of the bacteria may reach the sputum from the upper air-passages and that great numbers are usually present in decomposing sputum. Clinically, mixed infection is evidenced by fever.

Within the past few years much interest has centered in the so-called **"Much's granules."** These are Gram-positive but non-acid-fast granules which are apparently forms of the tubercle bacillus, since material containing them causes tuberculosis when injected into guinea-pigs. They may be present either alone or in company with the ordinary acid-fast form.

It is now fairly well established that Much's granules represent a less virulent form of the tubercle bacillus which is especially frequent in quiescent and mild chronic cases, and that they give place to the acid-fast forms when such cases become active. Their detection is therefore important, but it is not easy because of other granules—precipitated stain, micrococci, etc.—which may be mistaken for the true Much bodies. The following method, while somewhat complicated, reduces the chance for error to the minimum.

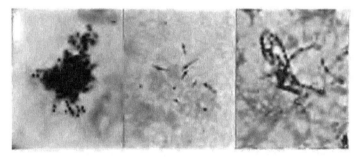

Fig. 22.—Much's granules in sputum stained by the method detailed in the text (× 1500).

Staining Method for Much's Granules.—1. To the twenty-four-hour amount of sputum add an equal volume of 0.6 per cent. sodium carbonate solution, shake thoroughly, and allow to stand in a warm place (preferably the incubator) for twenty-four hours. If it is not then completely homogeneous extend the time to forty-eight hours.

2. Centrifuge thoroughly, remove half of the supernatant fluid and mix an equal volume of 30 per cent. antiformin with the remaining half. Allow this to act for twenty minutes. It is imperative that the antiformin be fresh.

3. Centrifugalize, and make smears from the sediment.

4. Stain one smear by the Ziehl-Neelson method. This will demonstrate the ordinary tubercle bacilli, if present, and will also serve to show whether any cocci have been left undigested.

5. Immerse the remaining smears for forty-eight hours in a stain consisting of:

> Carbol-fuchsin.................... 3 parts
> Carbol-methyl violet.............. 1 part

This stain remains good for about two weeks. The carbol-methyl violet used for this purpose consists of 2 per cent. phenol, 9 parts; saturated alcoholic solution of methyl violet, 1 part. In order to avoid precipitates smears should stand on edge while in the stain.

6. Rinse gently in water.

7. Cover with Gram's iodin solution for five minutes, warming until steam rises.

8. Decolorize successively with 5 per cent. nitric acid for one minute, with 3 per cent. hydrochloric acid for ten seconds, and finally with a mixture of equal parts of acetone and 95 per cent. alcohol until color ceases to come off.

9. Dry, mount in cedar oil or balsam and examine with the oil-immersion lens.

Steps 1 and 2 in this method serve the double purpose of concentrating the sputum and of digesting any micrococci which may be present and which might be confused with Much's granules. One must be extremely cautious in interpreting isolated granules if any undigested cocci are found in the control slide. Partially digested cocci which take the color of the background will not cause confusion. The concentration and digestion may be omitted if desired, but results are then much less dependable.

Much's granules (Plate II, Fig. 4) are definite, clean-cut, round or oval bodies about 0.5 μ in diameter. They are thus about half the diameter of a staphylococcus. Ordinarily they are a deep purple, often with a tinge of red. They may lie singly or, more frequently, in rows of two to five. Connecting the granules can usually be seen a faint bluish or reddish band suggesting the body of a bacillus in or on which the granules lie. Isolated granules usually appear to lie at the end or in the middle of such a band, and unless the band is seen, they should not be accepted as true Much's granules.

(2) **Staphylococcus and Streptococcus** (see p. 516).— One or both of these organisms is commonly present in company with the tubercle bacillus in the sputum of advanced phthisis (Plate II, Figs. 2 and 3). They are often found in bronchitis, catarrhal pneumonia, and many other conditions. The streptococcus is a common cause of severe sore throat and tonsillitis.

(3) **Pneumococcus (Diplococcus of Fränkel).**—The pneumococcus is the causative agent in nearly all cases of croupous pneumonia, and is commonly found in large numbers in the rusty sputum of this disease. It is frequently met with in the sputum of catarrhal pneumonia, bronchitis, and tuberculosis. It is also an important factor in the causation of pleurisy, meningitis, otitis media, and other inflammations. It is frequently present in the saliva in health. Pneumococci are about the size of streptococci. They are ovoid in shape, and lie in pairs, end to end, often forming short chains. Each is

surrounded by a gelatinous capsule, which is its distinctive feature (Fig. 23).

The pneumococcus is closely related to the streptococcus, and it is sometimes extremely difficult to differentiate them even by culture methods (for which see p. 581). The morphology of the pneumococcus, the fact that it is Gram-positive, and the presence of a capsule are, however, generally sufficient for its recognition

FIG. 23.—*Diplococcus pneumoniæ* in the blood (× 1000) (Fränkel and Pfeiffer).

in smears from sputum or pus. The capsule is often seen as a halo around pairs of cocci in smears stained by the ordinary methods, particularly Gram's method, but to show it well special methods are required. There are numerous special methods of staining capsules which are applicable to other encapsulated bacteria, as well as to the pneumococcus, but few of them are satisfactory. Buerger's method can be recommended. It is especially useful with cultures upon serum media, but is applicable also to the sputum. Smith's and Rosenow's methods are easier of application, and apparently give uniformly good results. The sputum should be fresh—not more than three or four hours old.

Buerger's Method for Capsules.—1. Mix a few drops each of the sputum and blood-serum or egg-albumen solution (egg-albumen, distilled water, equal parts; shake, filter through cotton, and add about 0.5 per cent. phenol). Blood-serum can be obtained as described for the Widal test (see p. 606). Make thin smears from the mixture, and just as the edges begin to dry, cover with Müller's fluid (potassium dichromate, 2.5 Gm.; sodium sulphate, 1.0 Gm.; water, 100 c.c.) saturated with mercuric chlorid (ordinarily about 5 per cent.). Gently warm over a flame for about three seconds.

2. Rinse very quickly in water.

3. Flush once with alcohol.

4. Apply tincture of iodin for one to two minutes.

5. Thoroughly wash off the iodin with alcohol and dry in the air.

6. Stain about three seconds with weak anilin-gentian violet freshly made up as follows: Anilin oil, 10; water, 100; shake; filter; and add 5 c.c. of a saturated alcoholic solution of gentian violet.

7. Rinse off the stain with 2 per cent. solution of sodium chlorid, mount in this solution, and examine with an oil-immersion objective.

Buerger suggests a very useful variation as follows: After the alcohol wash and drying, the specimen is stained by Gram's method (see p. 572), counterstained with aqueous solution of fuchsin, washed, and mounted in water. The pneumococcus holds the purple stain, while all capsules take the pink counterstain.

W. H. Smith's Method.—1. Make thin smears of the sputum or other material, which should be as fresh as possible.

2. Fix in the flame in the usual manner.

3. Apply a 10 per cent. aqueous solution of phosphomolybdic acid (Merck) for four to five seconds.

4. Rinse in water.

5. Apply anilin-gentian violet (see p. 640), steaming gently for fifteen to thirty seconds.

6. Rinse in water.

7. Apply Gram's iodin solution, steaming gently for fifteen to thirty seconds.

8. Wash in 95 per cent. alcohol until the purple color ceases to come off.

9. Rinse in water.

10. Apply a 6 per cent. aqueous solution of eosin (Grübler, w. g.), and gently warm for one-half to one minute.

11. Rinse in water.

12. Wash in absolute alcohol.

13. Clear in xylol.

14. Mount in balsam.

This is essentially Gram's method (see p. 572), preceded by treatment with phosphomolybdic acid and followed by eosin. Gram-positive bacteria like the pneumococcus are deep purple; capsules are pink and stand out clearly.

When the method is applied to Gram-negative bacteria, steps five to nine inclusive are omitted; and between steps eleven and twelve the preparation is counterstained with Löffler's methylene blue, gently warming for fifteen to thirty seconds.

Rosenow's Method.—This is the same as Smith's with the exception that a 10 per cent. solution of tannic acid, applied while the film is still moist and allowed to act for ten to twenty seconds takes the place of the heat and phosphomolybdic acid in steps 2 and 3.

(4) **Bacillus of Friedländer (Bacillus mucosus capsulatus).**—In a small percentage of cases of pneumonia

this organism is found alone or in company with the pneumococcus. Its pathologic significance is uncertain. It is often present in the respiratory tract under normal conditions. Friedländer's bacilli are non-motile, encapsulated rods, sometimes arranged in short chains (Fig. 24). Very short individuals in pairs closely resemble pneumococci, from which they are distinguished by the fact that they are Gram-decolorizing.

FIG. 24.—Friedländer's bacillus in pus from pulmonary abscess (X about 1000). (Boston.)

(5) **Bacillus of Influenza.**—This is the etiologic factor in true influenza, although conditions which are clinically similar or identical may be caused by the pneu-

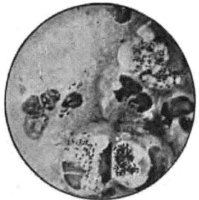

FIG. 25.—Bacillus of influenza; cover-glass preparation of sputum from a case of influenza, showing the bacilli in leukocytes; highly magnified (Pfeiffer).

mococcus, streptococcus, or *Micrococcus catarrhalis*. It is present, often in large numbers, in the nasal and bronchial secretions, and is also found in the local lesions following influenza. Chronic infection by influenza bacilli may be mistaken clinically for tuberculosis, and they should be searched for in all cases of obstinate chronic bronchitis.

Their recognition depends upon the facts that they are extremely small bacilli; that most of them lie within the pus-cells; that their ends stain more deeply than their centers, sometimes giving the appearance of minute diplococci; and that they are decolorized by Gram's method of staining (Figs. 25 and 212).

They are well stained by dilute fuchsin or by Pappenheim's pyronin-methyl green, but are more certainly recognized by Gram's method with the pyronin-methyl green for counterstain.

(6) **Bacillus pertussis.**—The whooping-cough bacillus is a minute, ovoid, Gram-negative bacillus which stains feebly with the ordinary dyes, and sometimes, though not usually, lies within pus cells. It can be demonstrated by the method given for the influenza bacillus.

(7) **Micrococcus catarrhalis.**—This organism is frequently present in the sputum in inflammatory conditions of the respiratory tract resembling influenza. It is sometimes present in the nasal secretions in health. It is a Gram-negative diplococcus, frequently intracellular, and can be distinguished from the meningococcus and gonococcus only by means of cultures (Fig. 26). The staining method recommended for the influenza bacillus is best. It grows readily on ordinary media.

2. Cells.—These include pus-corpuscles, epithelial cells, and red blood-corpuscles.

(1) **Pus-corpuscles** are present in every sputum, and at times the sputum may consist of little else. They are the polymorphonuclear leukocytes of the blood, and appear as rounded cells with several nuclei or one very irregular nucleus (Plate II, Fig. 3). They are often much disintegrated. Occasional lymphocytes are usu-

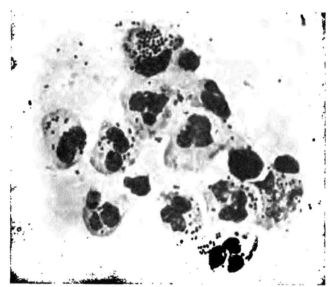

FIG. 26.—*Micrococcus catarrhalis* in smear from sputum (F. T. Lord; photo by L. S. Brown).

ally present. Their predominance is suggestive of early or mild tuberculosis.

Eosinophilic cells are quite constantly found in large numbers in the sputum of bronchial asthma near the time of the paroxysm, and constitute one of the most distinctive features of the sputum of this disease. They resemble ordinary pus-corpuscles, except that their cytoplasm is filled with coarse granules having a marked affinity for eosin. It is worthy of note that many of them, sometimes the majority, are mononuclear. Large numbers of free granules, derived from dis-integrated cells, are also found (Plate II, Fig. 2).

Ordinary pus-cells are easily recognized in sputum stained by any of the methods already given. For eosinophilic cells, some method which includes eosin must be used. A simple method is to stain the dried and fixed film two or three minutes with saturated solution of eosin, and then one-half to one minute with Löffler's methylene blue; nuclei and bacteria will be blue, eosinophilic granules bright red. Either Wright's or Jenner's stain (p. 313) will be found satisfactory.

(2) **Epithelial cells** may come from any part of the respiratory tract. A few are always present, since des-quamation of cells goes on constantly. Their recognition is important chiefly as an aid in deciding upon the source of the portion of the sputum in which they are found. In suspected lung conditions it is manifestly useless to study material from the nose only, yet this is not infrequently done. They have little diagnostic value, although a considerable excess would indicate a pathologic condition at the site of their origin. Any of the stains mentioned above will show them, and they can usually be identified in unstained sputum. In general, three forms are found:

(a) *Squamous Cells.*—Large, flat, polygonal cells with a comparatively small nucleus (Fig. 27, *i*). They come from the upper air-passages, and are especially numerous in laryngitis and pharyngitis. They are frequently studded with bacteria—most commonly diplococci.

(*h*) *Cylindric Cells from the Nose, Trachea, and Bronchi* (Fig. 27, *f*, *h*).—These are not usually abundant, and, as a rule, they are not identified because much altered from their original form, being usually round and swollen. When very fresh, they may retain their cylindric form, sometimes bearing cilia in active motion.

Fig. 27.—Different morphologic elements of the sputum (unstained): *a*, *b*, *c*, Pulmonary or alveolar epithelium—*a*, with normal lung pigment (carbon); *b*, with fat-droplets; *c*, with myelin globules; *d*, pus-corpuscles; *e*, red blood-corpuscles; *f*, cylindric beaker-shaped bronchial epithelial cells; *g*, free myelin globules; *h*, ciliated epithelium of different kinds from the nose, altered by coryza; *i*, squamous cells from the pharynx (after Bizzozero).

(*c*) *Alveolar Cells.*—Rather large, round, or oval cells, three to six times the diameter of a red corpuscle, with one or two round nuclei (Fig. 27). Their source is presumably the pulmonary alveoli. It is probable that many of the cells which have been included in this group are really large mononuclear leukocytes.

(3) **Red blood-corpuscles** may be present in small numbers in almost any sputum. When fairly constantly present in considerable numbers, they are suggestive of phthisis. The corpuscles, when fresh, can easily be recognized in unstained sputum or may be shown by any of the staining methods which include eosin. They are, however, commonly so much degenerated as to be unrecognizable and often only altered blood-pigment is left. Ordinarily, blood in the sputum is sufficiently recognized with the naked eye.

III. CHEMIC EXAMINATION

There is little to be learned from a chemic examination, and it is rarely undertaken. Recently, however, it has been shown that the presence or absence of albumin may have clinical significance. Albumin is almost constantly present in the sputum in pneumonia, pulmonary edema, and tuberculosis. It is usually absent in bronchitis. A test for albumin may, therefore, be of some value in distinguishing between bronchitis and tuberculosis. It is carried out as follows:

Method for Albumin in Sputum.—1. To 10 c.c. of the sputum add 30 c.c. of 1 per cent. acetic acid and shake until thoroughly mixed. This may be done in a stoppered bottle. Dilution and addition of acetic acid precipitates the mucus.

2. Filter through filter-paper.

3. Test the filtrate for albumin qualitatively and quantitatively, as described in the chapter upon the urine.

Active cases of phthisis, whether early or far advanced, generally show 0.2 per cent. or more albumin; slightly active cases, less than 0.2 per cent. The sputum must be fresh, otherwise a negative reaction may have changed to positive owing to disintegration of cells.

IV. THE SPUTUM IN DISEASE

Strictly speaking, any appreciable amount of sputum is abnormal. A great many healthy persons, however, raise a small quantity each morning, owing chiefly to the irritation of inhaled dust and smoke. Although not normal, this can hardly be spoken of as pathologic. It is particularly frequent in city dwellers and in those who smoke cigarettes to excess. In the latter the amount is sometimes so great as to arouse suspicion of tuberculosis. Such "normal morning sputum" generally consists of small, rather dense, mucoid masses, translucent-white, or, when due to inhaled smoke, gray in color. Microscopically, there are a few pus-corpuscles, and, usually, many alveolar cells, both of which may contain carbon particles. The alveolar cells commonly show myelin degeneration, and free myelin globules may be present in large numbers. Saprophytic bacteria may be present, but are not abundant.

1. Acute Bronchitis.—There is at first a small amount of tenacious, almost purely mucoid sputum, frequently blood streaked. This gradually becomes more abundant, mucopurulent in character, and yellowish or gray in color. At first the microscope shows a few leukocytes and alveolar and bronchial cells; later the leukocytes become more numerous. Bacteria are not usually abundant.

2. Chronic Bronchitis.—The sputum is usually abundant, mucopurulent, and yellowish or yellowish-

green in color. Nummular masses like those of tuberculosis are sometimes seen. Microscopically, there
are great numbers of leukocytes, often much disintegrated. Epithelium is not abundant. Bacteria of
various kinds, especially staphylococci, are usually
numerous.

In fibrinous bronchitis there are found, in addition,
fibrinous casts, usually of medium size.

In the chronic bronchitis accompanying long-continued passive congestion of the lungs, as in poorly
compensated heart disease, the sputum may assume a
rusty brown color, owing to presence of large numbers
of the "heart-failure cells" previously mentioned.

3. Bronchiectasis.—When there is a single large
cavity, the sputum is very abundant at intervals—
sometimes as high as a liter in twenty-four hours—and
has a very offensive odor. It is thinner than that of
chronic bronchitis, and upon standing separates into
three layers of pus, mucus, and frothy serum. It contains great numbers of miscellaneous bacteria.

4. Gangrene of the Lung.—The sputum is abundant, fluid, very offensive, and brownish in color. It
separates into three layers upon standing—a brown
deposit, a clear fluid, and a frothy layer. Microscopically, few cells of any kind are found. Bacteria are extremely numerous; among them may sometimes be
found an acid-fast bacillus probably identical with the
smegma bacillus. As before stated, elastic fibers are
usually present in large fragments.

5. Pulmonary Edema.—Here there is an abundant,
watery, frothy sputum, varying from faintly yellow or
pink to dark brown in color; a few leukocytes and

epithelial cells and varying numbers of red blood-corpuscles are found with the microscope.

6. Bronchial Asthma.—The sputum during and following an attack is scanty and very tenacious. Most characteristic is the presence of Curschmann's spirals, Charcot-Leyden crystals, and eosinophilic leukocytes.

7. Croupous Pneumonia.—Characteristic of this disease is a scanty, rusty red, very tenacious sputum, containing red corpuscles or altered blood-pigment, leukocytes, epithelial cells, usually many pneumococci, and often very small fibrinous casts. This sputum is seen during the stage of red hepatization. During resolution the sputum assumes the appearance of that of chronic bronchitis. When pneumonia occurs during the course of a chronic bronchitis, the characteristic rusty red sputum may not appear.

8. Pulmonary Tuberculosis.—The sputum is variable. In the earliest stages it may appear only in the morning, and is then scanty and almost purely mucoid, with an occasional yellow flake; or there may be only one very small mucopurulent mass. When the quantity is small, there may be no cough, the sputum reaching the larynx by action of the bronchial cilia. This is not well enough recognized by practitioners. A careful inspection of all the sputum brought up by the patient on several successive days, and a microscopic examination of all yellow portions, will not infrequently establish a diagnosis of tuberculosis when physical signs are negative. Intelligent coöperation of the patient is essential in such cases. Tubercle bacilli will sometimes be found in large numbers at this stage. Blood-streaked sputum is strongly suggestive of tuberculosis,

7

and is more common in the early stages than later. It usually indicates an advancing process.

The sputum of more advanced cases resembles that of chronic bronchitis, with the addition of tubercle bacilli and elastic fibers. Nummular masses—circular, "coin-like" disks, which sink in water—may be seen. Caseous particles containing immense numbers of the bacilli are common. Far-advanced cases with old cavities often show rather firm, spheric or ovoid grayish masses in a thin fluid—the so-called "globular sputum." These globular masses usually contain many tubercle bacilli. Considerable hemorrhages are not infrequent, and for some time thereafter the sputum may contain clots of blood or be colored brown.

CHAPTER II

THE URINE

Preliminary Considerations.—The urine is an extremely complex aqueous solution of various organic and inorganic substances. Most of the substances are either waste-products from the body metabolism or products derived directly from the foods eaten. Normally, the total amount of solid constituents carried off in twenty-four hours is about 60 Gm., of which the organic substances make up about 35 Gm. and the inorganic about 25 Gm.

The most important organic constituents are urea, uric acid, creatinin and ammonia. Urea constitutes about one-half of all the solids, or about 30 Gm. in twenty-four hours.

The chief inorganic constituents are the chlorids, phosphates, and sulphates. The chlorids, practically all in the form of sodium chlorid, make up about one-half of the inorganic substances, or about 13 Gm., in twenty-four hours.

Certain substances appear in the urine only in pathologic conditions. The most important of these are proteins, sugars, acetone, and related substances, bile, hemoglobin, and the diazo substances.

In addition to the substances in solution all urines contain various microscopic structures.

While, under ordinary conditions, the composition of

urine does not vary much from day to day, it varies greatly at different hours of the same day. It is evident, therefore, that the **collection of the specimen** is important and that *no quantitative test can be of value unless a sample of the mixed twenty-four-hour urine be used.* The patient should be instructed to void all the urine during the twenty-four hours into a clean vessel kept in a cool place, to mix it well, to measure the whole quantity, and to bring 8 or more ounces for examination. In order to avoid annoying misunderstandings, it is well to make these directions specific, telling him to empty the bladder at a specified time, say 8 a.m. and to discard this urine, to save all the urine voided up to 8 a.m. of the next day and at that time to empty the bladder whether he feels the need for it or not, and to add this final amount to the quantity collected. When it is desired to make only qualitative tests, as for albumin or sugar, a "sample" voided at random will answer. It should be remembered, however, that urine passed about three hours after a meal is most likely to contain pathologic substances. That voided first in the morning is least likely to contain them. To diagnose cyclic albuminuria samples obtained at various periods during the twenty-four hours must be examined.

The urine must be examined while fresh. **Decomposition** sets in rapidly, especially in warm weather, and greatly interferes with all the examinations. Decomposition may be delayed by adding 5 gr. of boric acid (as much of the powder as can be heaped upon a ten-cent piece) for each 4 ounces of urine. Formalin, in proportion of 1 drop to 4 ounces, is also an efficient preservative, but if larger amounts be used, it may give reactions for sugar and albumin, and is likely to cause a precipitate which greatly interferes with the microscopic examination. Thymol, toluol, and chloroform are likewise much used. The use of thymol is very convenient. A small lump, floating upon the surface, will preserve a bottle of urine for some days, but enough may dissolve to simulate the albumin reaction. The chief objection to toluol is the fact that it floats upon the surface, and the urine must be pipeted from beneath it. Chloroform is probably the least satisfactory. It reduces Fehling's solution; and it settles to the bottom in the form of globules which it is impossible to avoid when removing the sediment for microscopic examination. One of these preservatives may be placed in the vessel when collection of the twenty-four-hour sample is begun. Whenever possible the urine should be kept on ice.

Normal and **abnormal pigments,** which interfere with certain of the tests, can be removed by filtering the urine through animal charcoal, or precipitating with a solution of normal acetate of lead (sugar of lead) or with powdered lead acetate in substance and filtering.

Certain cloudy urines cannot be **clarified** by ordinary filtration through paper, particularly when the cloudiness is due to bacteria. Such urines can usually be rendered perfectly clear by adding a small amount of purified talc or infusorial earth, shaking well, and filtering.

A suspected fluid can be **identified as urine** by detecting any considerable quantity of urea in it (see p. 136). Traces of urea may, however, be met with in ovarian cyst fluid, while urine from very old cases of hydronephrosis may contain little or none.

The **frequency of micturition** is often suggestive in diagnosis. Whether it is unduly frequent can best be ascertained by asking the patient whether he has to get up at night to urinate. Increased frequency may be due to restlessness; to increased quantity of urine; to irritability of the bladder, usually an evidence of cystitis; to obstruction ("retention with overflow"); or to paralysis of the sphincter.

Clinical examination of the urine may conveniently be considered under five heads: I. General characteristics. II. Functional tests. III. Chemic examination. IV. Microscopic examination. V. The urine in disease.

I. GENERAL CHARACTERISTICS

1. Quantity.—The quantity passed in twenty-four hours varies greatly with the amount of liquids ingested, perspiration, etc. The normal may be taken as 1000 to 1500 c.c., or 35 to 50 ounces for an adult in this country. German writers give higher figures. For children the amount is somewhat greater in proportion to body weight.

The quantity is increased (polyuria) during absorption of large serous effusions and in many nervous conditions. It is usually much increased in chronic interstitial nephritis, diabetes insipidus, and diabetes mellitus. In these conditions a permanent increase in amount of urine is fairly constant—a fact of much value in diagnosis. In diabets mellitus the urine may, though rarely, reach the enormous amount of 50 liters.

The quantity is decreased (oliguria) in severe diarrhea; in fevers; in all conditions which interfere with circulation in the kidney, as poorly compensated heart disease;

in the parenchymatous forms of nephritis; and during
accumulation of fluid in the serous cavities. In uremia
the urine is usually very greatly decreased and may be
entirely suppressed (anuria).

Ordinarily, more urine is voided during the day than
during the night, the normal ratio being about 100 to 50
or 60. In certain diseases, notably arteriosclerosis and
cardiac and renal disease, conditions are reversed, and
the night urine (7 p.m. to 7 a.m.) equals or exceeds that
passed during the day.

2. Color.—This varies considerably in health, and
depends largely upon the quantity of urine voided, dilute urines being pale and concentrated urines, highly
colored. The usual color is yellow or reddish yellow,
due to the presence of several pigments, chiefly urochrome, which is yellow. Traces of hematoporphyrin,
uroerythrin, and urobilin are frequent. Uroerythrin
is chiefly responsible for the deep reddish tinge of urine
in acute fevers. Urobilin and hematoporphyrin have
clinical significance and are discussed later (see p. 184).
Acid urine is generally darker than alkaline. For
the sake of uniformity in recording the color, Vogel's
scale is widely used, the urine being filtered and examined by transmitted light in a glass 3 or 4 inches in
diameter. This scale uses nine colors: pale yellow,
light yellow, yellow, reddish yellow, yellowish red, red,
brownish red, reddish brown, and brownish black.

Color is sometimes greatly changed by abnormal
pigments. Blood-pigment gives a red or brown, smoky
color. Urine containing bile is yellowish or brown, with
a yellow foam when shaken. It may assume a greenish
hue after standing, owing to oxidation of bilirubin into

biliverdin. Ingestion of small amounts of methylene blue gives a pale green; large amounts give a marked greenish blue. Santonin produces a yellow; rhubarb, senna, cascara, and some other cathartics, a brown color; these change to red upon addition of an alkali, and if the urine be alkaline when voided, may cause suspicion of hematuria. A bright pink or red color appearing when the urine is alkalinized may be due to phenolphthalein. Thymol gives a yellowish green. Following poisoning from phenol and related drugs the urine may have a normal color when voided, but becomes olive green to brownish black upon standing. In susceptible individuals therapeutic doses of creosote, or absorption from carbolized dressings, may cause this change. Urine which contains melanin, as sometimes in melanotic tumors, and very rarely in wasting diseases, also becomes brown or black upon long standing. A similar darkening upon exposure to the air occurs in alkaptonuria (see p. 183). A milky color may be due to presence of chyle, or milk may have been added by a malingering patient.

A pale greenish urine with high specific gravity strongly suggests diabetes mellitus.

3. Transparency.—Freshly passed normal urine is clear. Upon standing, a faint cloud of mucus, leukocytes, and epithelial cells settles to the bottom—the so-called "nubecula." This is more abundant in women owing to vaginal cells and mucus. In urines of high specific gravity it may float near the middle of the fluid.

Abnormal cloudiness is usually due to presence of phosphates, urates, pus, blood, or bacteria. Epithelial cells and tube-casts are rarely present in sufficient number to produce more than a slight cloudiness although they may add to turbidity due to other causes.

Amorphous phosphates are precipitated in neutral or alkaline urine. They form a white cloud and sediment, which disappear upon addition of an acid.

Amorphous urates are precipitated only in acid urine. They form a white or pink cloud and sediment ("brick-dust deposit"), which disappear upon heating.

Pus resembles amorphous phosphates to the naked eye. Its nature is easily recognized with the microscope, or by adding a strong solution of caustic soda to the sediment, which is thereby transformed into a gelatinous mass (Donné's test).

Blood gives a reddish or brown, smoky color, and may be recognized with the microscope or by tests for hemoglobin.

Bacteria, when present in great numbers, give a uniform cloud, which cannot be removed by ordinary filtration. They are detected with the microscope.

The cloudiness of decomposing urine is due mainly to precipitation of phosphates and multiplication of bacteria.

4. Odor.—The characteristic aromatic odor has generally been attributed to volatile acids, but a newly discovered substance, called "urinod," has more recently been held responsible. The odor is most marked in concentrated urines. During decomposition the odor becomes ammoniacal. A fruity odor is sometines noted in diabetes, due probably to acetone. Urine which contains cystin may develop an odor of sulphureted hydrogen during decomposition.

Various articles of diet and drugs impart peculiar odors. Notable among these are asparagus, which gives a characteristic offensive odor, and turpentine, which imparts an odor somewhat suggesting that of violets.

5. Reaction.—Normally, the mixed twenty-four-hour urine is slightly acid in reaction. The acidity sometimes increases for a time after the urine is voided, the so-called "acid fermentation." The acidity was formerly held to be due wholly to acid phosphates, but Folin has shown that the acidity of a clear urine is ordinarily greater than the acidity of all the phosphates, the excess being due to free organic acids. Individual samples may be slightly alkaline, especially after a full meal; or they may be amphoteric, turning red litmus-paper blue and blue paper red, owing to presence of both alkaline and acid phosphates. The reaction is ordinarily determined by means of litmus-paper, which, however, is worthless unless of good quality. That put up in vials by Squibb can be recommended.

Acidity is increased after administration of certain drugs, by excess of protein in the diet, and whenever the urine is concentrated from any cause, as in fevers. A strongly acid urine may cause frequent micturition because of its irritation. This is often an important factor in the troublesome enuresis of children.

The urine always becomes alkaline upon long standing, owing to decomposition of urea with formation of ammonia.. If markedly alkaline when voided, it usually indicates such "ammoniacal decomposition" in the bladder, which is the rule in chronic cystitis, especially that due to paralysis or obstruction. Alkalinity due to ammonia, *volatile alkalinity*, can be recognized by the odor or by the fact that litmus-paper turned blue by the urine again becomes red upon gentle heating, or that the paper will turn blue when held in the steam over the boiling urine. A second form of alkalinity, *fixed alkalinity*, is due to alkaline salts, and is often observed during frequent vomiting, after the crisis of pneumonia, in various forms of anemia, during digestion of full meals, after abundant eating of fruits, and after administration of certain drugs, especially salts of vegetable acids.

Quantitative estimation of acidity of urine is not of much clinical value. When, however, it is desired to make it, the method of Folin will be found satisfactory. In every case the sample must be from the mixed twenty-four-hour urine and as fresh as possible.

Folin's Method.—Into a small flask measure 25 c.c. of the urine and add 1 or 2 drops 0.5 per cent. alcoholic solution of phenolphthalein and 15 or 20 Gm. of neutral potassium oxalate. Shake for a minute, and immediately titrate with decinormal sodium hydroxid, shaking meanwhile, until the first permanent pink appears. Read off from the buret the amount of decinormal sodium hydroxid solution added, and calculate the number of cubic centimeters which would be required for the entire twenty-four hours' urine. Most estimations run between 25 and 40 c.c. of decinormal solution for 100 c.c. of urine. Folin places the normal acidity for the twenty-four hour specimen at 554 to 669 c.c. of decinormal solution, but most other authors give lower figures. Much depends upon the diet.

6. Specific Gravity.—In a general way this varies inversely with the quantity of urine. The normal average is about 1.017 to 1.020. Samples of urine taken at random may go far above or below these figures, hence a sample of the mixed twenty-four-hour urine should always be used.

Pathologically, it may vary from 1.001 to 1.060. It is low in chronic interstitial nephritis, diabetes insipidus, and many functional nervous disorders. It is high in fevers and in parenchymatous disease of the kidney. In any form of nephritis a sudden fall without a corresponding increase in quantity of urine may foretell ap-

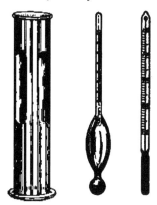

FIG. 28.—Squibb's urinometer with thermometer and cylinder.

proaching uremia. It is highest in diabetes mellitus. A high specific gravity when the urine is not highly colored, or when the quantity is above the normal, should lead one to suspect this disease. A normal specific gravity does not, however, exclude it.

The specific gravity is most conveniently estimated by means of the urinometer (Fig. 28). Squibb's urinometer is adjusted to give accurate readings at 22.5°C.; most other instruments, at 15°C. If the urine be brought to about the right temperature, a correction for

temperature will seldom be necessary in clinical work.
For accuracy, however, it is necessary to add 0.001 to
the urinometer reading for each 3°C. above the temperature for which the urinometer is standardized, and
to subtract 0.001 for each 3°C. below that point. Care
should be taken that the urinometer does not touch the
side of the tube, and that air-bubbles are removed from

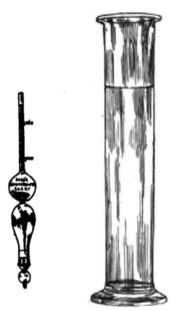

FIG. 29.—Saxe's urinopyknometer and jar for same.

the surface of the urine. Bubbles are easily removed
with a strip of filter-paper. With most instruments the
reading is taken from the bottom of the meniscus. A
long scale on the stem is desirable, because of the greater
ease of accurate reading. Many of the urinometers on
the market are too small to be of any real value.

One frequently wishes to ascertain the specific gravity of quantities of fluid too small to float a urinometer. A simple device for this purpose, which requires only about 3 c.c. and is very satisfactory in clinical work, has been designed by Saxe (Fig. 29). The urine is placed in the bulb at the bottom, the instrument is floated in distilled water, and the specific gravity is read off from the scale upon the stem.

7. Total Solids.—An estimation of the total amount of solids which pass through the kidneys in twenty-four hours is, in practice, one of the most useful of urinary examinations. The normal for a man of 150 pounds is about 60 Gm., or 950 gr. The principal factors which influence this amount are body weight (except with excessive fat), diet, exercise, and age, and these should be considered in making an estimation. After about the forty-fifth year it becomes gradually less; after the seventy-fifth it is about one-half the amount given.

In disease the amount of solids depends mainly upon the activity of metabolism and the ability of the kidneys to excrete. An estimation of the solids, therefore, furnishes an important clue to the functional efficiency of the kidneys. The kidneys bear much the same relation to the organism as does the heart; they cause no direct harm so long as they are capable of performing the work required of them. When, however, through either organic disease or functional inactivity, they fail to carry off their proportion of the waste-products of the body, some of these products must either be eliminated through other organs, where they cause irritation and disease, or be retained within the body, where they act as poisons. The great importance of these poisons in production of distressing symptoms and even organic disease is not well enough recognized by most practitioners. Disappearance of unpleasant and perplexing symptoms as the urinary solids rise to the normal under proper treatment is often most surprising.

When, other factors remaining unchanged, the amount of solids eliminated is considerably above the normal, increased destructive metabolism may be inferred.

The total solids can be estimated roughly, but accurately enough for most clinical purposes, by multiplying the last two figures of the specific gravity of the mixed twenty-four-hour urine by the number of ounces voided and to the product adding one-tenth of itself. This gives the amount in grains. If, for example, the twenty-four-hour quantity is 3 pints or 48 ounces, and the specific gravity is 1.018, the total solids would approximate 950 gr., as follows:

$$48 \times 18 = 864; \quad 864 + 86.4 = 950.4.$$

This method is especially convenient for the practitioner, because patients nearly always report the amount of urine in pints and ounces, and it avoids the necessity of converting into the metric system. Häser's method, which uses the metric system, is more widely used, but is less convenient. The last two figures of the specific gravity are multiplied by 2.33. The product is then multiplied by the number of cubic centimeters voided in twenty-four hours and divided by 1000. This gives the total solids in grams.

II. FUNCTIONAL TESTS

Within the past few years much thought has been devoted to methods of more accurately ascertaining the functional efficiency of the kidneys, especially of one kidney when removal of the other is under consideration. The most promising of the methods which have been devised are cryoscopy, electric conductivity, the phloridzin test, the methylene-blue test, and the phenolsulphonephthalein test. It is doubtful whether, except in the case of the last, these yield any more information than can be had from an intelligent consideration of the specific gravity and the twenty-four-hour quantity, together with a microscopic examination. They are most useful when the urines obtained from separate kidneys by segregation or ureteral catheterization are compared. Only the phenol-sulphonephthalein test will be given here. The reader is referred to larger works upon urinalysis for the others.

Phenolsulphonephthalein Test.—This test, which was offered by Rowntree and Geraghty in 1910, consists in the intramuscular injection of a solution of phenolsulphonephthalein, a drug which is eliminated only by the kidneys, and whose amount in the urine is easily estimated by colorimetric methods. The time of its first appearance in the urine and the quantity eliminated within a definite period are taken as a measure of the functional capacity of the kidneys. The test is harmless, comparatively simple, and apparently reliable. It will sometimes reveal a very serious degree of renal failure when twenty-four-hour quantity, total solids, and urea are practically normal.

Technic.—The original procedure, in which the patient was catheterized when the drug was injected and the catheter was left in place until the drug was detected in the urine, is now seldom followed. The catheter is still used if there be obstruction to the outflow of urine but ordinarily it is dispensed with and the procedure is as follows:

1. Give the patient 300 to 400 c.c. (about 2 glasses) of water to promote secretion of urine.

2. Twenty minutes afterward have him empty his bladder and discard the urine. Then, with a hypodermic syringe inject exactly 1 c.c. of the sterile phenolsulphone-phthalein solution[1] intramuscularly; preferably in the lumbar region.

3. In exactly one hour from the time of the injection, have the patient empty his bladder and save all the urine.

4. In two hours after the injection have the patient empty his bladder again, and save all the urine in a separate container. He should be under observation during the two-hour period, else it is difficult to make sure that he carries out his instructions exactly.

5. Estimate the output of phenolsulphonephthalein in each of the two portions of urine separately as described below.

Estimation of Output.—To each of the two portions of urine add sufficient sodium hydroxid solution to bring out the maximum purple-red color. Dilute each portion to exactly 1000 c.c. and estimate the amount of the drug contained in each by comparing the color with that of an alkalinized standard solution. The result is recorded in terms of the percentage of the amount injected.

In detail, this is done as follows:

---

[1] This solution may be obtained of any druggist. It is sold in 1-c.c. ampoules, sterilized ready for use; but it should be noted that these ampoules contain somewhat more than 1 c.c. hence one should not inject the entire contents.

8

1. Add 1 c.c. of the original phenolsulphonephthalein solution to about 800 c.c. water, alkalinize with sodium hydroxid and dilute to 1000 c.c. Since this contains the same amount of the drug as was injected, it may be rated as a 100 per cent. standard-color solution. No more than 100 c.c. of the standard solution will be needed, and there usually will be enough of the original solution left in the ampoule to make this amount.

2. Filter the diluted and alkalinized urine and place exactly 100 c.c. in one of two cylinder graduates whose corresponding graduations stand at the same height.

3. Into the other graduate pour the 100 per cent. standard solution, a little at a time, until the two cylinders show the same depth of color *when looked at from above* over a sheet of white paper. The height of the standard solution then indicates directly the excretion percentage. If, for example, the color of the first hourly portion was matched by 40 c.c. of the standard and that of the second by 20 c.c., then the excretion would be 40 per cent. and 20 per cent. for the one-hour portions and 60 per cent. for the two hours.

Another and probably a more accurate method requires the preparation of a series of standard dilutions representing various percentages and comparison of the color of the diluted urine with these, using test-tubes or cylinders of equal diameter and looking through them from the side. The tubes may be placed in an improvised frame with ground-glass back like that of the Sahli hemoglobinometer. The standard dilutions are easily made: for the 30 per cent. solution use 30 c.c. of the 100 per cent. standard and 70 c.c. water, etc.

In order to equalize the slight difference in color due to a highly colored urine, the standard color may be viewed through a faintly yellow-tinted piece of glass, or an amount of urine equal to that voided by the patient may be included in the standard solution. For those who do much work it is convenient to add a few drops of a solution of some yellow dyes such as Echtgelb Gor Tropaeolin OO.

For greater accuracy, more elaborate colorimeters are recommended. The simple and inexpensive Denison Laboratory instrument (see Fig. 32) is especially useful for this purpose. Results with this colorimeter are most dependable, when the unknown solution and the standard have nearly the same depth of color. It is therefore well to use a 50 per cent. standard solution for the phenol-sulphonephthalein estimation instead of the 100 per cent. standard above recommended. The colorimeter reading must then be divided by two.

When it is necessary to defer the color comparison for hours or days, the urine must be kept acid as the color fades in alkaline solution.

Under normal conditions the drug first appears in the urine in five to eleven minutes after the injection. Within the first hour 40 to 60 per cent. is eliminated; in the two hours, 60 to 85 per cent. Pathologically, the excretion may be reduced to a trace or even, in extreme cases, to none at all in the two hours.

Time has proved the great usefulness of this test in everyday practice, but it must be remembered that it is a test of functional capacity only, not a measure of the extent of anatomic changes in the kidney. Although it is true that these generally run more or less parallel, they do not always do so. The test is extremely valuable in diagnosis and prognosis of chronic nephritis where the phenolsulphonephthalein output runs fairly parallel with the course of the disease. In acute nephritis the result does not always agree with the clinical and pathological picture. Particularly is this true in the acute glomerulo-nephritis of scarlet fever where the excretion percentage may sometimes be fully up to the normal. Apparently the test speaks less definitely concerning glomerular changes than tubular.

III. CHEMIC EXAMINATION

The chemical constituents of the urine will be considered in two groups: those present normally, and those present in appreciable amount only under pathologic conditions.

Before discussing these in detail it is convenient at this place to include a general description of colorimetric and centrifugal methods, which have rather wide usefulness for quantitative estimations. Their application to individual susbtances will be given later.

Colorimetric Methods.—These combine comparative simplicity and great accuracy and are steadily growing in popularity.

In general they consist in treating the fluid under examination with such reagents as will produce a soluble colored compound with the substance to be estimated, and in comparing this color with that of a similar solution of known strength, upon the principle that the depth of color is directly proportionate to the amount of the substance present. Some preliminary treatment is usually necessary to remove interfering substances. Any device which will show the quantitative relationship between the colors is called a colorimeter.

The chief hindrances to the wide adoption of colorimetric methods for clinical purposes are the cost of the colorimeter and the difficulties in the way of preparing standard color solutions. Relatively stable standard solutions for many of the methods can be purchased with the instruments.

The **Hellige colorimeter** (Fig. 30) is one of the most satisfactory for general purposes. The solution under

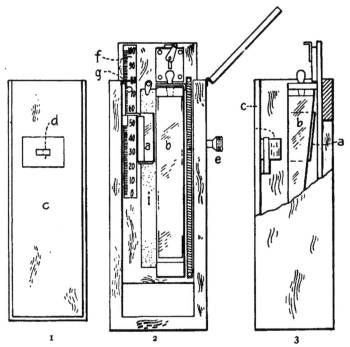

FIG. 30.—Hellige colorimeter. 1, sliding front, removed; 2, front view of interior; 3, side view with portion of side wall removed; *a*, glass trough for unknown solution; *b*, glass wedge for standard solution; *c*, sliding front; *d*, window; *e*, knerled head; *f*, scale; *g*, scale-pointer; *h*, double prism; *i*, ground glass back.

examination is placed in the box or trough (*a*), while the standard solution is placed in the wedge-shaped bottle (*b*), which can be moved up or down beside the trough. The front (*c*) is slipped into place and the two solutions are viewed through the window (*d*) behind which is a double prism (*h*) to bring the two colors close together. The wedge is moved up and down by means of the knerled head (*e*) until a point is reached where the two colors match. The figure on the scale (*f*) which then stands opposite the pointer (*g*) indicates the relation between the strengths of the two solutions. If the pointer stands at 40 then the unknown solution is 40 per cent. as strong as the known standard; if at 70, it is 70 per cent. as strong. In the older instruments the scale is reversed, the 100 mark being at the bottom. In both types the actual values are sometimes found by reference to a chart or "graph" which must be made for each standard solution. Hermetically sealed standard wedges for most of the tests, each accompanied by its appropriate graph can be purchased with the instrument.

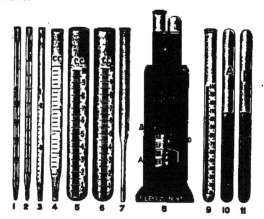

FIG. 31.—Kuttner micro-colorimeter, with pipets, graduated tubes and standard color tubes.

The **Kuttner colorimeter** (Fig. 31) is very similar to the Sahli hemoglobinometer, the chief differences being that the front is closed and the colors are viewed through a window supplied with a double prism like that of the Hellige. The unknown solution is diluted in the graduated tube until its color matches that of the standard, which is kept in a sealed tube. The pipets and test-tubes required for making the various tests are included. Standard color

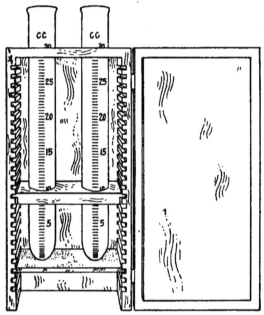

FIG. 32.—Denison Laboratory colorimeter, made from a slide box, blackened inside, and two 30-c.c. tubes which stand upon a ground-glass slide and are held in place by a wooden slide.

tubes for hemoglobin, blood sugar, and the phenolsul-phonephthalein kidney test are now supplied and the makers state that others are in preparation.

The **Denison Laboratory colorimeter**[1] (Fig. 32) is prob-

[1] Designed by the late A. R. Peebles, while director of the Denison Research Laboratory. Most of the colorimetric methods which are useful in blood and urine work have been modified for use with this colorimeter by R. C. Lewis and A. R. Peebles and will be published soon.

ably the simplest, most convenient and least expensive yet devised, and its accuracy equals or even exceeds that of the Hellige colorimeter. The instrument can be easily made by any one from a Pillsbury slide box and two graduated 30-c.c. test-tubes. These tubes are the right size, are carried in stock by most supply houses, and answer as well as specially graduated tubes. Equivalent graduations on the two tubes must stand at the same height.

To use the instrument the unknown solution is poured into one tube exactly to the 10-c.c. mark, and the standard solution is placed in the other, a little at a time by means of a medicine-dropper until the colors in the two tubes just match when looked at from above over a sheet of white paper or a small mirror, so placed that it reflects the light from a window. A small reflector can be placed in the bottom of the box at an angle of 45 degrees if desired but adds to the cost without commensurate advantage. When the two colors match, the height of the standard color solution—the reading being taken at the bottom of the meniscus—will indicate in percentage the relation between the strengths of the two solutions. If, for example, the top of the standard solution stands at the 7.5 c.c. mark, then the unknown solution is 75 per cent. as strong as the known standard. Readings are most accurate when the unknown solution and the standard have nearly the same depth of color.

This colorimeter can be strongly recommended for the phenolsulphonephthalein test (see p. 112) and for other estimations when one prepares his own standard solutions.

Centrifugal Methods.—As shown by Purdy, the centrifuge offers a means of making quantitative estimations of a number of substances in the urine. Results are easily and quickly obtained; and while the methods can lay no claim to accuracy, they will be found very useful in following the progress of a case from day to day when recourse to more elaborate methods is out of the question.

FIG. 33.—The Purdy electric centrifuge with four arms.

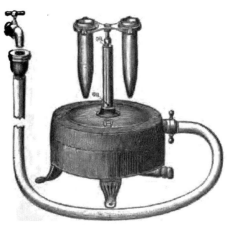

FIG. 34.—Water-motor centrifuge.

In general, the methods consist in precipitating the
substance to be estimated in a graduated centrifuge tube
by means of an appropriate reagent, and applying a definite
amount of centrifugal force for a definite length of time,
after which the percentage of precipitate is read off upon
the side of the tube. Interfering substances such as

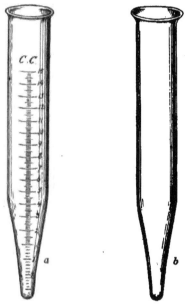

FIG. 35.—Purdy's tubes for the centrifuge: *a*, Percentage tube; *b*,
sediment tube.

albumin must be previously removed. Results are in
terms of *bulk of precipitate*, which must not be confused
with *percentage by weight*. The weight percentage can be
found by referring to Purdy's tables, given later; but in
following the progress of the same case from day to day
it suffices to compare the bulk of the precipitate, always
taking into consideration, of course, the twenty-four-hour
amount of urine.

To fulfil Purdy's requirements, upon which the tables are based, the centrifuge should have an arm with a radius of 6¾ inches when in motion, and should be capable of maintaining a speed of 1500 revolutions a minute. The electric centrifuge is to be recommended, although good work can be done with a water-power centrifuge or, after a little practice, with the hand centrifuge. A speed indicator is desirable with electric and water-motor machines, although one can learn to estimate the speed by the musical note. In general a four-arm centrifuge will be found most useful. Instead of the conical aluminum tube-shields usually supplied, it is well to get flat-bottomed shields with rubber cushions, because these permit the use of ordinary test tubes, which is a great convenience at times. When the centrifuge is in use, opposite tubes must carry the same weight, otherwise the machine will be quickly ruined. It is best to balance the filled tubes upon a scale, but it will usually suffice to fill them to the same height.

A. Normal Constituents

Of the large number of organic and inorganic substances normally present in the urine, only a few demand any consideration from the clinician. The following table, therefore, outlines the average composition from the clinical, rather than from the chemical, standpoint. Only the twenty-four-hour quantities are given, since they alone furnish an accurate basis for comparison. *The student cannot too soon learn that percentages mean little or nothing, excepting as they furnish a means of calculating the twenty-four-hour elimination.*

Although the conjugate sulphates are organic compounds, they are, for the sake of convenience, included with the inorganic sulphates in the following table.

COMPOSITION OF NORMAL URINE

	Grams in twenty-four hours	Approximate average
Water........................	1000–1500	1200
Total substances in solution...........	55–70	60
Inorganic substances..................	20–30	25
Chlorids (chiefly sodium chlorid)....	10–15	12.5
Phosphates (estimated as phosphoric		
acid), total..................	2.5–3.5	3
Earthy, ⅓ of total..............		1
Alkaline, ⅔ of total.............		2
Sulphates (estimated as sulphuric		
acid), total....................	1.5–3.0	2.5
Mineral, ⁹⁄₁₀ of total............		2.25
Conjugate, ¹⁄₁₀ of total..........		0.25
Includes indican..............		Trace
Ammonia......................	0.5–1.0	0.7
Organic substances...................	30–40	35
Urea..........................	25–35	30
Uric acid......................	0.4–1.0	0.7

Among constituents which are of little clinical importance, or are present only in traces, are:

Inorganic.—Iron, carbonates, nitrates, silicates, and fluorids.

Organic.—Creatinin, hippuric acid, purin bases, oxalic acid, volatile fatty acids, pigments, and acetone.

Variations in body weight, diet, and exercise cause marked fluctuations in the total solids and in individual substances.

1. Chlorids.—These are derived from the food, and are mainly in the form of sodium chlorid. The amount excreted normally is 10 to 15 Gm. in twenty-four hours. It is much affected by the diet, and is reduced to a minimum in starvation.

Excretion of chlorids is diminished in nephritis and

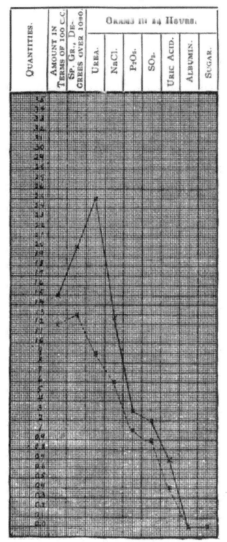

FIG. 36.—Graphic expression of quantities in the urine. Solid line, normal urine; dotted line, an example of pathologic urine in a case of cancerous cachexia (Saxe).

in fevers, especially in pneumonia and inflammations leading to the formation of large exudates. In nephritis the kidneys are less permeable to the chlorids, and it is possible that the edema is due largely to an effort of the body to dilute the chlorids which have been retained. Certainly an excess of chlorids in the food will in many cases increase both the albuminuria and the edema of nephritis. In fevers the diminution is due largely to decrease of food, though probably in some measure to impaired renal function. In pneumonia chlorids are constantly very low, and in some cases are absent entirely. Following the crisis they are increased. In inflammations leading to formation of large exudates—*e.g.*, pleurisy with effusion—chlorids are diminished because a considerable amount becomes "locked up" in the exudate. During absorption chlorids are liberated and appear in the urine in excessive amounts.

Diminution of chlorids is also sometimes observed in severe diarrhea, anemic conditions, and carcinoma of the stomach.

Detection of Chlorids.—The following simple test will show the presence of chlorids, and at the same time roughly indicate any pronounced alteration in amount:

To a few cubic centimeters of urine in a test-tube add a few drops of nitric acid to prevent precipitation of phosphates and then a few drops of silver nitrate solution of about 12 per cent. strength. A white, curdy precipitate of silver chlorid forms. If the urine merely becomes milky or opalescent, chlorids are markedly diminished.

Quantitative Estimation.—The well-known and reliable Volhard method has been simplified by Strauss, and this modification has recently been still further simplified by Bayne-Jones and by McLean and Selling, so that the method is now available for ordinary clinical work. The only difficulty is the preparation of solutions, and these can be purchased ready prepared. A much less accurate, though simple and very useful, method is afforded by the centrifuge.

1. **Simplified Volhard Method.**—As a rule albumin need not be removed. In an accurately graduated 50-c.c. cylinder place 5 c.c. of the urine and 10 c.c. of Solution No. 1. Mix by inverting several times. If a reddish color appears, add 3 drops of 10 per cent. potassium permanganate. After five minutes add Solution No. 2, a very little at a time, mixing after each addition, until a permanent red-brown color (best seen against a white background) appears. This is the end-point.

The solutions are so balanced that if the urine be chlorid-free the volume of fluid when the end-point is reached will be 35 c.c., and that for each gram per liter of chlorids in the urine the volume will be 1 c.c. less. Therefore, the difference between 35 c.c. and the height of the fluid at the end of the test gives directly the number of grams of chlorids per liter of urine, expressed as sodium chlorid. If, for example, the fluid reaches the 28-c.c. mark, $35 - 28 = 7$ Gm. of sodium chlorid per liter of urine.

A certified 50-c.c. graduated cylinder, with glass stopper, is required. The ordinary 50-c.c. graduate is inaccurate.

The **solutions** are as follows:

No. 1.—*Standard silver nitrate solution*:

Silver nitrate (C.P., anhydrous, crystallized).   29.055 Gm.;
Nitric acid (25 per cent.)..................   900 c.c.;
Ammonioferric alum (cold saturated solu-
    tion)................................   50 c.c.
Distilled water to.....................   1000 c.c.

No. 2.—*Ammonium sulphocyanate solution*:

Ammonium sulphocyanate................ .... 7 Gm.;

Distilled water.......................... 1000 c.c.

This solution is intentionally made too strong, and it must be standardized by diluting with distilled water until exactly 20 c.c. (and no less) will produce a red color when mixed with exactly 10 c.c. of Solution No. 1.

TABLE FOR THE ESTIMATION OF CHLORIDS AFTER CENTRIFUGATION

Showing the bulk-percentage of silver chlorid (AgCl) and the corresponding gravimetric percentages sodium chlorid (NaCl) and chlorin (Cl).— (Purdy.)

Bulk-percentage of AgCl.	Percentage NaCl.	Percentage Cl.	Bulk-percentage of AgCl.	Percentage NaCl.	Percentage Cl.
¼	0.03	0.02	8	1.04	0.63
½	0.07	0.04	8½	1.1	0.67
¾	0.1	0.06	9	1.17	0.71
1	0.13	0.08	9½	1.23	0.75
1¼	0.16	0.1	10	1.3	0.79
1½	0.19	0.12	10½	1.36	0.83
1¾	0.23	0.14	11	1.43	0.87
2	0.26	0.16	11½	1.49	0.91
2¼	0.29	0.18	12	1.50	0.95
2½	0.32	0.2	12½	1.62	0.99
2¾	0.36	0.22	13	1.69	1.02
3	0.39	0.24	13½	1.75	1.06
3¼	0.42	0.26	14	1.82	1.1
3½	0.45	0.28	14½	1.88	1.14
3¾	0.49	0.3	15	1.94	1.18
4	0.52	0.32	15½	2.01	1.22
4¼	0.55	0.34	16	2.07	1.26
4½	0.58	0.35	16½	2.14	1.3
4¾	0.62	0.37	17	2.2	1.34
5	0.65	0.39	17½	2.27	1.38
5½	0.71	0.43	18	2.33	1.42
6	0.78	0.47	18½	2.4	1.46
6½	0.84	0.51	19	2.46	1.5
7	0.91	0.55	19½	2.53	1.54
7½	0.97	0.59	20	2.59	1.58

Bulk-percentage to be read on the side of the tube.

2. **Centrifugal Method.**—Fill the graduated tube to the 10-c.c. mark with urine; add 15 drops strong nitric acid and then silver nitrate solution of 12 per cent. strength to the 15-c.c. mark. Mix by inverting several times. Let stand a few minutes for a precipitate to form, and then revolve in the centrifuge for three minutes at 1200 revolutions a minute. Each 0.1 c.c. of precipitate equals 1 per cent. by bulk. This may be converted into percentage by weight of chlorin or sodium chlorid by means of the table upon page 128.

2. **Phosphates** are derived largely from the food, only a small proportion resulting from metabolism. The normal daily output of phosphoric acid is about·2.5 to 3.5 Gm.

The urinary phosphates are of two kinds: *alkaline,* which make up two-thirds of the whole, and include the phosphates of sodium and potassium; and *earthy,* which constitute one-third, and include the phosphates of calcium and magnesium. Earthy phosphates are frequently thrown out of solution in neutral and alkaline urines, and as "amorphous phosphates" form a very common sediment. This sediment seldom indicates an excessive excretion of phosphoric acid. It is usually merely an evidence of diminished acidity of the urine, or of an increase in the proportion of phosphoric acid eliminated as earthy phosphates. This form of "phosphaturia" is most frequent in neurasthenia and hysteria. When the urine undergoes ammoniacal decomposition, some of the ammonia set free combines with magnesium phosphate to form ammoniomagnesium phosphate ("triple phosphate"), which is only slightly soluble in alkaline urine and is deposited in typical crystalline form (see p. 211).

9

Excretion of phosphates is *increased* by a diet rich in nucleins; in active metabolism; in certain nervous and mental disorders; in leukemia; and in phosphatic diabetes, an obscure disturbance of metabolism (not related to diabetes mellitus) which is associated with an increase in the output of phosphates up to 10 Gm. or more in twenty-four hours. Phosphates are *decreased* in chronic diseases with lowered metabolism; in hepatic cirrhosis and acute yellow atrophy; in pregnancy, owing to developing fetal bones; and in nephritis, owing to kidney impermeability.

TABLE FOR THE ESTIMATION OF PHOSPHATES AFTER CENTRIFUGATION

Showing bulk-percentages of uranyl phosphate ($H[UO_2]PO_4$) and the corresponding gravimetric percentages of phosphoric acid (P_2O_5).— (Purdy.)

Bulk-percentage of $H(UO_2)PO_4$.	Percentage P_2O_5.	Bulk-percentage of $H(UO_2)PO_4$.	Percentage P_2O_5.
½	0.02	11	0.14
1	0.04	12	0.15
1½	0.045	13	0.16
2	0.05	14	0.17
2½	0.055	15	0.18
3	0.06	16	0.19
3½	0.065	17	0.2
4	0.07	18	0.21
4½	0.075	19	0.22
5	0.08	20	0.23
6	0.09	21	0.24
7	0.1	22	0.25
8	0.11	23	0.26
9	0.12	24	0.27
10	0.13	25	0.28

Bulk-percentage to be read from graduation on the side of the tube.

Quantitative estimation does not furnish much of definite clinical value. The centrifugal method is the most convenient.

Purdy's Centrifugal Method.—Take 10 c.c. urine in the graduated tube, add 2 c.c. of 50 per cent. acetic acid, and 3 c.c. of 5 per cent. uranium nitrate solution. Mix; let stand a few minutes, and revolve for three minutes at 1200 revolutions a minute. Each 0.1 c.c. of precipitate is 1 per cent. by bulk. The corresponding percentage of phosphoric acid by weight is found by consulting the table on page 130.

3. Sulphates.—The urinary sulphates are derived partly from the food, especially meats, and partly from body metabolism. The normal output of sulphuric acid is about 1.5 to 3 Gm. daily. It is increased in conditions associated with active metabolism, and in general may be taken as a rough index of protein metabolism.

Quantitative estimation of the total sulphates yields little of clinical value.

Purdy's Centrifugal Method.—Take 10 c.c. urine in the graduated tube and add 5 c.c. barium chlorid solution (barium chlorid, 4 parts; concentrated hydrochloric acid, 1 part; and distilled water, 16 parts). Mix; let stand a few minutes, and revolve for three minutes at 1200 revolutions a minute. Each 0.1 c.c. of precipitate is 1 per cent. by bulk. The percentage by weight of sulphuric acid is calculated from the table on page 132.

About nine-tenths of the sulphuric acid is in combination with various mineral substances, chiefly sodium, potassium. calcium, and magnesium (*mineral* or *preformed sulphates*). One-tenth is in combination with

certain aromatic substances, which are mostly products
of protein putrefaction in the intestine, but are derived in part from destructive metabolism (*conjugate*
or *ethereal sulphates*). Among these aromatic substances
are indol, phenol, and skatol. By far the most important of the conjugate sulphates and representative of
the group is potassium indoxyl sulphate.

TABLE FOR THE ESTIMATION OF SULPHATES AFTER
CENTRIFUGATION

*Showing the bulk-percentages of barium sulphate ($BaSO_4$) and the cor-
responding gravimetric percentages of sulphuric acid (SO_3).—(Purdy.)*

Bulk-percentage of $BaSO_4$.	Percentage SO_3.	Bulk-percentage of $BaSO_4$.	Percentage SO_3.
1/8	0.04	2 1/4	0.55
1/4	0.07	2 1/2	0.61
3/8	0.1	2 3/4	0.67
1/2	0.13	3	0.73
5/8	0.16	3 1/4	0.79
3/4	0.19	3 1/2	0.85
7/8	0.22	3 3/4	0.91
1	0.25	4	0.97
1 1/4	0.31	4 1/4	1.03
1 1/2	0.37	4 1/2	1.09
1 3/4	0.43	4 3/4	1.15
2	0.49	5	1.21

Bulk-percentage to be read from graduation on the side of the tube.

Potassium indoxyl sulphate, or **indican,** is derived
from indol. Indol is absorbed and oxidized into indoxyl, which combines with sulphuric acid and potassium and is thus excreted. Under normal conditions
the amount in the urine is small. It is increased by a
meat diet.

Unlike the other ethereal sulphates, which are derived in part from metabolism, indican originates practically wholly from putrefactive processes. It alone, therefore, and not the total ethereal sulphates, can be taken as an index of such putrefaction. A pathologic increase is called indicanuria. It is noted in:

(a) *Diseases of the Small Intestine.*—This is by far the most common source. Intestinal obstruction gives the largest amounts of indican. It is also much increased in intestinal indigestion—so-called "biliousness;" in inflammations, especially in cholera and typhoid fever; and in paralysis of peristalsis, such as occurs in peritonitis. Simple constipation and diseases of the *large* intestine alone do not so frequently cause indicanuria.

(b) *Diseases of the stomach* associated with deficient hydrochloric acid, as chronic gastritis and gastric cancer. Diminished hydrochloric acid favors intestinal putrefaction.

(c) *Diminished Flow of Bile.*—Since the bile serves as a stimulant to peristalsis and in several ways retards putrefaction, a diminished flow from any cause favors occurrence of indicanuria.

(d) *Decomposition of exudates* anywhere in the body, as in empyema, bronchiectasis, and large tuberculous cavities.

Detection of indican depends upon its decomposition and oxidation of the indoxyl set free into indigo-blue. This change sometimes takes place spontaneously in decomposing urine, causing a dirty blue color. Crystals of indigo (see Fig. 49) may then be found both in the sediment and the scum.

Obermayer's Method.—Take a test-tube about one-third full of the urine and add an equal volume of Obermayer's reagent and a few cubic centimeters of chloroform. Mix by inverting a few times; avoid shaking violently. If indican be present in excess, the chloroform, which sinks to the bottom, will assume an indigo-blue color. It will take up the indigo more quickly if the urine be warm. The depth of color indicates the comparative amount of indican if the same proportions of urine and reagents are always used, but one should bear in mind the total amount of urine voided. The indican in normal urine may give a faint blue by this method. Urine of patients taking iodids gives a reddish-violet color, which disappears upon addition of a few drops of strong sodium hyposulphite solution and shaking. Occasionally, owing to slow oxidation, indigo-red will form instead of indigo-blue. This gives a color like that due to iodids, but it does not disappear when treated with sodium hyposulphite. Bile-pigments, which interfere with the test, must be removed if present (see p. 101).

Obermayer's reagent consists of strong hydrochloric acid (sp. gr., 1.19), 1000 c.c., and ferric chlorid, 2 Gm. This makes a yellow, fuming liquid which keeps well.

4. Urea.—From the standpoint of physiology urea is the most important constituent of the urine. It is the principal waste-product of metabolism, and constitutes about one-half of all the solids excreted—about 20 to 35 Gm. in twenty-four hours. It represents 85 to 90 per cent. of the total nitrogen of the urine, and its quantitative estimation is a simple, though not very accurate, method of ascertaining the state of nitrogenous excretion. This is true, however, only in normal individuals upon average mixed diet. Upon a low protein diet it may fall to 60 per cent. of the total nitrogen. Under pathologic conditions, the proportion of nitrogen distributed among the various nitrogen-containing substances undergoes great variation. The only accurate index of protein metabolism is, therefore, the total output of nitrogen, which can be estimated by the Kjeldahl method or one of its modifications such as the new direct Nesslerization method of Folin and Denis. The whole subject of "nitrogen partition" (distribution of nitrogen among the nitrogen-containing bodies) and "nitrogen equilibrium" (relation of excretion to intake) is an important one, but is out of the province of this book, since as yet it concerns the physiologic chemist more than the clinician.

It may be helpful to state here, however, that upon a mixed diet the nitrogen of the urine is distributed about as follows: urea nitrogen, 86.9 per cent.; ammonia nitrogen, 4.4 per cent.; creatinin nitrogen, 3.6 per cent.; uric acid nitrogen, 0.75 per cent.; "undetermined nitrogen," chiefly in amino-acids, 4.3 per cent.

Normally, the amount is greatly influenced by exercise and diet. It is increased by copious drinking of water and administration of ammonium salts of organic acids.

Pathologically, urea is increased in fevers, in diabetes when acidosis is not marked, and especially during resolution of pneumonia and absorption of large exudates. As above indicated, when other factors are equal, the amount of urea indicates the activity of metabolism. In deciding whether in a given case an increase of urea is due to increased metabolism the relation between the amounts of urea and of the chlorids is a helpful consideration. Upon a mixed diet the amount of urea is normally about twice that of the chlorids. If the proportion is much increased above this, increased tissue destruction may be inferred, since other conditions which increase urea also increase chlorids.

In general, a pathologic decrease in amount of urea is due either to lessened formation within the body or to diminished excretion. *Decreased formation* of urea occurs in diseases of the liver with destruction of liver substance, such as marked cirrhosis, carcinoma, and acute yellow atrophy. The state of acidosis likewise decreases formation of urea, because nitrogen which would otherwise be built into urea is eliminated in the form of ammonia (see p. 145). *Retention* of urea occurs in most cases of nephritis. In acute nephritis the amount of urea in the urine is markedly decreased, and a return to normal denotes improvement. In the early stages of chronic nephritis, when diagnosis is difficult, it is usually normal. In the late stages, when diagnosis is comparatively easy, it is decreased. Hence estimation of urea is of little help in the diagnosis of this disease, and is of no value whatever when, as is so frequently the case, a small quantity of urine taken at random is used. When, however, the diagnosis is established, estimations made at frequent intervals under the same conditions of diet and exercise are of much value, *provided a sample of the mixed twenty-four-hour urine be used.* A steady decline is a very bad prognostic sign, and a sudden marked diminution is usually a forerunner of uremia.

The presence of urea can be shown by allowing a few

drops of the fluid partially to evaporate upon a slide, and adding a small drop of pure, colorless nitric acid or saturated solution of oxalic acid. Crystals of urea nitrate or oxalate (Fig. 37) will soon appear and can be recognized with the microscope.

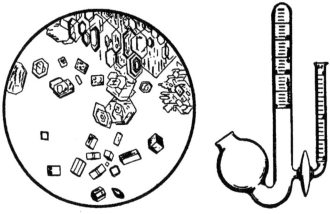

FIG. 37.—Crystals of nitrate of urea (upper half) and oxalate of urea (lower half) (after Funke).

FIG. 38. — Doremus-Hinds' ureometer without foot.

Quantitative Estimation.—The hypobromite method, which has long been used in clinical work, is very simple, but is notoriously inaccurate. The new urease methods are much more accurate.

1. **Hypobromite Method.**—This depends upon the fact that urea is decomposed by sodium hypobromite with liberation of nitrogen. The amount of urea is calculated from the volume of nitrogen set free. Of the many forms of apparatus devised for this purpose, that of Doremus-Hinds (Fig. 38) is probably the most convenient.

Pour some of the urine into the smaller tube of the apparatus, then open the stop-cock and quickly close it so as to fill its lumen with urine. Rinse out the larger tube with water and fill it and one-half of the bulb with 25 per cent. caustic soda solution. Add to this 1 c.c. of bromin by means of a medicine-dropper and mix well. This prepares a fresh solution of sodium hypobromite with excess of caustic soda, which serves to absorb the carbon dioxid set free in the decomposition of urea. When handling bromin, keep an open vessel of ammonia near to neutralize the irritant fumes.

Pour the urine into the smaller tube, and then turn the stop-cock so as to let as much urine as desired (usually 1 c.c.) run slowly into the hypobromite solution. When bubbles have ceased to rise, read off the height of the fluid in the large tube by the graduations upon its side. This gives the amount by weight of urea in the urine added, from which the amount excreted in twenty-four hours can easily be calculated. If the urine contains much more than the normal amount, it should be diluted.

This method has fallen into disrepute largely because of inconstant results, and because it gives more nearly the total nitrogen than the urea. According to Robinson and Müller the discrepancies are due to insufficient mixing of urine and hypobromite and can be obviated by gentle shaking after the first vigorous reaction is over. Results are then constant, but too high, owing to decomposition of other nitrogenous constituents.

To avoid handling pure bromin, which is disagreeable, Rice's solutions may be employed:

(a) Bromin.............................. 31 Gm.;
    Potassium bromid..................... 31 Gm.;
    Distilled water...................... 250 c.c.;

(b) Sodium hydroxid..................... 100 Gm.;
    Distilled water...................... 250 c.c.

Equal parts of these solutions are mixed and used for the test. The bromin solution must be kept in a tightly stoppered bottle or it will rapidly lose strength.

2. **Urease Methods.**—These are based upon the conversion of urea into ammonium carbonate by urease, a ferment first extracted by Takeuchi from the soy bean in 1909. The urea is estimated from the amount of ammonium carbonate produced by the fermentation. There are several clinical methods, two of which are here given in detail. In the first, the urine after fermentation is titrated with decinormal hydrochloric acid in the presence of the indicator, methyl orange. It is not entirely accurate, but is much superior to the hypobromite method. In the second, the ammonia is determined by direct Nesslerization. This method is sufficiently accurate for the most exacting work but is too complicated for use in a physician's office laboratory. It is included here because of the growing importance of exact estimations of the various nitrogenous substances in blood and urine.

Neither albumin nor sugar nor any other substance likely to be present in body fluids interferes with the action of urease.

Marshall's Urease Method.—1. Into each of two 200-c.c. flasks measure 5 c.c. of the urine and about 100 c.c. of water, and to one add 1 c.c. of a 10 per cent. solution of urease.[1]

2. Overlay the fluid in each flask with about 1 c.c. of toluol, insert corks and let stand over night at room temperature (or for three hours in the incubator at 37°C.).

3. At the end of this time, titrate the contents of each flask to a distinct pink color with decinormal hydrochloric

---

[1] It is more convenient to use the 0.025-Gm. tablets sold by Hynson, Westcott and Dunning, Baltimore. One of these is crushed and dissolved in 5 c.c. of water, and the whole of this solution is used for the test.

acid, using a few drops of 0.5 per cent. methyl orange solution as indicator.

4. Find the difference between the number of cubic centimeters of decinormal acid used in the two titrations and multiply this by the factor 0.06 to obtain the percentage of urea in the urine. From the percentage calculate the twenty-four-hour elimination.

Urease Method of Folin and Denis.—This method is comparatively simple if one has the reagents at hand. All except the soy bean meal suspension can be purchased ready prepared, and for this, the tablets mentioned in the foot-note on page 139 may be substituted. Ammonia-free distilled water must be used throughout.

Reagents Required.—(a) One per cent. suspension of soy bean flour. The "soja bean meal" which is sold as a food for diabetics and is obtainable from any wholesale drug house may be used. Rub up 5 Gm. of the meal to a uniform paste with about 15 c.c. water, and gradually add enough water to make 400 c.c. To this add 100 c.c. alcohol. The suspension remains good for about two days.

(b) Nessler's reagent. Dissolve 7.5 Gm. of potassium iodid in 50 c.c. warm water and add 10 Gm. of mercuric iodid. Add about 40 c.c. water, filter and dilute to 100 c.c. This is a stock solution. The Nessler's solution to be used in this method, consists of 30 c.c. of the stock solution, 20 c.c. of 10 per cent. sodium hydroxid and 50 c.c. water.

(c) Standard ammonium sulphate solution. In exactly 1000 c.c. distilled water dissolve 4.716 Gm. Kahlbaum's[1] C.P. ammonium sulphate which has been dried for an hour at 110°C. before weighing. Take 10 c.c. of this solution and dilute to exactly 200 c.c. Twenty cubic

[1] Most ammonium sulphate contains pyridine bodies which interfere with the Nesslerization. In the future it should be possible to obtain satisfactory ammonium sulphate of American make.

centimeters of this final solution contains exactly 0.001 Gm. of nitrogen.

(d) Meta-phosphoric acid, 25 per cent. aqueous solution made without heat. This deteriorates after two or three days.

(e) Merck's blood charcoal.

Method.—1. Place exactly 1 c.c. of the urine in a 100-c.c. flask, add 10 or 15 c.c. of the 1 per cent. soy meal suspension, stopper the flask and let stand for one hour at room temperature or fifteen minutes in a water bath at 50°C.

2. Add 25 c.c. water and 1 c.c. fresh 25 per cent. meta-phosphoric acid. Mix well.

3. Add about 1 Gm. Merck's blood charcoal, make up to 100 c.c. and filter.

4. To 10 c.c. of the filtrate add about 60 c.c. distilled water and 15 c.c. Nessler's solution, and make up to 100 c.c. with distilled water.

5. Make up a standard solution representing 0.001 Gm. ammonia nitrogen as follows: To 20 c.c. of the standard ammonium sulphate solution add 15 c.c. Nessler's solution, and make up to 100 c.c. with distilled water.

6. Find the strength of the unknown solution by comparing it with the standard in a colorimeter. The quantity of nitrogen in the unknown represents the urea-nitrogen plus the ammonia nitrogen in 0.1 c.c. urine, and must be multiplied by 1000 to find the amount in 100 c.c. urine. If, for example, the reading on one of the newer Hellige instruments is 70, then the unknown solution (representing 0.1 c.c. urine) contains 0.0007 Gm. combined urea and ammonia nitrogen, and 100 c.c. urine contains 0.7 Gm.

7. In another sample of urine estimate ammonia nitrogen alone for 100 c.c. urine (see p. 147) and subtract this from the figure obtained above. The remainder is the urea nitrogen for 100 c.c. of urine. To express this in terms of urea multiply by 2.14.

This method, slightly modified, is also applicable to estimation of urea in blood, ammonia in urine (see p. 147) and total nitrogen in urine.

5. Uric acid is the most important of a group of substances, called *purin bodies*, which are derived chiefly from the nucleins of the food, *exogenous uric acid*, and from metabolic destruction of the nuclei of the body, *endogenous uric acid*. The daily output of uric acid is about 0.4 to 1 Gm. The amount of the other purin bodies together is about one-tenth that of uric acid. Excretion of these substances is greatly increased by a diet rich in nucleins, as sweetbreads and liver.

Uric acid exists in the urine in the form of urates, chiefly of sodium and potassium, which in concentrated urines are readily thrown out of solution and constitute the familiar sediment of "amorphous urates." This, together with the fact that uric acid is frequently deposited as crystals, constitutes its chief interest to the practitioner. It is a very common error to consider these deposits as evidence of excessive excretion.

Pathologically, the greatest increase of uric acid occurs in leukemia, where there is extensive destruction of leukocytes, in diseases with active destruction of the liver and other organs rich in nuclei and during absorption of a pneumonic exudate. There is generally an increase during x-ray treatment. Uric acid is decreased before an attack of gout and increased for several days after it, but its etiologic relation is still uncertain. An increase is also noted in acute fevers.

Quantitative Estimation of Purin Bodies.—There is no accurate method which is simple enough for clinical purposes. Of clinical methods, the two given here are most satisfactory. They are based upon the same principle: precipitation and removal of phosphates, and then precipitation of purin bodies with silver nitrate which is strongly ammoniated in order to hold silver chlorid in solution. The amount of purin bodies is calculated from the bulk of the silver-purin, which in Cook's method is thrown down by the centrifuge, and in Hall's is allowed to settle for twenty-four hours. The urine must be albumin-free.

1. **Cook's Method.**—In a centrifuge tube take 10 c.c. urine and add about 1 Gm. (about 1 c.c.) sodium carbonate and 1 or 2 c.c. strong ammonia. Shake until the soda is dissolved. The earthy phosphates will be precipitated. Centrifugalize thoroughly and pour off all the clear fluid into a graduated centrifuge tube. To this fluid add 2 c.c. ammonia and 2 c.c. ammoniated silver nitrate solution. Let stand a few minutes, and revolve in the centrifuge until the bulk of precipitate *remains constant*. Each 0.1 c.c. of sediment represents 0.001176 Gm. purin bodies.

Ammoniated silver nitrate solution is prepared by dissolving 5 Gm. of silver nitrate in 100 c.c distilled water, and adding ammonia until the solution clouds and again becomes clear.

2. **Hall's Method.**—The instrument is shown in Fig. 39. Close the stop-cock, introduce 90 c.c. urine and 20 c.c. of the magnesia solution, and mix by inverting a few times. Open the stop-cock and let the instrument stand for about ten minutes, or until the precipitated phosphates have settled into the lower chamber. Then close the stop-cock, and pour in ammoniated silver nitrate solution until the level of the fluid reaches the 100-c.c. mark. Mix well, and if any white precipitate of silver chlorid persists, bring it into solution by adding a few drops of ammonia. Stand

the instrument in the dark for twenty-four hours and read off the bulk of the precipitate. The corresponding percentage of purin nitrogen is found by reference to a

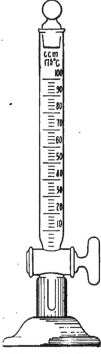

table which accompanies the instrument. Albumin must be removed before making the test.

The *magnesia mixture* is prepared by dissolving 10 Gm. of magnesium chlorid in 75 c.c. of water and adding 10 Gm. of ammonium chlorid and 100 c.c. strong ammonium hydroxid. If a precipitate forms, it is dissolved by further addition of ammonia. Add water to bring the volume to 200 c.c. and finally add 10 Gm. of finely powdered talcum.

The *ammoniated silver nitrate solution* used in Hall's method consists of silver nitrate, 1 Gm.; ammonium hydroxid, 100 c.c.; talcum, 5 Gm.; distilled water, 100 c.c.

Quantitative Estimation of Uric Acid.—Ruhemann's method, while far from accurate, will probably answer for clinical work. The estimation is, however, seldom of any clinical value.

FIG. 39.—Hall's purinometer.

Ruhemann's Method for Uric Acid.—The urine must be slightly acid. By means of a pipet fill Ruhemann's tube (Fig. 40) to the mark *S* with the indicator, carbon disulphid, so that the lowest part of the meniscus is on a level with the mark, as indicated in Fig. 40. Next add Ruhemann's reagent until the base of the upper arch of the meniscus is level with the mark *J*. The carbon disulphid

will assume a violet color. Add the urine, a
small quantity at a time, closing the tube
with the glass stopper and shaking vigorously after each addition, until the disulphid
loses every trace of its violet color and becomes pure white. This completes the test.
Toward the end the reagent should be
added a very little at a time, and the
shaking should be prolonged in order not
to pass the end-point. The figure in the
right-hand column of figures corresponding
to the top of the fluid gives the amount of
uric acid in parts per thousand. The
presence of diacetic acid interferes with the
test, as do also, to some extent, bile and
albumin. Diacetic acid can be driven off
by boiling; bile-pigment and albumin are
removed as described elsewhere (see pp.
101 and 166).

Ruhemann's reagent consists of iodin, 0.5
Gm.; potassium iodid, 1.25 Gm.; absolute
alcohol, 7.5 Gm.; glycerol, 5 Gm.; distilled
water to 100 c.c.

6. Ammonia.—A small amount of
ammonia, combined with hydrochloric,
phosphoric, and sulphuric acids, is always
present. Estimated as NH₃, the normal
average is about 0.7 Gm. in twenty-four
hours. This represents 4 to 5 per cent.
of the total nitrogen of the urine, ammonia
standing next to urea in this respect.

Under ordinary conditions, most of
the ammonia which results from the

FIG. 40.—Ruhemann's uricometer.

metabolic processes is transformed into urea. When, however, acids are present in excess, either from ingestion of mineral acids or from abnormal production of acids within the body (as in fevers, diabetes, pernicious vomiting of pregnancy, delayed chloroform-poisoning, etc.), ammonia combines with them and is so excreted, urea being correspondingly decreased. It is thus that the body protects itself against acid intoxication. A marked increase of ammonia is, therefore, very important as an *index of the tendency to acidosis*, particularly that associated with the presence of diacetic and oxybutyric acids.

In diabetes mellitus ammonia elimination may reach 4 or 5 Gm. daily. It is likewise markedly increased in pernicious vomiting of pregnancy, but *not in nervous vomiting;* and in conditions in which the power to synthesize urea is interfered with, notably cirrhosis and other destructive diseases of the liver and conditions associated with deficient oxygenation. Certain drugs have a marked influence upon ammonia elimination; thus, fixed alkalies and salts of organic acids diminish it, while inorganic acids such as hydrochloric increase it.

Quantitative Estimation.—The urine must be fresh, since decomposition increases the ammonia. The formalin method is satisfactory for clinical work though subject to some inaccuracies. When carried out without use of lead acetate, it includes amino-acids with the ammonia, hence gives figures that are too high. The Folin and Denis method gives ammonia only and is accurate. The difference between the figures obtained by the two methods therefore represents amino-acids.

Ronchese-Malfatti Formalin Method.—This depends upon the fact that when formalin is added to the urine the ammonia combines with it, forming hexamethylenamin. The acids with which the ammonia was combined are set free, and their quantity, ascertained by titration with sodium hydroxid, indicates the amount of ammonia.

Take 10 c.c. of the urine in a beaker or evaporating dish, add 50 c.c. water and 10 drops of 0.5 per cent. alcoholic solution of phenolphthalein. Neutralize by adding a weak caustic soda or sodium carbonate solution until a permanent pink color appears. To 5 c.c. formalin add 15 c.c. water and neutralize in the same way. Pour the formalin into the urine. The pink color at once disappears, owing to liberation of acids. Now add decinormal sodium hydroxid solution from a buret until the pink color just returns. Each cubic centimeter of the decinormal solution used in this titration corresponds to 0.0017 Gm. of NH_3. This must be multiplied by 10 to obtain the percentage from which the twenty-four-hour elimination of ammonia is calculated.

The method is more complicated, but distinctly more accurate, when carried out as suggested by E. W. Brown: Treat 60 c.c. of urine with 3 Gm. of basic lead acetate, stir well, let stand a few minutes, and filter. This removes certain interfering nitrogenous substances. Treat the filtrate with 2 Gm. neutral potassium oxalate, stir well, and filter. Take 10 c.c. of the filtrate, add 50 c.c. water and 15 Gm. neutral potassium oxalate, and proceed with the ammonia estimation as above outlined.

Method of Folin and Denis.—The reagents used are the same as those already given for the similar urea method. (See page 140.)

1. To 10 c.c. of urine in a small flask add 1 c.c. of 25 per cent. meta-phosphoric acid, 9 c.c. distilled water, and 2 Gm. Merck's blood charcoal. Shake for at least one minute and filter.

2. Transfer 2 c.c. of the filtrate to a 100-c.c. flask, add about 70 c.c. distilled water and 15 c.c. Nessler's solution, make up to 100 c.c. with distilled water and mix well.

3. Make up a standard solution consisting of 20 c.c. standard ammonium sulphate solution (representing 0.001 Gm. ammonia nitrogen), 15 c.c. Nessler's solution and water to make exactly 100 c.c.

4. Find the amount of nitrogen in the unknown solution by comparing it with the standard in a colorimeter. This amount then represents the ammonia nitrogen in 1 c.c. of urine. Multiplied by 100, it gives the percentage of ammonia nitrogen. To transform these figures into terms of NH_3 multiply by 1.214.

7. Amylase.—A small quantity of starch-digesting ferment—derived chiefly from the pancreas—can be detected in the urine of healthy persons. According to Brown under normal conditions the twenty-four-hour urine will digest 1500 to 12,000 c.c. of 1 per cent. starch solution in one-half hour at 38°C.; the normal amount of amylase is therefore said to be 1500 to 12,000 units. It is somewhat influenced by the diet.

Amylase is diminished in pancreatic disease and in nephritis with deficient renal permeability. It is increased in simple obstruction of the pancreatic duct, although as the pancreas becomes involved in the pathologic process the amount diminishes. The estimation of urinary amylase is therefore important in suspected disease of the pancreas, particularly when considered in connection with the pancreatic ferments of the feces. It has also been proposed as a test of renal function but does not promise much in this field.

Estimation of Amylase.—1. Obtain the twenty-four-hour urine, which must be kept in a cool place and may be preserved by addition of an ounce of toluol. It should be examined without delay.

2. Dilute the urine to 3000 c.c. and mix well.

3. Proceed exactly as for fecal amylase, steps 1 to 5, except that a 0.1 per cent. starch solution must be substituted for the 1 per cent. solution recommended for the feces, and a weaker iodin solution must be used. One part of Gram's iodin solution diluted with 4 parts of water will answer.

The normal falls between tube 8 (1500 units) and tube 11 (12000 units).

B. Abnormal Constituents

Those substances which appear in the urine only in pathologic conditions are of much more interest to the clinician than are those which have just been discussed. Among them are: proteins, sugars, the acetone bodies, bile, urobilin, hemoglobin, hematoporphyrin and the diazo substances. The detection of drugs in the urine will also be discussed under this head.

1. Proteins.—Of the proteins which may appear in the urine, serum-albumin and serum-globulin are the most important. Mucin, proteose, and a few others are found occasionally, but are of less interest.

(1) Serum-albumin and Serum-globulin.—These two proteins constitute the so-called "urinary albumin." They usually occur together, have practically the same significance, and both respond to all the ordinary tests for "albumin."

Their presence, or *albuminuria*, is probably the most important pathologic condition of the urine. It is either *accidental* or *renal*. The physician can make no greater mistake than to regard all cases of albuminuria as indicating kidney disease.

Accidental or *false albuminuria* is due to admixture with the urine of albuminous fluids, such as pus, blood, and vaginal discharge. The microscope will usually reveal its nature. It occurs most frequently in pyelitis, cystitis, and chronic vaginitis, and the quantity is usually small.

Renal albuminuria refers to albumin which has passed from the blood into the urine through the walls of the kidney tubules or the glomeruli.

Albuminuria sufficient to be recognized by clinical methods probably never occurs as a physiologic condition, the so-called *physiologic albuminuria* appearing only under conditions which must be regarded as abnormal. Among these may be mentioned excessive muscular exertion in those unaccustomed to it; excessive ingestion of proteins; prolonged cold baths; and childbirth. In these conditions the albuminuria is slight and transient.

There are certain other forms of albuminuria which have still less claim to be called physiologic, but which are not always regarded as pathologic. Among these are *cyclic albuminuria*, which regularly recurs at a certain period of the day, and *orthostatic* or *postural albuminuria*, which appears only when the patient is standing. They are rare and of obscure origin, and occur for the most part in neurasthenic subjects during adolescence. It is noteworthy in this connection that nephritis sometimes begins with a cyclic albuminuria.

In pathologic conditions and in most, at least, of the "functional" conditions just enumerated, renal albuminuria may be referred to one or more of the following causes. In nearly all cases it is accompanied by tube-casts.

(a) *Changes in the blood* which render its albumin more diffusible, as in severe anemias, purpura, and scurvy. Here the albumin is small in amount.

(b) *Changes in circulation in the kidney*, either anemia or congestion, as in excessive exercise, chronic heart disease, and pressure upon the renal veins. The quantity of albumin is usually, but not always, small. Its presence is constant or temporary, according to the cause. Most of the causes, if continued, will produce organic changes in the kidney.

(c) *Organic Changes in the Kidney.*—These include the inflammatory and degenerative changes commonly grouped together under the name of nephritis, and also renal tuberculosis, neoplasms, and cloudy swelling due to irritation of toxins and drugs. The amount of albumin eliminated in these conditions varies from minute traces to 20 Gm., or even more, in the twenty-four hours, and, except in acute processes, bears little relation to the severity of the disease. In acute and chronic parenchymatous nephritis the quantity is usually very large. In chronic interstitial nephritis it is small—frequently no more than a trace. It is small in cloudy swelling from toxins and drugs, and variable in renal tuberculosis and neoplasms. In amyloid disease of the kidney the quantity is usually small, and serum-globulin may be present in especially large proportion, or even alone. Roughly distinctive of serum-globulin is the appearance of an opalescent cloud when a few drops of the urine are dropped into a glass of distilled water.

Detection of albumin depends upon its precipitation by chemicals or coagulation by heat. There are many tests, but none is entirely satisfactory, because other substances as well as albumin are precipitated. The most common source of error is mucin. When any considerable amount of mucin is present it can be removed by acidifying with acetic acid and filtering. Urine voided early in the evening or a few hours after a meal is most likely to contain albumin.

It is very important that urine to be tested for albumin be rendered clear by filtration or centrifugation. This is too often neglected in routine work. When ordinary methods do not suffice, it can usually be cleared by shaking up with a little purified talc, infusorial earth or animal charcoal and filtering. This will remove a part of the albumin by adsorption, but the remainder is more easily detected. If the urine is alkaline, sufficient acetic acid should be added to make it acid to litmus. Vaughan has recently called attention to the fact that if bacteria be abundant in an alkaline urine, some of the bacterial proteins may go into solution and give the tests for albumin.

Technic of Ring or Contact Tests.—Since this simple and widely useful method of testing is best known in connection with the detection of albumin a general description is given at this place.

Take a few cubic centimeters of the heavier fluid in a conical test-glass, hold the glass in an inclined position, and run the lighter fluid gently down the inside of the glass by means of a medicine-dropper so that it will form a layer on top of the other without mixing. In the case of the urine, which must be filtered before testing, it may be run in directly from the stem of the funnel by touching this against the wall of the test glass. If the test be positive a sharply defined white or colored ring will appear where the two fluids come into contact. According to its color the ring is seen most clearly if viewed against a white or a black background, as the case may be; and one side of the test-glass may be painted half white, half black, for this purpose.

In the writer's experience this is the most satisfactory technic. The common practice of taking the reagent in a narrow test-tube and pouring the urine in on top of it from a bottle is much inferior. Boston brings the fluids into contact in a glass pipet which is immersed first in the lighter fluid and then (after wiping the outside of the pipet) in the heavier. This is convenient for the routine testing of a large number of urines, but cannot be recommended for accuracy, owing to the small diameter of the column of fluid. Substitution of a medicine-dropper in place of the pipet renders Boston's method more convenient but no more accurate. For those who do only a little testing the "horismascope" (Fig. 41) will be found very convenient and satisfactory. The instrument is, however, fragile and somewhat expensive.

The albumin tests here given are widely used and can be recommended for clinical purposes. They make no distinction between serum-albumin and serum-globulin. As a rule the most sensitive tests are not the most useful for clinical purposes. The writer prefers Purdy's heat test and Roberts' ring test for routine testing, but other workers will have other favorites. The extremely sensitive trichloracetic and sulphosalicylic acid tests are used only in special cases.

1. **Trichloracetic Acid Test.**—The reagent consists of a saturated aqueous solution of trichloracetic acid to which

magnesium sulphate is added to saturation. A simple saturated solution of the acid may be used, but addition of magnesium sulphate favors precipitation of globulin, and, by raising the specific gravity, makes the test easier to apply.

The test is carried out by the "ring" or "contact" method just described. If albumin be present, a white, cloudy ring will appear where the two fluids come in contact.

This is an extremely sensitive test, but, unfortunately, both mucin and proteoses respond to it; urates, when abundant, may give a confusing white ring, and the reagent is comparatively expensive. It is not much used in routine work except as a control to the less sensitive tests.

2. **Sulphosalicylic acid** in 20 per cent. aqueous solution may be used in the same way as the trichloracetic acid reagent. It is fully as sensitive and is somewhat more reliable in that urates and resins are not precipitated. This may also be applied by adding a few drops of the reagent to a few cubic centimeters of the urine in a test-tube and obtaining a white cloud in the presence of albumin, or by adding a bit of the sulphosalicylic acid in the solid state. The last is especially convenient for the practitioner who wishes to make albumin tests at the bedside.

3. **Roberts' Test.**—The reagent consists of pure nitric acid, 1 part, and saturated aqueous solution of magnesium sulphate, 5 parts. It is applied by the "ring" or "contact" method above described.

Albumin gives a white ring, which varies in density with the amount present and when traces only are present, may not appear for two or three minutes. A similar white ring may be produced by primary proteose, thymol, and resinous drugs. White rings or cloudiness in the urine *above* the zone of contact may result from excess of urates or mucus. Colored rings near the junction of the fluids may be produced by iodids, urinary pigments, bile, or indican, but these are not so frequent as with Heller's test.

Roberts' test is one of the best for routine work, although the various rings are apt to be confusing to the inexperienced. It is more sensitive than Heller's test, of which

FIG. 41.—Horismascope: adding the reagent.

it is a modification, and has the additional advantage that the reagent is not so corrosive.

4. **Ulrich's test** avoids the somewhat confusing colored rings. The reagent consists of saturated solution of common salt, 98 c.c.; glacial acetic acid, 2 c.c. It must be perfectly clear. Boil a few cubic centimeters of this fluid in a test-tube, and immediately overlay with the urine as in the preceding tests. Albumin and globulin give a white ring at the zone of contact.

5. **Purdy's Heat Test.**—Take a test-tube two-thirds full of urine, add about one-sixth its volume of saturated solution of sodium chlorid, and 5 to 10 drops of 50 per cent. acetic acid. Mix, and boil the upper inch, holding the tube with the fingers near the bottom. A white cloud in the heated portion shows the presence of albumin. A faint cloud is best seen when viewed against a black background at a distance of two or three feet.

This is a valuable test for routine work. It is simple, sufficiently accurate for clinical purposes, and has practically no fallacies. *Addition of the salt solution, by raising the specific gravity, prevents precipitation of mucin.* Bence-Jones' protein may produce a white cloud, which disappears upon boiling and reappears upon cooling.

6. **Heat and Nitric Acid Test.**—This is one of the oldest of the albumin tests, and, if properly carried out, one of the best. Boil about 5 c.c. of filtered urine in a test-tube and add 1 to 3 drops of concentrated nitric acid. The tube may be held with a test-tube clamp or simply with a strip of muslin, the center of which is folded once around the neck of the tube. A white cloud or flocculent precipitate (which usually appears during the boiling, but if the quantity be very small only after addition of the acid) denotes the presence of albumin. A similar white precipitate, which disappears upon addition of the acid, is due to earthy phosphates. The acid should not be added before boiling, and the proper amount should always be used; otherwise, part of the albumin may fail to be precipitated or may be transformed to acid-albumin and redissolved.

A decided advantage of this test is the fact that it allows a rough estimation of the amount of albumin from the volume of the sediment after standing over night. When the entire fluid solidifies the albumin amounts to 2 to 3 per cent. Sediments reaching to one-half, one-third, one-fourth and one-tenth the height of the column of urine correspond respectively to about 1, 0.5, 0.25, and 0.1 per cent. albumin. When there is only a slight cloudiness the albumin does not exceed 0.01 per cent.

Quantitative Estimation.—The gravimetric, which is the most reliable method, is too elaborate for clinical work. The three which follow are simple and are very widely used but none is entirely satisfactory.

1. **Esbach's Method.**—The urine must be clear, of acid reaction, and not concentrated. Always filter before testing, and, if necessary, add acetic acid and dilute with water, making allowance for the dilution in the final calculation. Esbach's tube (Fig. 42) is essentially a test-tube with a mark U near the middle, a mark R near the top, and graduations ½, 1, 2, 3, etc., near the bottom. Fill the tube to the mark U with urine and to the mark R with the reagent. Close with a rubber stopper, invert slowly several times, and set aside in a cool place. At the end of twenty-four hours read off the height of the precipitate. This gives the amount of albumin in *grams per liter, and must be divided by 10 to obtain the percentage.*

Fig. 42.—Esbach's albuminometer, improved form.

Lenk advises addition of a small quantity of powdered charcoal, pumice or kaolin after adding Esbach's reagent. This hastens sedimentation which is complete in ten minutes to half an hour. Andresen uses 0.1 to 0.2 Gm. of barium sulphate.

Esbach's reagent consists of picric acid, 1 Gm., citric acid, 2 Gm., and distilled water, to make 100 c.c.

2. **Tsuchiya's Method.**—This is carried out in the same manner as the Esbach method, using the following reagent:

Phosphotungstic acid..................... 1.5 Gm.;
Alcohol (96 per cent.)..................... 95.0 c.c.;
Concentrated hydrochloric acid........... 5.0 c.c.

The urine should be diluted to a specific gravity not exceeding 1.008. The method is said to be much more accurate than the original Esbach method, particularly with small quantities of albumin, but in the writer's work this has not proved to be true.

3. **Purdy's Centrifugal Method.**—This is detailed in the table on opposite page. Since 10 c.c. of urine were used, each 0.1 c.c. of precipitate is 1 per cent. by bulk.

(2) **Mucin.**—Traces of the substances (mucin, mucoid, nucleo-protein, etc.) which are loosely classed under this name are present in normal urine; increased amounts are observed in irritations and inflammations of the mucous membrane of the urinary tract. They are of interest chiefly because they may be mistaken for albumin in most of the tests. If the urine be diluted with water and acidified with acetic acid, the appearance of a white cloud indicates the presence of mucin.

Mucin and mucoid are glyco-proteins, and upon boiling with an acid or alkali, as in Fehling's test, yield a carbohydrate substance which reduces copper.

(3) **Bence-Jones' Protein.**—The protein known by this name was originally classed as an albumose, but its protein nature is now well established. It was formerly regarded as practically pathognomonic of multiple myeloma but has recently been found in a number of cases of chronic leukemia, of both lymphatic and myelogenous types and in osteomalacia.

To detect Bence-Jones' protein the urine is slightly acidified with acetic acid and gently heated in a water-bath.

PURDY'S QUANTITATIVE METHOD FOR ALBUMIN IN URINE (CENTRIFUGAL)

Table showing the relation between the volumetric and gravimetric percentage of albumin obtained by means of the centrifuge with radius of six and three-quarter inches; rate of speed, 1500 revolutions per minute; time, three minutes.

Volumetric percentage by centrifuge.	Percentage by weight of dry albumin.	Volumetric percentage by centrifuge.	Percentage by weight of dry albumin.	Volumetric percentage by centrifuge.	Percentage by weight of dry albumin.
1/4	0.005	13 1/2	0.281	31 1/2	0.656
1/2	0.01	14	0.292	32	0.667
3/4	0.016	14 1/2	0.302	32 1/2	0.677
1	0.021	15	0.313	33	0.687
1 1/4	0.026	15 1/2	0.323	33 1/2	0.698
1 1/2	0.031	16	0.333	34	0.708
1 3/4	0.036	16 1/2	0.344	34 1/2	0.719
2	0.042	17	0.354	35	0.729
2 1/4	0.047	17 1/2	0.365	35 1/2	0.74
2 1/2	0.052	18	0.375	36	0.75
2 3/4	0.057	18 1/2	0.385	36 1/2	0.76
3	0.063	19	0.396	37	0.771
3 1/4	0.068	19 1/2	0.406	37 1/2	0.781
3 1/2	0.073	20	0.417	38	0.792
3 3/4	0.078	20 1/2	0.427	38 1/2	0.801
4	0.083	21	0.438	39	0.813
4 1/4	0.089	21 1/2	0.448	39 1/2	0.823
4 1/2	0.094	22	0.458	40	0.833
4 3/4	0.099	22 1/2	0.469	40 1/2	0.844
5	0.104	23	0.479	41	0.854
5 1/2	0.111	23 1/2	0.49	41 1/2	0.865
6	0.125	24	0.5	42	0.875
6 1/2	0.135	24 1/2	0.51	42 1/2	0.885
7	0.146	25	0.521	43	0.896
7 1/2	0.156	25 1/2	0.531	43 1/2	0.906
8	0.167	26	0.542	44	0.917
8 1/2	0.177	26 1/2	0.552	44 1/2	0.927
9	0.187	27	0.563	45	0.938
9 1/2	0.198	27 1/2	0.573	45 1/2	0.948
10	0.208	28	0.583	46	0.958
10 1/2	0.219	28 1/2	0.594	46 1/2	0.969
11	0.229	29	0.604	47	0.979
11 1/2	0.24	29 1/2	0.615	47 1/2	0.99
12	0.25	30	0.625	48	1.0
12 1/2	0.26	30 1/2	0.635		
13	0.271	31	0.646		

Test.—Three cubic centimeters of 10 per cent. solution of ferrocyanid of potassium and 2 cubic centimeters of 50 per cent. acetic acid are added to 10 cubic centimeters of the urine in the percentage tube and *stood aside for ten minutes*, then placed in the centrifuge and revolved at rate of speed and time as stated at head of the table. If albumin is excessive, dilute the urine with water until volume of albumin falls below 10 per cent. Multiply result by the number of dilutions employed before using the table.

If this substance be present, the urine will begin to be turbid at about 40°C. and a precipitate will form at about 60°C. As the boiling-point is reached the precipitate wholly or partially dissolves. It reappears upon cooling. It may easily be overlooked in the presence of albumin.

(4) **Proteoses.**—These are intermediate products in the digestion of proteins and are frequently, although incorrectly, called albumoses. Two groups are generally recognized: *primary proteoses*, which are precipitated upon half-saturation of their solutions with ammonium sulphate; and *secondary proteoses*, which are precipitated only upon complete saturation.

The secondary proteoses have been observed in the urine in febrile and malignant diseases and chronic suppurations, during resolution of pneumonia, and in many other conditions, but their clinical significance is indefinite. In pregnancy, albumosuria may be due to absorption of amniotic fluid.

Primary proteoses are rarely encountered in the urine.

The proteoses are not coagulable by heat, but are precipitated by such substances as trichloracetic acid, sulphosalicylic acid, and phosphotungstic acid. The primary proteoses alone are precipitated by concentrated nitric acid.

Proteoses may be detected by acidifying the urine with acetic acid, boiling, filtering while hot to remove mucin, albumin, and globulin, and testing the filtrate by the trichloracetic acid test. As above indicated, the nitric acid test, and half and complete saturation with ammonium sulphate, will separate the two groups.

2. Sugars.—Various sugars may at times be found in the urine. Dextrose is by far the most common, and is the only one of much clinical importance. Levulose, lactose, and some others are occasionally met.

(1) **Dextrose** (**Glucose**).—Traces of glucose, too small to respond to the ordinary tests, are present in the urine in health. Its presence in appreciable amount constitutes "glycosuria" and is almost uniformly a result of hyperglycemia.

Transitory glycosuria is unimportant, and may occur in many conditions, as after general anesthesia and administration of certain drugs, in pregnancy, and following shock and head injuries. Recently attention has been directed to glycosuria following strong emotions (anger, fear, anxiety) due, according to Cannon, to increased adrenal secretion. The urine of a considerable percentage of a class of students will give positive tests for sugar following a long and hard examination. The possibility that a trace of sugar found in a patient's urine after a physical examination may be due to his anxiety must be kept in mind. Glycosuria may also occur after eating excessive amounts of carbohydrates (alimentary glycosuria). The "assimilation limit" varies with different individuals and with different conditions of exercise. It also depends upon the kind of carbohydrate. The normal for glucose is about 100 to 150 Gm. When more than this amount is taken at one time some of it will be excreted in the urine. Excretion lasts for a period of four or five hours.

Persistent glycosuria has been noted in brain injuries involving the floor of the fourth ventricle. As a rule, however, persistent glycosuria is diagnostic of diabetes mellitus, of which disease it is the essential symptom. The amount of glucose eliminated in diabetes is usually

11

considerable, and is sometimes very large, reaching 500 grams, or even more, in twenty-four hours, but it does not bear any uniform relation to the severity of the disease. Glucose may, on the other hand, be almost or entirely absent temporarily and in mild cases it may appear only at certain hours of the day.

Detection of Dextrose.—If albumin be present in more than traces, it must be removed by boiling and filtering.

1. **Haines' Test.**—Take about 4 c.c. of Haines' solution in a test-tube, boil, examine carefully for a precipitate, and, if none is present, add 6 or 8 drops of urine. A heavy yellow or red precipitate, which settles readily to the bottom, shows the presence of sugar. Neither precipitation of phosphates, as a light, flocculent sediment, nor simple decolorization of the reagent should be mistaken for a positive reaction.

This is one of the best of the copper tests, all of which depend upon the fact that in strongly alkaline solutions glucose reduces cupric hydrate to cuprous hydrate (yellow) or cuprous oxid (red). They are somewhat inaccurate, because they make no distinction between glucose and less common forms of sugar; because certain normal substances, when present in excess, especially mucin, uric acid, and creatinin, may reduce copper, and because many drugs— *e.g.*, chloral, chloroform, copaiba, acetanilid, benzoic acid, morphin, sulphonal, salicylates—are eliminated as copper-reducing substances. To minimize these fallacies *dilute the urine, if it be concentrated; do not add more than the specified amount of urine, and do not boil after the urine is added*. If chloroform has been used as a preservative, it should be removed by boiling the urine before making the test.

Haines' solution is prepared as follows: Completely dissolve 2 Gm. pure copper sulphate in 16 c.c. distilled water, and add 16 c.c. pure glycerin; mix thoroughly, and add 156 c.c. liquor potassæ. The solution keeps well.

2. **Fehling's Test.**—Two solutions are required—one containing 34.64 Gm. pure crystalline copper sulphate in 500 c.c. distilled water; the other, 173 Gm. Rochelle salt and 100 Gm. potassium hydroxid in 500 c.c. distilled water. Mix equal parts of the two solutions in a test-tube, dilute with 3 or 4 volumes of water, and boil. Add the urine a little at a time, heating, but not boiling, between additions. In the presence of glucose a heavy red or yellow precipitate will appear. The quantity of urine should not exceed that of the reagent. The fallacies mentioned under Haines' test apply equally to this.

3. **Benedict's Test.**—This new test promises to displace all other reduction tests for glucose. The reagent is said to be ten times as sensitive as Haines' or Fehling's, and not to be reduced by uric acid, creatinin, chloroform, or the aldehyds. It consists of:

Copper sulphate (pure crystallized)........ 17.3 Gm.;
Sodium or potassium citrate............... 173.0 Gm.;
Sodium carbonate (crystallized).......... 200.0 Gm.;
   (or 100 Gm. of the anhydrous salt).
Distilled water, to make................. 1000.0 c.c.

Dissolve the citrate and carbonate in 700 c.c. of water, with the aid of heat, and filter. Dissolve the copper in 100 c.c. of water and pour slowly into the first solution, stirring constantly. Cool, and make up to one liter. The reagent keeps indefinitely. *It can not be used for quantitative estimations.*

Take about 5 c.c. of this reagent in a test-tube, and add 8 *or* 10 *drops* (*not more*) of the urine. Heat to vigorous boiling, keep at this temperature for one or two minutes, and allow to cool slowly. In the presence of glucose the entire body of the solution will be filled with a precipitate, which may be red, yellow, or green in color. When traces only of glucose are present, the precipitate may appear only upon cooling. In the absence of glucose, the solution remains clear or shows only a faint, *bluish* precipitate, due to urates.

4. **Phenylhydrazin Test.**—*Kowarsky's Method.*—The following directions include certain modifications which have

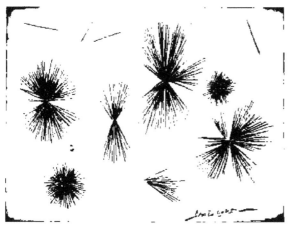

FIG. 43.—Crystals of phenylglucosazone from diabetic urine—Kowarsky's test (× 500).

been worked out by C. S. Bluemel in the writer's laboratory: In a wide test-tube take 5 drops pure phenylhydrazin, 10 drops glacial acetic acid, and 1 c.c. saturated solution of sodium chlorid. A curdy mass results. Add 3 or 4 c.c. of the urine and 4 or 5 c.c. of water. Boil vigorously for two or three minutes. The annoying bumping can be reduced or obviated by shaking continually, or, much better, by placing in the test-tube a number of pieces of glass tubing, varying in length from 1½ to 3 inches, so as to produce an organ-pipe effect. The volume of fluid remaining after boiling should be 2 to 3 c.c. Set aside to cool, or if the glass tubes were used pour the fluid into another hot test-tube and allow to cool. Examine the sediment with the microscope, using a two-thirds objective. If glucose be present, characteristic crystals of phenyl-glucosazone will be seen. These are yellow, needle-like crystals arranged mostly in clusters or in sheaves (Fig. 43). When traces only of glucose are present, the crystals may not appear for one-half hour or more. The best crystals are obtained when the fluid is cooled very slowly. It must not be agitated during cooling. The test-tubes and pieces of tubing can be cleaned when necessary by boiling in a solution of caustic soda or acetic acid.

This is an excellent test for clinical work. Bluemel finds that when applied as above directed, with the tubing to prevent bumping, it will readily detect 0.025 per cent. of glucose in urine, the crystals appearing in three to four hours. The test has practically no fallacies excepting levulose, which is a fallacy for all the ordinary tests. Other carbohydrates which are capable of forming crystals with phenylhydrazin are extremely unlikely to do so when the test is applied directly to the urine. Even if not used routinely, this test should always be resorted to when the copper tests give a positive reaction in doubtful cases.

5. **Fermentation Test.**—This is simple and reliable, but owing to the time required it is not much used in routine work, except as an aid in distinguishing dextrose from other forms of sugar. It is carried out in the same manner as the quantitative test (see p. 169). A home-made device which answers well for the purpose is shown in Fig. 44.

Quantitative Estimation.—In quantitative work Fehling's solution, for so many years the standard, has been largely displaced by Benedict's quantitative solution, which appears to be more exact and more satisfactory than any other titration method available for sugar work. The older method is still preferred by some and both are therefore given.

Should the urine contain much glucose, it must be

diluted before making any quantitative test, allowance being made for the dilution in the subsequent calculation. Albumin, if present, must be removed by acidifying a considerable quantity of urine with acetic acid, boiling, and filtering. Any water lost during the boiling should be replaced before filtering.

A rough but sometimes useful approximation of the amount of sugar in the urine of a diabetic patient can be made by estimating the total solids (see p. 110), subtracting what may be regarded as normal for the individual and regarding the remainder as sugar.

Fig. 44.—Simple device for fermentation test for dextrose.

1. **Fehling's Method.**—Take 10 c.c. of Fehling's solution (made by mixing 5 c.c. each of the copper and alkaline solutions described on page 163) in a flask or beaker, add 3 or 4 volumes of water, boil, and add the urine very slowly from a buret until the solution is completely decolorized, heating but not boiling after each addition.

Fehling's solution is of such strength that the copper in 10 c.c. will be reduced by exactly 0.05 Gm. of glucose. Therefore, the amount of urine required to decolorize the test solution contains just 0.05 Gm. glucose, and the percentage is easily calculated.

The chief objection to Fehling's method is the difficulty of determining the end-point. The use of an "outside indicator," however, obviates this. When reduction is thought to be complete, a few drops of the solution are filtered through a fine-grained filter-paper on to a porcelain plate, quickly acidified with acetic acid, and mixed with a drop of 10 per cent. potassium ferrocyanid. Immediate appearance of a red-brown color shows the presence of unreduced copper.

A somewhat simpler application of this method, which is accurate enough for most clinical purposes, is as follows: Take 1 c.c. of Fehling's solution in a large test-tube, dilute with about 5 c.c. of water, heat to boiling, and, while keeping the solution hot but not boiling, add the urine drop by drop from a medicine-dropper until the blue color is entirely gone. Toward the end add the drops very slowly, not more than 4 or 5 a minute. Divide 10 by the number of drops required to discharge the blue color; the quotient will be the percentage of glucose. If 20 drops were required, the percentage would be $10 \div 20 = 0.5$ per cent. It is imperative that the drops be of such size that 20 of them will make 1 c.c. Test the dropper with urine, not water, and hold it always at the angle which will give the right sized drop. If the drops are too large, draw out the tip of the dropper; if too small, cut off the tip.

2. **Benedict's Method.**—The following modification of his copper solution has been offered by Benedict for quantitative estimations.

The reagent consists of:

Copper sulphate (pure crystallized)............	18.0 Gm.;
Sodium carbonate (crystallized)................	200.0 Gm.;
(or 100 Gm. of the anhydrous salt).	
Sodium or potassium citrate..................	200.0 Gm.;

Potassium sulphocyanate...................... 125.0 Gm.;
Potassium ferrocyanid solution (5 per cent.)...... 5.0 c.c.;
Distilled water, to make...................... 1000.0

With the aid of heat dissolve the carbonate, citrate, and sulphocyanate in about 700 c.c. of the water and filter. Dissolve the copper in 100 c.c. of water and pour slowly into the other fluid, stirring constantly. Add the ferro-cyanid solution, cool, and dilute to 1000 c.c. Only the copper need be accurately weighed. This solution is of such strength that 25 c.c. are reduced by 0.05 Gm. glucose. It keeps well.

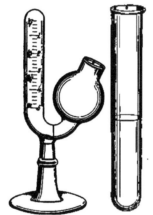

FIG. 45.—Einhorn's saccharimeter.

To make a sugar estimation, take 25 c.c. of the reagent in a small flask, add 10 to 20 Gm. of sodium carbonate crystals (or one-half this weight of the anhydrous salt) and a small quantity of powdered pumice-stone or talcum. Heat to boiling, and add the urine rather rapidly from a buret until a chalk-white precipitate forms and the blue color of the reagent begins to fade. After this point is reached, add the urine a few drops at a time until the last trace of blue just disappears. This end-point is easily recognized. During the whole of the titration the mixture must be kept vigorously boiling. Loss by evaporation must be made up by adding water. The quantity of urine required to discharge the blue color contains exactly 0.05 Gm. glucose, and the percentage contained in the original sample is easily calculated.

3. **Fermentation Method.**—This is convenient and satisfactory, its chief disadvantage being the time required. It depends upon the fact that glucose is fermented by yeast with evolution of CO_2. The amount of gas evolved is an index of the amount of glucose. No preservative must have been added. Einhorn's saccharimeter (Fig. 45) is the simplest apparatus.

The urine must be so diluted as to contain not more than 1 per cent. of glucose. A fragment of fresh yeast-cake about the size of a split-pea is mixed with a definite quantity of the urine measured in the tube which accompanies the apparatus. The exact amounts of yeast and urine are unimportant. It should form an emulsion free from lumps or air-bubbles. The long arm of the apparatus and about half the bulb are then filled with the mixture, all bubbles being carefully discharged by tipping the instrument with the thumb over the opening, and the instrument is stood in a warm place. At the end of fifteen to twenty-four hours fermentation will be complete, and the percentage of glucose can be read off upon the side of the tube. The result must then be multiplied by the degree of dilution. Since yeast itself sometimes gives off gas, a control test must be carried out with normal urine and the amount of gas evolved must be subtracted from that of the test. A control should also be made with a known glucose solution to make sure that the yeast is active. It has recently been shown that yeast can split off carbon dioxid from amino-acids, so that the results with the fermentation method are likely to be a little high.

The method is not applicable to urine which is undergoing ammoniacal fermentation.

4. Roberts' Differential Density Method.—While this method gives only approximate results, it is convenient, and requires no special apparatus but an accurate urinometer. Mix a quarter of an yeast-cake with about 100 c.c. of urine. Take the specific gravity and record it. Set the urine in a warm place for twenty-four hours or until fermentation is complete. Then cool to the temperature at which the specific gravity was originally taken, and take it again. The difference between the two readings gives the number of grains of sugar per ounce, and this, multiplied by 0.234, gives the *percentage* of sugar. If the original reading is 1.035, and that after fermentation is 1.020, the urine contains 1.035 − 1.020 = 15 gr. of sugar per fluidounce; and the percentage equals 15 × 0.234 = 3.5.

(2) **Levulose, or fruit sugar,** is seldom present in urine except in association with dextrose, and has about the same significance. According to von Noorden, its appearance in diabetes indicates an advanced case. Its name is derived from the fact that it rotates polarized light to the left.

The normal assimilation limit for levulose is about 100 Gm. This fact is used in the Strauss test of the functional capacity of the liver. One hundred grams of levulose are given upon an empty stomach, and the subsequent appearance of levulose in the urine is taken as evidence of deficiency of the glycogenic function. The degree of the hepatic derangement is measured by a quantitative estimation.

Detection of Levulose.—Levulose responds to all the tests above given for dextrose. It may be distinguished from dextrose by the following:

Borchardt's Test.—Mix about 5 c.c. each of the urine and 25 per cent. hydrochloric acid (concentrated HCl, 2 parts; water, 1 part) in a test-tube and add a few crystals of resorcinol. Heat to boiling and boil for not more than one-half minute. In the presence of levulose a red color appears. Cool in running water, pour into a beaker, and render slightly alkaline with solid sodium or potassium hydroxid. Return to the test-tube, add 2 or 3 c.c. of acetic ether, and shake. If levulose be present, the ether will be colored yellow. A similar yellow color will follow administration of rhubarb and senna.

If indican be present the test must be modified as follows: Perform Obermayer's test and extract the indican with chloroform. Reduce the acidity of the indican-free urine by adding one-third its volume of water, add a few crystals of resorcinol, and proceed with Borchardt's test.

Quantitative Estimation of Levulose.—The methods are the same as for dextrose (see p. 165). Twenty-five cubic centimeters of Benedict's quantitative solution are reduced by 0.053 Gm. levulose.

(3) **Lactose,** or **milk-sugar,** is sometimes present in the urine of nursing women and in that of women who have recently miscarried. It is of interest chiefly because it may be mistaken for glucose. *It reduces copper, but does not ferment with yeast.* In strong solution it can form crystals with phenylhydrazin, but is extremely unlikely to do so when the test is applied directly to the urine.

(4) **Maltose** and **cane-sugar** are of little or no clinical mportance. Maltose has been found along with dextrose in diabetes. It reduced copper, 0.074 Gm. being equivalent to 25 c.c. of Benedict's solution. Cane-sugar

(sucrose) is sometimes added to the urine by malingering patients. It does not reduce copper.

(5) **Pentoses.**—These sugars are so named because the molecule contains 5 atoms of carbon. Vegetable gums form their chief source. They reduce copper strongly but slowly, and give crystals with phenyl-hydrazin, but do not ferment with yeast.

Pentosuria is uncommon. It has been noted after ingestion of large quantities of pentose-rich substances, such as cherries, plums, and fruit-juices, and is said to be fairly constant in habitual use of morphin. It sometimes accompanies glycosuria in diabetes. An obscure chronic form of pentosuria without clinical symptoms has been observed. The pentose excreted in these cases is believed to be optically inactive arabinose, although recent work indicates that ribose is present in some cases at least.

Bial's Orcinol Test.—Dextrose is first removed by fermentation. About 5 c.c. of Bial's reagent are heated in a test-tube, and after removing from the flame the urine is added drop by drop, not exceeding 20 drops in all. The appearance of a green color denotes pentose.

The reagent consists of:

Hydrochloric acid (30 per cent.).......... 500 c.c.;
Ferric chlorid solution (10 per cent.)....... 25 drops;
Orcinol............................... 1 Gm.

3. Acetone Bodies.—This is a group of closely related substances—acetone, diacetic acid, and beta-oxybutyric acid—whose chief source is in abnormal katabolism of fats. Formerly beta-oxybutyric acid was held to be the mother substance, but it is now believed that diacetic acid is first formed and that the others are derived from it. In general, their presence in the urine is a sign of acidosis or "acid-intoxication." When the disturbance is mild, acetone occurs alone; as it becomes more marked, diacetic acid is also found, and finally beta-oxybutyric acid appears.

(1) **Acetone.**—Minute traces, too small for the ordinary tests, may be present in the urine under normal conditions. Larger amounts are not uncommon when the intake of carbohydrates is limited and in fevers, gastro-intestinal disturbances, and certain nervous disorders. A notable degree of acetonuria has likewise been observed in pernicious vomiting of pregnancy and in eclampsia.

Acetonuria is practically always observed in acid intoxication, and, together with diaceturia, constitutes its most significant diagnostic sign. A similar or identical toxic condition, always accompanied by acetonuria and often fatal, is now recognized as a not infrequent late effect of anesthesia, particularly of chloroform anesthesia. This postanesthetic toxemia is more likely to appear, and is more severe, when the urine contains any notable amount of acetone before operation, which suggests the importance of routine examination for acetone in surgical cases.

Acetone is present in considerable amounts in many cases of diabetes mellitus, and is always present in severe cases. Its amount is a better indication of the severity of the disease than is the amount of sugar. A progressive increase is a grave prognostic sign. It can be diminished temporarily by more liberal allowance of carbohydrates in the diet, by addition of certain vegetables to the diet (because of their content of alkaline salts), and by administration of bicarbonates.

Acetonuria from any cause is apt to be especially marked in children, and this doubtless plays an important part in acute and chronic diseases of childhood, especially in those requiring a restricted diet. In fact, the urine of a considerable percentage of young children shows acetone under normal conditions.

FIG. 46.—A simple distilling apparatus.

According to Folin, acetone is usually present in only small amounts in the above mentioned conditions, the substance shown by the usual tests, particularly after distillation of the urine, being really diacetic acid. In this connection, Frommer's test is to be recommended, since it does not require distillation, and does not react to diacetic acid unless too great heat is applied.

Detection of Acetone.—The urine may be tested directly, but it is much better to distil it after adding a little phosphoric or hydrochloric acid to prevent foaming, and to test the first few cubic centimeters of distillate. A simple distilling apparatus is shown in Fig. 46. The test-tube may be attached to the delivery tube by means of a two-hole rubber cork as shown, the second hole serving as air vent, or, what is much less satisfactory, it may be tied in place with a string. Should the vapor not condense well, the test-tube may be immersed in a glass of cold water. If a sufficiently long delivery tube be used (2 feet) there will be little difficulty about condensation.

When diacetic acid is present, a considerable proportion will be converted into acetone during distillation.

Owing to the marked and variable loss of acetone through the lungs a quantitative estimation is not of much value. After the existence of an acidosis has been established by the detection of acetone bodies, it is better to rely upon an estimation of ammonia as a measure of its severity.

1. **Gunning's Test.**—To about five cubic centimeters of urine or distillate in a test-tube add 5 drops of strong ammonia and then Lugol's solution in sufficient quantity to produce a black cloud which does not immediately disappear. This cloud will gradually clear up and, if acetone be present, iodoform, usually crystalline, will separate out. The iodoform can be recognized by its odor, especially upon heating (there is danger of explosion if the mixture be heated before the black cloud disappears), or by detection of the crystals microscopically. The latter, alone, is dependable, unless one has an acute

sense of smell.   The odor of iodin, which is also present,
is often confusing.   Iodoform crystals are yellowish six-
pointed stars or six-sided plates (Fig. 47).

This modification of Lieben's test is less sensitive than the
original, but is sufficient for all clinical work; it has the advantage that alcohol does not cause confusion, and especially
that the sediment of iodoform is practically always crystalline.   When it is applied directly to the urine, phosphates are precipitated and may form large feathery, star-

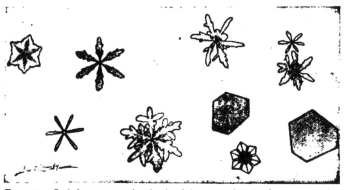

FIG. 47.—Iodoform crystals obtained in several tests for acetone by
Gunning's method (× about 600).

shaped crystals which are confusing to the inexperienced
(see Fig. 56).   Albumin prevents formation of the crystals,
and when it is present, the urine must be distilled for the
test.

2. **Lange's Test.**—This is a modification of the well-
known Legal test.   It is more sensitive and gives a sharper
end-reaction.   To a small quantity of urine add about one-
twentieth its volume (1 drop for each 1 c.c.) of glacial acetic
acid and a few drops of fresh concentrated aqueous solution
of sodium nitroprussid, and gently run a little ammonia
upon its surface.   If acetone be present, a reddish-purple

ring will form within a few minutes at the junction of the two fluids.

Lange's test is even more sensitive to diacetic acid than to acetone. For this reason, **Rothera's test,** which is more sensitive to acetone, is to be preferred: To 5 or 10 c.c. of urine add about a gram of ammonium sulphate, and 2 or 3 drops of fresh concentrated sodium nitroprussid solution and overlay with strong ammonia. A permanganate colored ring shows the presence of acetone.

3. **Frommer's Test.**—This test has proved very satisfactory in the hands of the writer. The urine need not be distilled. Alkalinize about 10 c.c. of the urine with 2 or 3 c.c. of 40 per cent. caustic soda solution, add 10 or 12 drops of 10 per cent. alcoholic solution of salicylous acid (salicyl aldehyd), heat the upper portion to about 70°C. (it should not reach the boiling-point), and keep at this temperature five minutes or longer. In the presence of acetone an orange color, changing to deep red, appears in the heated portion A yellow to brown color may appear in the absence of acetone.

The test can be made more definite by adding the caustic soda in substance (about 1 Gm.), and before it goes into solution adding the salicyl aldehyd and warming the lower portion.

(2) **Diacetic (aceto-acetic) acid** occurs in the same conditions as acetone, but has more serious significance. In diabetes its presence is a grave symptom and often forewarns of approaching coma. It rarely or never occurs without acetone.

Detection.—The urine must be fresh. If a preservative must be used, toluene is best.

Gerhardt's Test.—To a few cubic centimeters of the urine add solution of ferric chlorid (about 10 per cent.)

12

drop by drop until the phosphates are precipitated; filter and add more of the ferric chlorid. If diacetic acid be present, the urine will assume a Bordeaux-red color which disappears upon boiling. Several minutes boiling are required; simply bringing the fluid to the boiling-point will not suffice. A red or violet color which does not disappear upon boiling may be produced by other substances, as phenol, salicylates, and antipyrin. Whenever the reaction is doubtful the urine should be distilled and the distillate tested for acetone.

The test is somewhat more definite if applied by the contact or "ring" method (see p. 152).

(3) **Oxybutyric acid** has much the same significance as diacetic acid, but is of more serious import.

Hart's Test.—Remove acetone and diacetic acid by diluting 20 c.c. urine with 20 c.c. water, adding a few drops of acetic acid, and boiling down to 10 c.c. To this add 10 c.c. water, mix, and divide between two test-tubes. To one tube add 1 c.c. of hydrogen peroxid, warm gently, and cool. This transforms β-oxybutyric acid to acetone. Now apply Lange's test for acetone (see p. 176) to each tube. A positive reaction in the tube to which hydrogen peroxid has been added shows the presence of β-oxybutyric acid in the original sample of urine.

4. Bile.—The pigment of bile has its origin in the never-ceasing destruction of red blood-corpuscles within the body.

The significance of bile in the urine is practically the same as that of bile-staining of the tissues, known as icterus or jaundice. Small amounts of bile may, however, be found in the urine when the disturbance is not severe enough to produce recognizable jaundice or, in other cases, before the jaundice supervenes. The usual cause of icterus is some obstruction to the outflow of bile from the liver, which may be in the nature of foreign bodies or new growths inside or outside of the bile-passages, or inflammatory swelling of the walls with narrowing of the lumen. Jaundice may also occur when there is excessive destruction of red blood-corpuscles from any cause. This leads to excessive formation of a bile which is more inspissated than normal and thus tends to block the bile-capillaries. Another, less frequent, cause is rapid destruction of liver cells as in acute yellow atrophy and phosphorus poisoning. Strong emotion has also been known to cause jaundice in some obscure way. Both bile-pigment and bile acids may be found. They generally occur together. but the pigment is not infrequently present alone.

Of the several pigments bilirubin, alone, occurs in freshly voided urine, the others (biliverdin, bilifuscin, etc.) being produced from this by oxidation as the urine stands. The acids, which occur chiefly as sodium salts, are almost never present without the pigments, and are, therefore, seldom tested for clinically. Crystals of bilirubin (hematoidin) (Fig. 49, 4) may be deposited in heavily bile-charged urine.

Detection of Bile-pigment.—Bile-pigment gives the urine a greenish-yellow, yellow, or brown color, which upon shaking is imparted to the foam. Cells, casts, and other structures in the sediment may be stained brown or yellow. This, however, should not be accepted as proving the presence of bile without further tests.

1. **Smith's Test.**—Overlay the urine with tincture of iodin diluted with nine times its volume of alcohol. An

emerald-green ring at the zone of contact shows the presence of bile-pigments. It is convenient to use a conical test-glass, one side of which is painted white.

2. **Gmelin's Test.**—This consists in bringing slightly yellow nitric acid into contact with the urine. A play of colors, of which green and violet are most distinctive, denotes the presence of bile-pigment. Blue and red may be produced by indican and urobilin. Colorless nitric acid will become yellow upon standing in the sunlight. The test may be applied in various ways: by overlaying the acid with the urine; by bringing a drop of each together upon a porcelain plate; by filtering the urine through thick filter-paper, and touching the paper with a drop of the acid; and, probably best of all, by precipitating with lime-water, filtering, and touching the precipitate with a drop of the acid. In the last method bilirubin is carried down as an insoluble calcium compound which concentrates the pigment and avoids interfering substances.

Detection of Bile Acids.—Hay's test is simple, sensitive, and fairly reliable, and will, therefore, appeal to the practitioner. It depends upon the fact that bile acids lower surface tension. Other tests require isolation of the acids for any degree of accuracy.

Hay's Test.—Upon the surface of the urine, *which must not be warm*, sprinkle a little finely powdered sulphur ("flowers of sulphur"). If it sinks at once, bile acids are present to the amount of 0.01 per cent. or more; if only after gentle shaking, 0.0025 per cent. or more. If it remains floating, even after gentle shaking, bile acids are absent. It is said that urobilin when present in large amount also reduces surface tension.

5. Hemoglobin.—The presence in the urine of hemoglobin or pigments directly derived from it, accompanied by few, if any, red corpuscles, constitutes *hemoglobinuria*. It is a comparatively rare condition, and must be distinguished from *hematuria*, or *blood* in the urine, which is common. In both conditions chemic tests will show hemoglobin, but in the latter the microscope will reveal the presence of red corpuscles. Urines which contain notable amounts of hemoglobin have a reddish or brown color, and may deposit a sediment of brown, granular pigment.

Hemoglobinuria occurs when there is such extensive destruction of red blood-cells within the body that the liver cannot transform all the hemoglobin set free into bile-pigment. The most important examples are seen following extensive burns in poisoning, as by mushrooms and potassium chlorate, in scurvy and purpura, in malignant malaria (blackwater fever), and in the obscure condition known as "paroxysmal hemoglobinuria." This last is characterized by the appearance of large quantities of hemoglobin at intervals, usually following exposure to cold, the urine remaining free from hemoglobin between the attacks.

Detection.—Teichmann's test may be applied to the precipitate after boiling and filtering, but this is not very satisfactory, and the guaiac or benzidin test will be found more convenient in routine work. For further discussion of blood tests, including spectroscopic methods, see page 364.

Guaiac Test.—Mix a few cubic centimeters each of "ozonized" turpentine and a fresh 1 : 60 alcoholic solution of guaiac. The guaiac solution may be freshly prepared by dissolving a pocket-knife-pointful of powdered guaiac in 4 or 5 c.c. of alcohol. Carry out the test by the "ring" or "contact" method described on page 152, stratifying the guaiac-turpentine mixture upon the urine. A bright blue ring will appear at the zone of contact within a few minutes if hemoglobin be present. The guaiac should be kept in an amber-colored bottle. Fresh turpentine can be "ozonized" by allowing it to stand a few days in an open vessel in the sunlight. Instead of turpentine, hydrogen peroxid may be used.

This test is very sensitive, and a negative result proves the absence of hemoglobin. Positive results are not conclusive, because numerous other substances—few of them likely to be found in the urine—may produce the blue color. That most likely to cause confusion is pus, but the blue color produced by it disappears upon heating and will appear equally well if the turpentine be omitted. The thin film of copper often left in a test-tube after testing for sugar may give the reaction, as may also the fumes from an open bottle of bromin. Most sources of error can be avoided by extracting the hemoglobin with ether as described on page 364.

Benzidin Test.—The reagents employed are hydrogen peroxid and a saturated solution of benzidin in glacial acetic acid. The benzidin labeled "For blood tests" should be employed. The reagents are mixed in equal parts and a few cubic centimeters are added to a like amount of the urine. A blue color appears in the presence of hemoglobin. The test has the same fallacies as the guaiac test, but is more sensitive and in general more satisfactory.

The benzidin test may be simplified by use of the tablets recently put upon the market by the firm of E. R. Squibb & Sons. These contain benzidin and sodium perborate. A tablet is thoroughly moistened with the fluid to be tested and is then touched with a drop of glacial acetic acid, the appearance of a blue color indicating blood. The test is less delicate than those given above.

Spectroscopic Method.—This is discussed on page 366. It is more reliable than the preceding tests but less sensitive. Render the urine slightly acid, dilute if very highly colored and examine with a small direct-vision spectroscope. The usual bands seen are those of oxyhemoglobin and met-hemoglobin.

To detect traces of blood proceed as described in section 2 on page 368. This method will easily detect 2 drops of blood in 8 ounces of urine.

6. Alkapton Bodies.—The name "alkaptonuria" has been given to a condition in which the urine turns reddish brown to brownish black upon standing and strongly reduces copper (but not bismuth), owing to the presence of certain substances which result from imperfect protein metabolism. The chief of these is homogentisic acid. The change of color takes place quickly when fresh urine is alkalinized, hence the name, *alkapton bodies.*

Alkaptonuria is unaccompanied by other symptoms, and has little clinical importance. Only a few cases, mostly congenital, have been reported. The change in color of the urine and the reduction of copper with no reduction of bismuth nor fermentation with yeast would suggest the condition.

7. Melanin.—Urine which contains melanin likewise darkens upon exposure to the air, assuming a dark brown or black color. This is due to the fact that the substance is eliminated as a chromogen—melanogen—which is later converted into the pigment. It does not reduce copper.

Melanuria occurs in most, but not all, cases of melanotic tumor. Its diagnostic value is lessened by the fact that it has been observed in other wasting diseases.

Tests for Melanin.—1. Addition of ferric chlorid gives a gray precipitate which blackens on standing.

2. Bromin water causes a yellow precipitate which gradually turns black.

8. Hematoporphyrin is an iron-free pigment derived from hemoglobin. Its formation within the body is not well understood. Normally it appears in the urine only in slight traces. An increase has been observed in a variety of conditions, notably in organic liver diseases, in lead-poisoning, and, especially, during continued use of sulphonal, trional and tetronal. In the presence of abnormal amounts the urine may have a dark red or "port wine" color, which, however, appears to be due in part to the simultaneous presence of related but little-known pigments.

Hematoporphyrin does not respond to the guaiac or the hemin test and is best detected with the spectroscope. Treat 100 c.c. of the urine with 20 c.c. of 10 per cent. sodium hydroxid solution. The pigment will be carried down with the phosphates. Filter (or centrifugalize), wash the precipitate with water then with alcohol, and finally dissolve in about 5 c.c. of alcohol to which 5 to 10 drops of concentrated hydrochloric acid have been added. Examine the acid solution for the absorption bands of acid hematoporphyrin (Fig. 142).

9. Urobilin.—Traces of this pigment, too small for detection by the ordinary tests, are present under normal conditions. It is now regarded as identical with hydrobilirubin, the principal coloring matter of the feces. It is excreted as a chromogen, *urobilinogen*, which is changed into urobilin through the action of light within a few hours after the urine is voided. A great excess gives the urine a dark brown color suggesting the presence of bile, but does not color the foam as does bile. Small amounts cause no perceptible change in color.

The mode of formation of urobilin is not yet clearly understood. According to the generally accepted view it is a decomposition product of bilirubin, formed chiefly in the intestine through the action of bacteria. Upon the other hand, the formation of small amounts in the liver itself under certain conditions cannot be denied. Urobilinogen is first formed. Under normal conditions a portion of this is absorbed from the intestine, carried to the liver in the portal blood, and there reconverted into bilirubin. When the liver cells are deranged, this transformation into bilirubin does not take place and urobilinogen reaches the general circulation and is excreted by the kidneys. The remainder in the intestine, changed largely into urobilin, passes out with the feces. The pigment and the chromogen have exactly the same significance in the urine, and the name "urobilin" is commonly used to cover both.

Owing to the many unknown factors it is impossible to ascribe definite clinical significance to urobilinuria. Certain statements, however, seem justified. Whenever, owing to excessive destruction of blood-corpuscles, there is excessive formation of bilirubin and hence an increase of urobilin in the feces, there is also a marked increase in the urine. With this exception, urobilinuria usually points toward functional incapacity of the liver. Its recognition is very simple and has considerable practical usefulness, as for example in the diagnosis of hepatic cirrhosis; in judging the amount of damage done to the liver parenchyma by poisons and the chronic congestion of poorly compensated heart disease; and in differentiating anemias associated with excessive blood destruction (*e.g.*, pernicious anemia) from those due to other causes (carcinoma, hemorrhage). Urobilinuria is frequent in acute infectious diseases, especially in scarlet fever and pneumonia, and usually means either hemolysis or damage to the liver. In severe nephritis urobilin may fail to be excreted even when other conditions favor its appearance in the urine. It is nearly or entirely absent in obstructive jaundice.

To be of value, tests for urobilin should be made upon several successive days, because for some unknown reason it may be absent for a day or two, and it is advisable to make a simultaneous study of the urobilin of the feces.

1. **Ehrlich's Test for Urobilinogen.**—To a few cubic centimeters of the urine in a test-tube add a few crystals of para-dimethyl-amino-benzaldehyd and make definitely acid with hydrochloric acid. In the presence of pathologic amounts of urobilinogen a cherry-red color appears. This is best seen by viewing the tube from the top over a sheet of white paper. Normal amounts will cause the red color only when the urine is heated.

2. **Schlesinger's Test for Urobilin.**—To about 5 c.c. of the urine in a test-tube add a few drops of Lugol's solution to transform the chromogen into the pigment. Now add 4 or 5 c.c. of a saturated alcoholic solution of zinc acetate

or zinc chlorid and filter. A greenish fluorescence, best seen when the tube is viewed in bright sunlight against a black background and when the light is concentrated upon it with a lens, shows the presence of urobilin. The fluorescence becomes more marked after an hour or two. Bile-pigment, if present, should be previously removed by adding about one-fifth volume of 10 per cent. calcium chlorid solution and filtering.

Quantitative Estimation.—The indirect although clinically satisfactory method of Wilber and Addis which is given in detail on page 432 may be applied to the urine as follows:

1. To a 10-c.c. portion of the mixed twenty-four-hour urine, which has been preserved with a crystal of thymol and kept in darkness, add 10 c.c. of a saturated alcoholic solution of zinc acetate and filter.

2. To 10 c.c. of the filtrate (representing 5 c.c. of urine) add 1 c.c. of Ehrlich's reagent, the formula for which is as follows:

> Para-dimethyl-amino-benzaldehyd........... 10 gm.
> Concentrated hydrochloric acid.............. 75 c.c.
> Water................................... 75 c.c.

3. Let stand in the dark for one to three hours, not longer.

4. Examine with a small direct-vision spectroscope and dilute with tap water until absorption bands disappear. Calculate the total dilution value for the twenty-four-hour quantity of urine as described for urobilin in feces, *basing the calculation upon the 5 c.c. of urine used.*

Example.—If the twenty-four-hour urine amounts to 1200 c.c. and it is necessary to dilute the 10-c.c. filtrate to 50 c.c. to get rid of the absorption bands (supposing that they disappear together), then the combined dilution value for urobilin and urobilinogen in the 5 c.c. of urine would be 10 + 10 = 20; and the twenty-four-hour value would be 20 × 240 = 4800.

10. Diazo Substances.—Certain imperfectly known substances sometimes present in the urine give a characteristic color reaction—the "diazo-reaction" of Ehrlich—when treated with diazo-benzol-sulphonic acid and ammonia. This reaction has much clinical value, *provided its limitations be recognized*. It is at best an empirical test and must be interpreted in the light of clinical symptoms. Although it has been met with in a considerable number of diseases, its usefulness is practically limited to typhoid fever, tuberculosis, and measles.

(1) **Typhoid Fever.**—Practically all cases give a positive reaction, which varies in intensity with the severity of the disease. It is so constantly present that it is sometimes said to be "negatively pathognomonic:" if negative upon several successive days *at a stage of the disease when it should be positive*, typhoid is almost certainly absent. Upon the other hand, a reaction when the urine is highly diluted (1:50 or more) has much positive diagnostic value, since this dilution prevents the reaction in most conditions which might be mistaken for typhoid; but it should be noted that mild cases of typhoid may not give it at this dilution. Ordinarily the diazo- appears a little earlier than the Widal reaction—about the fourth or fifth day—but it may be delayed. In contrast to the Widal, it begins to fade about the end of the second week, and soon thereafter entirely disappears. An early disappearance is a favorable sign. It reappears during a relapse, and thus helps

to distinguish between a relapse and a complication, in which it does not reappear.

(2) **Tuberculosis.**—The diazo-reaction has been obtained in many forms of the disease. It has little or no diagnostic value. Its continued presence in pulmonary tuberculosis is, however, a grave prognostic sign, even when the physical signs are slight. After it once appears it generally persists more or less intermittently until death, the average length of life after its appearance being about six months. The reaction is often temporarily present in mild cases during febrile complications, and has then no significance.

(3) **Measles.**—A positive reaction is usually obtained in measles, and may help to distinguish this disease from German measles, in which it does not occur. It generally appears before the eruption and remains about five days.

Technic.—Although the test is really a very simple one, careful attention to technic is imperative. Many of the early workers were very lax in this regard. Faulty technic and failure to record the stage of the disease in which the tests were made have probably been responsible for the bulk of the conflicting results reported.

Certain drugs often given in tuberculosis and typhoid interfere with the reaction or prevent it. The chief are creosote, tannic acid and its compounds, opium and its alkaloids, salol, phenol, and the iodids. The reagents are:

(1) Sulphanilic acid..................... 1 Gm.;
   Concentrated hydrochloric acid...... 10 c.c.;
   Water........................... 200 c.c.;
(2) Sodium nitrite..................... 0.5 Gm.;
   Water........................... 100 c.c.
(3) Strong ammonia.

Mix 100 parts of (1) and 1 part of (2).[1] In a test-tube·
take equal parts of this mixture and the urine, and pour 1 or
2 c.c. of the ammonia upon its surface. If the reaction be
positive, a garnet ring will form at the junction of the two
fluids; and, upon shaking, a distinct pink color will be imparted to the foam. The color of the foam is the essential
feature. If desired, the mixture may be well shaken before
the ammonia is added: the pink color will then instantly
appear in that portion of the foam which the ammonia has
reached, and can be readily seen. The color varies from
eosin-pink to deep crimson, depending upon the intensity
of the reaction. *It is a pure pink or red; any trace of yellow
or orange denotes a negative reaction.* A doubtful reaction
should be considered negative.

Substitutes for the Diazo-reaction.—The two following tests, which have been offered as simple and satisfactory substitutes for the diazo, have found rather
wide acceptance. They are supposed to be positive in
the same classes of cases as the diazo and to have the
same clinical significance, but are claimed to be more
reliable.

1. **Weis's Urochromogen Test.**—In a test-tube mix 2 c.c.
of urine and 4 c.c. distilled water, and add 3 drops of 1 : 1000
aqueous solution of potassium permanganate. The appearance of a yellow color denotes a positive reaction. The
color is best judged by comparison with a tube of diluted
urine to which no permanganate has been added, the two
tubes being viewed from the top over a sheet of white
paper. The color of a genuine reaction is a canary yellow. A yellow color, usually not so bright, and tending
more toward brown, may be produced by urobilin and

---

[1] These proportions are recommended by Greene, and are now generally used. Ehrlich used 40 parts of (1) and 1 part of (2).

other substances, but these false reactions fade quickly, usually within thirty seconds, while the color of a true reaction remains a longer time.

Weis believes the diazo-reaction to be due principally to urochromogen, which, because of the effect of certain toxins upon metabolism, fails of conversion into urochrome; and he has offered this permanganate reaction as a more satisfactory test, both for urochromogen and for an antecedent substance which has the same significance as urochromogen, but which the diazo fails to detect. This test has been studied chiefly in its relation to prognosis in tuberculosis, in which it appears to have about the same value as the diazo, with the differences that it is more frequently noted and is less intermittent in a given case and probably has less serious import.

2. **Russo's Methylene-blue Test.**—To 5 c.c. of the urine in a test-tube add 5 drops of 1:1000 aqueous solution of methylene-blue, and mix. An emerald- or mint-green color, in which there must be no trace of blue, denotes a positive reaction. There is considerable difficulty in judging the color.

Since this test was offered in 1905, it has been condemned by many workers and extolled by others. In the writer's opinion it is worthless for the purpose for which it was offered. At most, it is a very rough quantitative test for urochrome.

11. Drugs.—The effect of various drugs upon the of the urine has been mentioned (see p. 103). Most color poisons are eliminated in the urine, but their detection is more properly discussed in works upon toxicology. A few drugs which are of interest to the practitioner, and which can be detected by comparatively simple methods, are mentioned here.

Acetanilid and Phenacetin.—The urine is evaporated by gentle heat to about half its volume, boiled for a few minutes with about one-fifth its volume of strong hydrochloric acid, and shaken out with ether. The ether is evaporated, the residue dissolved in water, and the following test applied: To about 10 c.c. are added a few cubic centimeters of 3 per cent. phenol, followed by a weak solution of chromium trioxid (chromic acid) drop by drop. The fluid assumes a red color, which changes to blue when ammonia is added. If the urine is very pale, extraction with ether may be omitted.

Antipyrin.—This drug gives a dark-red color when a few drops of 10 per cent. ferric chlorid are added to the urine. The color does not disappear upon boiling, which excludes diacetic acid.

Arsenic.—*Reinsch's Test.*—Add to the urine in a test-tube or small flask about one-seventh its volume of hydrochloric acid, introduce a piece of bright copper-foil about ⅛ inch square, and boil for several minutes. If arsenic be present, a dark-gray film is deposited upon the copper. The test is more delicate if the urine be concentrated by slow evaporation. This test is well known and is widely used, but is not so reliable as the following:

Gutzeit's Test.—In a large test-tube place a little arsenic-free zinc, and add 5 to 10 c.c. pure dilute hydrochloric acid and a few drops of iodin solution (Gram's solution will answer), then add 5 to 10 c.c. of the urine. At once cover the mouth of the tube with a filter-paper cap moistened with saturated aqueous solution of silver nitrate (1 : 1) If arsenic be present, the paper quickly becomes lemon yellow, owing to formation of a com-

pound of silver arsenid and silver nitrate, and turns
black when touched with a drop of water.   To make
sure that the reagents are arsenic-free, the paper cap
may be applied for a few minutes before the urine is
added.

Atropin will cause dilatation of the pupil when a few
drops of the urine are placed in the eye of a cat or
rabbit.

Bromids can be detected by acidifying about 10 c.c. of
the urine with dilute sulphuric acid, adding a few drops
of fuming nitric acid and a few cubic centimeters of
chloroform, and shaking.   In the presence of bromin the
chloroform, which settles to the bottom, assumes a
yellow color.

Chloral hydrate appears in the urine chiefly as urochloralic acid, which reduces the copper solutions used
for sugar tests.   To detect it, evaporate about 500 c.c.
of the urine to about one-fourth its volume, make
decidedly acid with hydrochloric acid, add about 50 c.c.
of ether, shake thoroughly, and separate the ether.
Now evaporate the ether and dissolve the residue in a
little water.   If urochloralic acid be present this
aqueous solution will respond to Fehling's test.

Hexamethylenamin.—Interest in this drug centers
chiefly in its value as a urinary antiseptic, which depends upon its decomposition with liberation of formaldehyd.   According to a number of recent workers
formaldehyd can be detected in the urine of only about
50 per cent. of patients who are taking hexamethylenamin.   A test for formaldehyd is, therefore, necessary
in order to know whether the object in administering
the drug is being accomplished.

13

Rimini-Burnam Test for Formaldehyd.—To about 10 c.c. of the urine add successively 3 drops of 0.5 per cent. solution of phenylhydrazin hydrochlorid, 3 drops of 5 per cent. solution of sodium nitroprussid, and a few drops of a saturated solution of sodium hydroxid. The last is allowed to trickle down the inside of the tube; and if formaldehyd be present a purplish-black color, changing to green and then to yellow, will appear as it mingles with the urine.

Iodin from ingestion of iodids or absorption from iodoform dressings is tested for in the same way as the bromids, the chloroform assuming a pink to reddish-violet color; or Obermayer's reagent may be used in the same way as described for indican (see p. 134). To detect traces, a large quantity of urine should be rendered alkaline with sodium carbonate and greatly concentrated by evaporation before testing.

Lead.—No simple method is sufficiently sensitive to detect the traces of lead which occur in the urine in chronic poisoning. Of the more sensitive methods, that of Arthur Lederer is probably best suited to the practitioner:

It is essential that all apparatus used be lead-free. Five hundred cubic centimeters of the urine are acidified with 70 c.c. pure sulphuric acid, and heated in a beaker or porcelain dish. About 20 to 25 Gm. of potassium persulphate are added a little at a time. This should decolorize the urine, leaving it only slightly yellow. If it darkens upon heating, a few more crystals of potassium persulphate are added, the burner being first removed to prevent boiling over; if it becomes cloudy, a small amount of sulphuric acid is added. It is then boiled until it has evaporated to 250 c.c. or less. After cooling, an equal volume of alcohol is added, and the mixture allowed to stand in a cool place for four or five hours, during which time all the lead will be precipitated as insoluble sulphate.

The mixture is then filtered through a small, close-grained filter-paper (preferably an ashless, quantitative filter-paper), and any sediment remaining in the beaker or dish is carefully washed out with alcohol and filtered. A test-tube is placed underneath the funnel; a hole is

FIG. 48.—A simple hydrogen sulphid generator.

punched through the tip of the filter with a small glass rod, and all the precipitate (which may be so slight as to be scarcely visible) washed down into the test-tube with a jet of distilled water from a wash-bottle, using as little water as possible. Ten cubic centimeters will usually suffice. This fluid is then heated, adding crystals of sodium acetate until it becomes perfectly clear. It now contains all the lead of the 500 c.c. urine in the form of lead acetate. It is allowed to cool, and hydrogen sulphid gas is passed through it for about five minutes. The slightest yellowish-brown discoloration indicates the presence of lead. A very slight discoloration can be best seen when looked at from above. For comparison, the gas may be passed through a test-tube containing an equal amount of distilled water. The quantity of lead can be determined by comparing the discoloration with that produced by passing the gas through lead acetate (sugar of lead) solutions of known strength. One gram of lead acetate crystals contains 0.54 Gm. of lead. Hydrogen sulphid is easily prepared in the simple apparatus shown in Fig. 48. A small quantity of iron sulphid is placed in the test-tube; a little dilute hydrochloric acid is added; the cork is replaced; and the delivery tube is inserted to the bottom of the fluid to be tested.

Mercury.—Traces can be detected in the urine for a considerable time after the use of mercury compounds by ingestion or inunction.

About a liter of urine is acidified with 10 c.c. hydrochloric acid, and a small piece of copper-foil or gauze is introduced. This is gently heated for an hour, and allowed to stand for twenty-four hours. The metal is then removed, and washed successively with very dilute sodium hydroxid solution, alcohol, and ether. When dry, it is placed in a long, slender test-tube, and the lower portion of the tube is heated to redness. A tube with a constriction in its upper portion is better. If mercury be present, it will volatilize and condense in the upper portion of the tube as small, shining globules which can be seen with a hand-magnifier or low power of the microscope. If, now, a crystal of iodin be dropped into the tube and gently heated, the mercury upon the side of the tube is changed first to the yellow iodid, and later to the red iodid, which are recognized by their color.

Morphin.—Add sufficient ammonia to the urine to render it distinctly ammoniacal, and shake thoroughly with a considerable quantity of pure acetic ether. Separate the ether and evaporate to dryness. To a little of the residue in a watch-glass or porcelain dish add a few drops of formaldehyd-sulphuric acid, which has been freshly prepared by adding 1 drop of formalin to 1 c.c. pure concentrated sulphuric acid. If morphin be present, this will produce a purple-red color, which changes to violet, blue violet, and finally nearly pure blue.

Phenol.—As has been stated, the urine following phenol-poisoning turns olive green and then brownish black upon standing. Tests are of value in recognizing poisoning from ingestion and in detecting absorption from carbolized dressings.

The urine is acidulated with hydrochloric acid and distilled. To the first few cubic centimeters of distillate is added 10 per cent. solution of ferric chlorid drop by drop. The presence of phenol causes a deep amethyst-blue color, as in Uffelmann's test for lactic acid (see p. 402).

Phenolphthalein, which is now widely used as a cathartic, gives a bright pink color when the urine is rendered alkaline.

Quinin—A considerable quantity of the urine is rendered alkaline with ammonia and extracted with ether; the ether is evaporated, and a portion of the residue dissolved in about 20 drops of dilute alcohol. The alcoholic solution is acidulated with dilute sulphuric acid, 1 drop of an alcoholic solution of iodin (tincture of iodin diluted about ten times) is added, and the mixture is warmed. Upon cooling, an iodin compound of quinin (herapathite) will separate out in the form of a microcrystalline sediment of green plates.

The remainder of the residue may be dissolved in a little dilute sulphuric acid. This solution will show a characteristic blue fluorescence when quinin is present.

Resinous drugs cause a white precipitate like that of albumin when strong nitric acid is added to the urine. This is dissolved by alcohol.

Salicylates, salol, aspirin, and similar drugs give a bluish-violet color, which does not disappear upon heating, upon addition of a few drops of 10 per cent. ferric chlorid solution. When the quantity of salicylates is small, the urine may be acidified with hydrochloric acid and extracted with ether, the ether evaporated, and the test applied to an aqueous solution of the residue.

Tannin and its compounds appear in the urine as gallic acid, and the urine becomes greenish black (inky, if much gallic acid be present) when treated with a solution of ferric chlorid.

IV. MICROSCOPIC EXAMINATION

A careful microscopic examination will often reveal structures of great diagnostic importance in urine which seems perfectly clear, and from which only very slight sediment can be obtained with the centrifuge. Upon the other hand, cloudy urines with abundant sediment are often shown by the microscope to contain nothing of clinical significance.

Since the nature of the sediment soon changes, the urine must be examined while fresh, preferably within six hours after it is voided. When possible it should be kept on ice. The sediment is best obtained by means of the centrifuge. If a centrifuge is not available, the urine may be allowed to stand in a conical test-glass for six to twenty-four hours after adding some preservative (see p. 100).

A small amount of the sediment should be transferred to a slide by means of a pipet. It is very important to do this properly. The best pipet is a simple glass tube 7 or 8 inches long which has been drawn out at one end to a tip with a 1 or 1.5-mm. opening. The centrifuge tube containing the sediment is held on a level with the eye, the larger end of the pipet is closed with the index-finger, which must be dry, and the tip is carried down into the sediment. By carefully loosening the finger, but not entirely removing it, a small amount of the sediment is then allowed to run slowly into the pipet. Slightly rotating the pipet will aid in accomplishing this, and at the same time will serve to loosen any structures which cling to the bottom of the tube. After wiping off the urine which adheres to the outside, a drop from the pipet is placed upon a clean slide. A hair is then placed in the drop and a large cover-glass applied. The correct size of the drop can be learned only by experience. It should not be so large as to float the cover-glass about, nor so small as to leave unoccupied space beneath the cover. Many workers use no cover. This offers a thicker layer and larger area of urine, the chance of finding scanty structures being proportionately increased. It has the disadvantage that any jarring of

the room (as by persons walking about) sets the microscopic field into vibratory motion and makes it impossible to see anything clearly; and, since it does not allow
satisfactory use of high-power objectives, one cannot
examine details as carefully as one often wishes to do.
It is true that a cover can be applied later, but any
structure which one has found with the low power and
wishes to study with the high is sure to be lost when the
cover is applied. A large cover-glass (about 22 mm.
square) with a hair beneath it avoids these disadvan-
tages, and gives enough urine to find any structures
which are present in sufficient number to have clinical
significance, provided other points in the technic have
been right. It is best, however, to examine several
drops; and, when the sediment is abundant, drops from
the upper and lower portions should be examined
separately.

In examining urinary sediments microscopically no
fault is so common, nor so fatal to good results, as im-
proper illumination (see Fig. 6), and none is so easily
corrected. The light should be central and very subdued for ordinary work, but oblique illumination, obtained by swinging the mirror a little out of the optical
axis, will be found helpful in identifying certain delicate
structures like hyaline casts. The 16-mm. objective
should be used as a finder, while the 4-mm. is reserved
for examining details. An experienced worker will rely
almost wholly upon the lower power.

It is well to emphasize that *the most common errors
which result in failure to find important structures, when
present, are: (a) lack of care in transferring the sediment to*

the slide, (b) too strong illumination, and (c) too great magnification.

In order to distinguish between similar structures it is often necessary to watch the effect upon them of certain reagents. This is especially true of the various unorganized sediments. They very frequently cannot be identified from their form alone. With the structures still in focus, a drop of the reagent may be placed at one edge of the cover-glass and drawn underneath it by the suction of a piece of blotting-paper touched to the opposite edge; or, better, a small drop of the reagent and of the urine may be placed close together upon a slide and a cover gently lowered over them. As the two fluids mingle, the effect upon various structures may be seen.

Urinary sediments may be studied under three heads: A. Unorganized sediments. B. Organized sediments. C. Extraneous structures.

A. Unorganized Sediments

In general, these have little diagnostic or prognostic significance. Most of them are substances normally present in solution, which have been precipitated either because present in excessive amounts, or, more frequently, because of some alteration in the urine (as in reaction, concentration, etc.) which may be purely physiologic, depending upon changes in diet or habits. Various substances are always precipitated during decomposition, which may take place either within or without the body. Unorganized sediments may be classified according to the reaction of the urine in which they are *most likely* to be found. This classification is useful, but is not accurate, since the characteristic sediments of acid urine may remain after the urine has become alkaline, while the alkaline sediments may be precipitated in a urine which is still acid.

In acid urine: Uric acid, amorphous urates, sodium urate, calcium oxalate, leucin and tyrosin, cystin, and fat-globules. Uric acid, the urates, and calcium oxalate

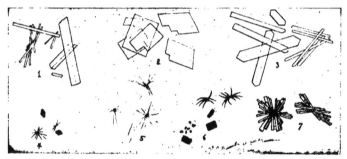

FIG. 49.—Unusual urinary crystals (drawn from various authors): 1, Calcium sulphate (colorless); 2, cholesterol (colorless); 3, hippuric acid (colorless); 4, hematoidin (brown); 5, fatty acids (colorless); 6, indigo (blue); 7, sodium urate (yellowish).

are the common deposits of acid urines; the others are less frequent, and depend less upon the reaction of the urine.

In alkaline urine: Phosphates, calcium carbonate, and ammonium urate.

Other crystalline sediments (Fig. 49) which are rare and require no further mention are: Calcium sulphate, cholesterol, hippuric acid, hematoidin, fatty acids, and indigo.

The following brief table will aid the student in identifying the chemical sediments which one meets every day:

PLATE III.

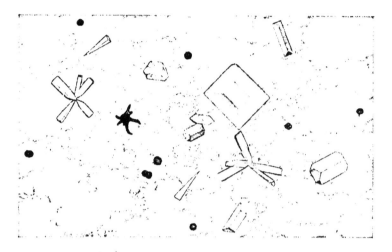

Fig. 1.—Common sediments of alkaline urine: Triple phosphate crystals, calcium phosphate crystals, ammonium urate crystals, and amorphous phosphates. X 150.

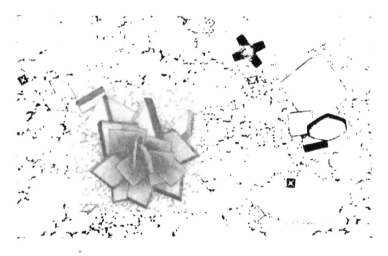

Fig. 2.—Common sediments of acid urine: Uric-acid crystals, calcium oxalate crystals, and amorphous urates. X 150.

	In acid urine	In alkaline urine
Yellow crystals.	Uric acid—dissolve in KOH.	Ammonium urate—dissolve in HCl.
Colorless crystals.	Calcium oxalate—dissolve in HCl.	Phosphate crystals—dissolve in acetic acid.
Amorphous material.	Urates—dissolve with heat.	Amorphous phosphates—dissolve in acetic acid.

1. In Acid Urine.—(1) **Uric-acid Crystals.**—These crystals are the red grains—"gravel" or "red sand"—which are often seen adhering to the sides and bottom of a vessel containing urine. Microscopically, they are yellow or reddish-brown crystals, which differ greatly in size and shape. The color is due to urinary pigments, chiefly uroerythrin. The most characteristic forms (Plate III and Fig. 50) are "whetstones;" roset-like clusters of prisms and whetstones; and rhombic plates, which have usually a paler color than the other forms and are sometimes colorless. A very rare form is a colorless hexagonal plate resembling cystin. Recognition of the crystals depends less upon their shape than upon their color, the reaction of the urine, and the facts that they are soluble in caustic soda solution and insoluble in hydrochloric or acetic acid. When ammonia is added, they dissolve and crystals of ammonium urate appear.

A deposit of uric-acid crystals has no significance unless it occurs before or very soon after the urine is voided. Every urine, if kept acid, will in time deposit its uric acid. Factors which favor an early deposit are high acidity, diminished urinary pigments, and excessive excretion of uric acid. The chief clinical interest of the crystals lies in their tendency to form calculi, owing to the readiness with which they collect about any solid object. Their presence in the freshly voided urine in

FIG. 50.—Forms of uric acid: 1, Rhombic plates; 2, whetstone forms; 3, 3, quadrate forms; 4, 5, prolonged into points; 6, 8, rosets; 7, pointed bundles; 8, barrel forms precipitated by adding hydrochloric acid to urine (Ogden).

clusters of crystals suggests stone in the kidney or bladder, especially if blood is also present (see Fig. 82).

It was formerly believed that the uric acid stone is the most common form of renal calculus, but from a recent study of a series of calculi Kahn and Rosenbloom believe that the great majority are composed of calcium oxalate although all contain a trace of uric acid.

(2) **Amorphous Urates.** These are chiefly urates of sodium and potassium which are thrown out of solution as a yellow or red "brick-dust" deposit. In pale urines this sediment is almost white. It disappears upon heating. A deposit of amorphous urates is very common in concentrated and strongly acid urines, especially in cold weather, and has no clinical significance. Under the microscope it appears as fine yellowish granules, sometimes almost colorless (Plate III). Often they are so abundant as to obscure all other structures. In such cases the urine should be warmed before examining. The granules have a tendency to collect upon tube-casts, strands of mucus, and other structures. Amorphous urates are readily soluble in caustic soda solutions. When treated with hydrochloric or acetic acid, they slowly dissolve and rhombic crystals of uric acid appear in ten to twenty minutes.

Rarely, sodium urate occurs in crystalline form—slender prisms, arranged in fan- or sheaf-like structures (see Fig. 49).

(3) **Calcium Oxalate.**—Characteristic of calcium oxalate are colorless, glistening, octahedral crystals, giving the appearance of small squares crossed by two intersecting diagonal lines—the so-called "envelope crystals" (see Fig. 77). They vary greatly in size, being sometimes so small as to seem mere points of light with medium-power objectives. Unusual forms, which, however, seldom occur except in conjunction with the octahedra, are colorless dumb-bells, spheres, and variations of the octahedra (Fig. 51). The spheres might be mistaken for globules of fat or red blood-corpuscles. Crystals of calcium oxalate are insoluble in acetic acid

or caustic soda. They are dissolved by strong hydrochloric acid, and recrystallize as octahedra upon addition of ammonia. They are sometimes encountered in
alkaline urine.

The crystals are commonly found in the urine after
ingestion of vegetables rich in oxalic acid, as tomatoes,

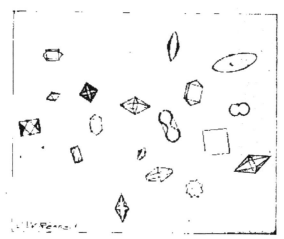

FIG. 51.—Various forms of calcium oxalate crystals from urine.
The majority are the typical octahedra seen in different positions
(× 450).

spinach, asparagus, and rhubarb. They have no definite significance pathologically. They often appear in
digestive disturbances, in neurasthenia, and when the
oxidizing power of the system is diminished. When
abundant, they are generally associated with a little
mucus; and, in men, frequently with a few spermatozoa.
Their chief clinical interest lies in their tendency to
form calculi, and their presence in fresh urine, together
with evidences of renal or cystic irritation, should be

viewed with suspicion, particularly if they are clumped in small masses.

(4) **Leucin and Tyrosin.**—These substances are cleavage products of the protein molecule. They are of comparatively rare occurrence in the urine and generally appear together. In general, their presence indicates autolysis of tissue proteins. Clinically, they are seen most frequently in severe fatty destruction of the liver, such as occurs in acute yellow atrophy and phosphorus-poisoning. Crystals are deposited spontaneously only when the substances are present in large amount. Usually they will be deposited when the urine is evaporated to a small volume on a water-bath. It is best, however, to separate them from the urine as follows:

Treat 500 to 1000 c.c. of urine, which has been freed from albumin, with neutral, then with basic, lead acetate until a precipitate no longer forms. Filter, remove excess of lead with hydrogen sulphid (see p. 196), and filter again. Concentrate to a syrup on a water-bath. Extract repeatedly with small quantities of absolute alcohol to remove urea. Treat the residue with hot dilute alcohol to which a little ammonia has been added. Filter and evaporate the filtrate to a small volume and let stand for the leucin and tyrosin to separate out. The leucin can be separated from the tyrosin by boiling with glacial acetic acid. Leucin dissolves, leaving the tyrosin, and can again be recovered by evaporating the acetic acid.

The crystals cannot be identified from their morphology alone, since other substances, notably calcium phosphate (see Fig. 57) and ammonium urate, may take similar or identical forms. It is, therefore, necessary to try out their solubility in various reagents or to apply special tests.

Leucin crystals (Fig. 52) as they appear in the urine do not represent the pure substance. They are slightly yellow, oily-looking spheres, many of them with radial and concentric striations. Some may be merged together in clusters. They are not soluble in hydrochloric acid nor in ether.

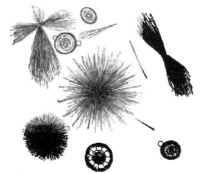

FIG. 52.—Leucin spheres and tyrosin needles (Stengel).

Tyrosin crystallizes in very fine needles, which may appear black and which are usually arranged in sheaves, with a marked constriction at the middle (Fig. 53). It is soluble in ammonia and hydrochloric acid, but not in acetic acid.

Mörner's Test for Tyrosin.—To a small quantity of the crystals in a test-tube add a few cubic centimeters of Mörner's reagent (formalin, 1 c.c.; distilled water, 45 c.c.; concentrated sulphuric acid, 55 c.c.). Heat gently to the boiling-point. A green color shows the presence of tyrosin.

(5) **Cystin crystals** are colorless, highly refractive, rather thick, hexagonal plates with well-defined edges.

They lie either singly or superimposed to form more or less irregular clusters (Fig. 54). Uric acid sometimes takes this form and must be excluded. Cystin is soluble in hydrochloric acid, insoluble in acetic; it is readily soluble in ammonia and recrystallizes upon addition of acetic acid.

FIG. 53.—Tyrosin crystals from urine (× 450).

Cystin is one of the amino-acids formed in decomposition of the protein molecule, and is present in traces in normal urine. Crystals are deposited only when the substance is present in excessive amount. Their presence is known as *cystinuria*. It is a rare condition due to an obscure abnormality of protein metabolism and usually continues throughout life. The amount of cystin can be greatly diminished by a low-protein diet, and the formation of crystals can in some measure be prevented by administration of sodium carbonate. There are rarely any symptoms save those referable

14

to renal or cystic calculus, to which the condition strongly predisposes.

(6) **Fat-globules.**—Fat appears in the urine as highly refractive globules of various sizes, frequently very small. These globules are easily recognized from the fact that they are stained black by osmic acid and orange or red by Sudan III. The stain may be applied

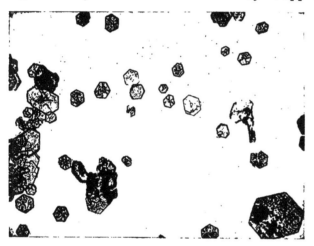

Fig. 54.—Cystin crystals from urine of patient with cystin calculus
(× 200).

upon the slide, as already described (see p. 201). Osmic acid should be used in 1 per cent. aqueous solution; Sudan III, in saturated solution in 70 per cent. alcohol, to which one-half its volume of 10 per cent. formalin may advantageously be added.

Fat in the urine is usually a contamination from unclean vessels, oiled catheters, etc. A very small amount may be present after ingestion of large quantities of cod-liver oil or other fats. In fatty degeneration of the

kidney, as in phosphorus-poisoning and chronic parenchymatous nephritis, fat-globules are commonly seen,
both free in the urine and embedded in cells and tube-
casts. Fat-droplets are common in pus-corpuscles and
in degenerating cells of any kind.

In *chyluria*, or admixture of chyle with the urine as a
result of rupture of a lymph-vessel, minute droplets of
fat are so numerous as to give the urine a milky appearance. The droplets are smaller than those of milk,
which is sometimes added by malingerers. The fluid
is often blood-tinged. The condition is best recognized
by shaking up with ether, which, when separated, leaves
the urine comparatively clear. If, then, the ether be
evaporated a fatty residue remains. Chyluria occurs
most frequently as a symptom of infection by filaria
(see p. 462), the larvæ of which can usually be
found in the milky urine. In other cases the etiology
is obscure.

2. In Alkaline Urine.—(1) **Phosphates.**—While
most common in alkaline urine, phosphates are sometimes deposited in amphoteric or feebly acid urines.
The usual forms are: (*a*) Ammoniomagnesium phosphate crystals; (*b*) acid calcium phosphate crystals,
and (*c*) amorphous phosphates. All are readily soluble
in acetic acid.

(*a*) *Ammoniomagnesium Phosphate Crystals.*—They
are the common "triple phosphate" crystals, which are
generally easily recognized (Figs. 55, 56 and 83, and
Plate III). They are colorless, except when bile stained.
Their usual form is some modification of the prism,
with oblique ends. Most typical are the well-known
"coffin-lid" and "hip-roof" forms. The long axis

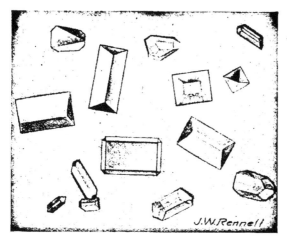

FIG. 55.—Prismatic forms of triple phosphate crystals, from urine (× 450).

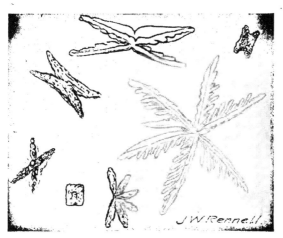

FIG. 56.—Triple phosphate crystals: forms produced by rapid precipitation and by partial solution of prisms (× 450).

of the hip roof crystal is often so shortened that it resembles the envelope crystal of calcium oxalate. It does not, however, have the same luster; this, and its solubility in acetic acid, will always prevent confusion.

When rapidly deposited, as by artificial precipitation, triple phosphate often takes feathery, star-, or leaf-like forms (Fig. 56). These gradually develop into the more common prisms. X-forms may be produced by partial solution of prisms.

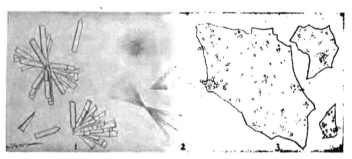

FIG. 57.—Crystals of calcium phosphate: 1, Common form (copied from Rieder's Atlas); 2, needles resembling tyrosin (drawn from nature); 3, large, irregular plates (from nature).

(b) *Dicalcium Phosphate Crystals.*—In feebly acid, amphoteric, or feebly alkaline urines acid calcium phosphate, wrongly called "neutral calcium phosphate," is not infrequently deposited in the form of colorless prisms arranged in stars and rosets (Fig. 57, 1). Because of the shape of the crystals it is sometimes called "stellar phosphate." The individual prisms are usually slender, with one beveled, wedge-like end, but are sometimes needle-like. They may sometimes take forms resembling tyrosin (Fig. 57, 2), calcium sulphate, or hippuric acid, but are readily distinguished by their solubility in acetic acid.

Calcium phosphate often forms large, thin, irregular, usually granular, colorless plates (Fig. 57, 3) which should be easily recognized, although small plates might be mistaken for squamous epithelial cells. These crystals most frequently form a scum upon the surface of the urine. They are regarded by some as magnesium phosphate.

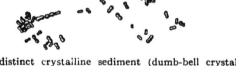

FIG. 58.—Indistinct crystalline sediment (dumb-bell crystals) of calcium carbonate. Similar crystals are sometimes formed by calcium oxalate and calcium sulphate (after Funke).

(c) *Amorphous Phosphates.*—The earthy phosphates are thrown out of solution in most alkaline and many amphoteric urines as a white, amorphous sediment, which may be mistaken for pus macroscopically. Under the microscope the sediment is seen to consist of numerous colorless granules, distinguished from amorphous urates by their color, their solubility in acetic acid, and the reaction of the urine.

The various phosphatic deposits frequently occur together. They are sometimes due to excessive excretion of phosphoric acid, but usually merely indicate that the urine has become, or is becoming, alkaline (see Phosphates, p. 129).

(2) **Calcium Carbonate** may sometimes be mingled with the phosphatic deposits, usually as amorphous

granules, or, more rarely, as colorless spheres and dumb-bells (Fig. 58), which are soluble in acetic acid with gas formation.

(3) **Ammonium Urate Crystals.**—This is the only urate deposited in alkaline urine. It forms opaque yellow crystals, usually in the form of spheres (Plate III. and Fig. 83), which are often covered with fine or coarse

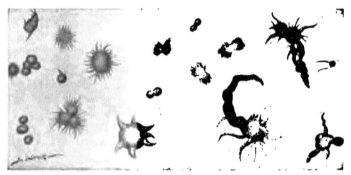

FIG. 59.—Crystals of ammonium urate (one-half of the forms copied from Rieder's Atlas; the others, from nature).

spicules— "thorn-apple crystals." Sometimes dumb-bells, compact sheaves of fine needles, and irregular rhizome forms are seen (Fig. 59). Upon addition of acetic acid they dissolve, and rhombic plates of uric acid appear.

These crystals occur only when free ammonia is present. They are generally found along with the phosphates in decomposing urine and have no clinical significance.

B. ORGANIZED SEDIMENTS

The principal organized structures in urinary sediments are: Tube-casts; epithelial cells; pus-corpuscles;

red blood-corpuscles; spermatozoa; bacteria, and animal parasites. They are much more important than the unorganized sediments just considered.

1. Tube-casts.—These interesting structures are albuminous casts of the uriniferous tubules. Their presence in the urine (known as *cylindruria*) probably always indicates some pathologic change in the kidney, although this change may be very slight or transitory. Large numbers may be present in temporary irritations and congestions. *They do not in themselves, therefore, imply organic disease of the kidney.* They rarely occur in urine which does not contain, or has not recently contained, albumin.

While it is not possible to draw a sharp dividing line between the different varieties, casts may be classified as follows:

(1) Hyaline casts.
    (*a*) Narrow.
    (*b*) Broad.
(2) Waxy casts.
(3) Fibrinous casts.
(4) Granular casts.
    (*a*) Finely granular.
    (*b*) Coarsely granular.
(5) Fatty casts.
(6) Casts containing organized structures.
    (*a*) Epithelial casts.
    (*b*) Blood-casts.
    (*c*) Pus-casts.
    (*d*) Bacterial casts.

As will be seen later, practically all varieties are modifications of the hyaline.

The significance of the different varieties is more readily understood if one considers their mode of formation. Albuminous material, the source and nature of which are not definitely known, but which are doubtless not the same in all cases, probably enters the lumen of a uriniferous tubule in a fluid or plastic state. The material has been variously thought to be an exudate from the blood, a pathologic secretion of the renal cells, and a product of epithelial degeneration. In the tubule it hardens into a cast which, when washed out by the urine, retains the shape of the tubule, and contains within its substance whatever structures and débris were lying free within the tubule or were loosely attached to its wall. If the tubule be small and has its usual lining of epithelium, the cast will be narrow; if it be large or entirely denuded of epithelium, the cast will be broad. *A cast, therefore, indicates the condition of the tubule in which it is formed, but does not necessarily indicate the condition of the kidney as a whole.* In any particular case of kidney disease several forms or even all may be found. Their number and the preponderance of certain forms will, as is shown later, furnish a clue to the nature of the pathologic process but further than this one cannot go with certainty. One cannot rely upon the casts for accurate diagnosis of the histologic changes in the kidney.

At times during the course of a nephritis the urine is suddenly flooded with great numbers of tube-casts. Such "showers" may be of serious import but are not necessarily so. In some cases they may result from a clearing out of the plugged renal tubules coincident with improvement and increased flow of urine.

The search for casts must be carefully made. The urine must be fresh, since hyaline casts soon dissolve when it becomes alkaline. It should be thoroughly centrifugalized. When the sediment is abundant, casts, being light structures, will be found near the top of the sediment. In cystitis, where casts may be entirely hidden by the pus, the bladder should be irrigated to remove as much of the pus as possible and the next urine examined. In order to prevent solution of the casts the urine, if alkaline, must be rendered acid by previous administration of boric acid or other drugs. Heavy sediments of urates, blood, or vaginal cells may likewise obscure casts and other important structures. The last can be avoided by catheterization. Urates can be dissolved by gently warming before centrifugalizing, care being taken not to heat enough to coagulate the albumin. The aluminum shield of the centrifuge tube may also be heated. Blood can be destroyed by centrifugalizing, pouring off the supernatant urine, filling the tube with water, adding a few drops of dilute acetic acid, mixing well, and again centrifugalizing; this process being repeated until the blood is completely decolorized. Too much acetic acid will dissolve hyaline casts.

In searching for casts the low-power objective should invariably be used, although a higher power may occasionally be desirable in studying details as, for example, in distinguishing between an epithelial and a pus-cast. The casts are perhaps most frequently found near the edge of the cover-glass. Their cylindric shape can be best seen by slightly moving the cover-glass while observing them, or by pressing upon one edge of the cover with a needle, thus causing them to roll. This little manipulation should be practised until it can be done satisfactorily. It will prove useful in many examinations.

Various methods of staining casts so as to render them more conspicuous have been proposed. They offer no special advantage to one who understands how to use the substage mechanism of his microscope. The "negative-staining" method is as good as any. It consists simply in adding a little India-ink to the drop of urine on the slide. Casts, cells, etc., will stand out as colorless structures on a dark background.

(1) **Hyaline Casts.**—Typically, these are colorless, homogeneous, semitransparent, cylindric structures,

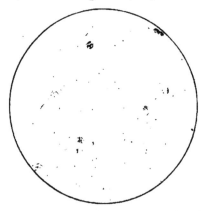

FIG. 60.—Hyaline casts showing fat-droplets and leukocytes (obj. 4 mm.) (Boston).

with parallel sides and usually rounded ends. Not infrequently they are more opaque or show a few granules or an occasional cell or oil-globule, either adhering to them or contained within their substance. Generally they are straight or curved; less commonly, convoluted.

Their length and breadth vary greatly: they are sometimes so long as to extend across several fields of a medium-power objective, but are usually much shorter; in breadth they vary from one to seven or eight times the diameter of a red blood-corpuscle (see Figs. 6, 60, 61, and 66).

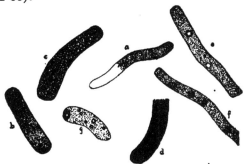

FIG. 61.—Various kinds of casts: *a*, Hyaline and finely granular cast; *b*, finely granular cast; *c*, coarsely granular cast; *d*, brown granular cast; *e*, granular cast with normal and abnormal blood cells adherent; *f*, granular cast with renal cells adherent; *g*, granular cast with fat and a fatty renal cell adherent (Ogden).

Hyaline casts are the least significant of all the casts, and occur in many slight and transitory conditions. Small numbers are common following ether anesthesia, in fevers, after excessive exercise, and in congestions and irritations of the kidney. They are always present, and are usually stained yellow when the urine contains much bile. While they are found in all organic diseases of the kidney, they are most important in chronic interstitial nephritis. Here they are seldom abundant, but their constant presence is the most reliable urinary sign of the disease. Small areas of chronic interstitial change are probably responsible for the few hyaline casts so frequently found in the urine of elderly persons.

Very broad hyaline casts commonly indicate complete desquamation of the tubular epithelium, such as occurs in the late stages of nephritis.

(2) **Waxy Casts.**—Like hyaline casts, these are homogeneous when typical, but frequently contain a few granules or an occasional cell. They are much more opaque than the hyaline variety, and are usually shorter and broader, with irregular, broken ends, and sometimes appear to be segmented. They are grayish or colorless, and have a dull, waxy look, as if cut from paraffin (Figs. 62 and 81). They are sometimes composed

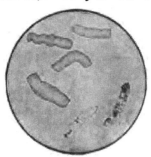

FIG. 62.—Waxy casts (upper part of figure). Fatty and fat-bearing casts (lower part of figure) (from Greene's " Medical Diagnosis").

of material which gives the amyloid reactions. All gradations between hyaline and waxy casts may be found. Waxy casts are found in most advanced cases of nephritis, where they are an unfavorable sign. They are perhaps most abundant in amyloid disease of the kidney, but are not distinctive of the disease, as is sometimes stated.

(3) **Fibrinous Casts.**—Casts which resemble waxy casts, but have a distinctly yellow color, as if cut from beeswax, are often seen in acute nephritis. They are

called fibrinous casts, but the name is inappropriate, as they are not composed of fibrin. They are often classed with waxy casts, but should be distinguished, as their significance is much less serious.

(4) **Granular Casts.**—These are merely hyaline casts in which numerous granules are embedded (Figs. 61, 63, and 66).

Finely granular casts contain many fine granules, are usually shorter, broader, and more opaque than the hyaline variety, and are more conspicuous. Their color is grayish or pale yellow.

FIG. 63.—Granular and fatty casts and two compound granule cells (Stengel).

Coarsely granular casts contain larger granules and are darker in color than the finely granular, being often dark brown, owing to presence of altered blood-pigment. They are usually shorter and more irregular in outline, and more frequently have irregularly broken ends.

(5) **Fatty Casts.**—Small droplets of fat may at times be seen in any variety of cast. Those in which the droplets are numerous are called fatty casts (Figs. 62, 63 and 81). The fat-globules are not difficult to recognize.

Staining with osmic acid or Sudan III (see p. 210) will remove any doubt as to their nature.

The granules and fat-droplets seen in casts are products of epithelial degeneration. Granular and fatty casts, therefore, always indicate partial or complete disintegration of the renal epithelium. The finely granular variety is the least significant, and is found when the epithelium is only moderately affected. Coarsely granular, and especially fatty casts, if present in considerable numbers, point toward a serious parenchymatous nephritis.

(6) **Casts Containing Organized Structures.**—Cells and other structures are frequently seen adherent to a cast or embedded within it (see Figs. 60 and 61). When numerous, they give name to the cast.

(a) *Epithelial casts* contain epithelial cells from the renal tubules. The cells vary in size and are often flattened, oval, or elongated. They may be recognized as epithelial cells by irrigating with dilute acetic acid, which usually brings out the nucleus clearly. Epithelial casts always imply desquamation of epithelium, which rarely occurs except in parenchymatous inflammations (see Figs. 80 and 81). When the cells are well preserved they point to acute nephritis.

(b) *Blood-casts* contain red blood-corpuscles, usually much degenerated (Figs. 64, 65, and 80). They always indicate hemorrhage into the tubules, which is most common in acute nephritis or an acute exacerbation of a chronic nephritis.

(c) *Pus-casts* (see Fig. 82), composed almost wholly of pus-corpuscles, are uncommon, and point to a chronic suppurative process in the kidney. Casts containing a few pus-corpuscles, either alone or in combination with epithelial or red blood cells are common. In these the pus-cells have no special significance.

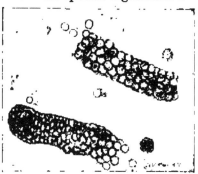

FIG. 64.—Two blood-casts, one containing a leukocyte; six free red blood-cells; and two renal epithelial cells. From the urine of a child with acute nephritis (× 300).

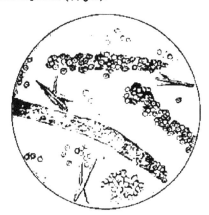

FIG. 65.—Red blood-corpuscles and blood-casts (courtesy of Dr. A. Scott) (obj. 4 mm.) (Boston).

(d) True *bacterial casts* are rare. They indicate a septic condition in the kidney. Bacteria may permeate a cast after the urine is voided.

Structures Likely to be Mistaken for Casts.—(1)
Mucous Threads.—Mucus frequently appears in the
form of long strands which slightly resemble hyaline
casts (Fig. 66). They are, however, more ribbon-like,
have less well-defined edges, and usually show faint
longitudinal striations. Their ends taper to a point or
are split or curled upon themselves, and are never evenly
rounded, as is commonly the case with hyaline casts.

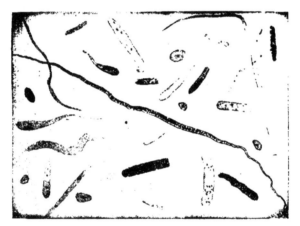

FIG. 66.—Hyaline and granular casts, mucous threads, and cylindroids.
There are also a few epithelial cells from the bladder (Wood).

Such threads form a part of the nubecula of normal
urine, and are especially abundant when calcium oxalate
crystals are present. When there is an excess of mucus,
as in irritations of the urinary tract, every field may be
filled with an interlacing meshwork.

Mucous threads are microscopic and should not be
confused with urethral shreds or "gonorrheal threads,"
which are macroscopic, 0.5 to 1 cm. long, and consist of

15

a matrix of mucus in which many epithelial and pus-cells are embedded.

(2) **Cylindroids.**—This name is sometimes given to the mucous threads just described, but is more properly the applied to certain peculiar structures more nearly allied to casts. They resemble hyaline casts in structure, but differ in being broader at one end and tapering to a slender tail, which is often twisted or curled upon itself (Fig. 66). They frequently occur in the urine along with hyaline casts, especially in irritations of the kidney, and have practically the same significance.

(3) **Masses of amorphous urates, or phosphates, or very small crystals** (Fig. 67), which accidentally take a

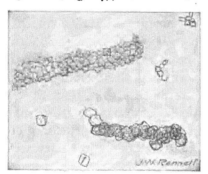

FIG. 67.—Two pseudo-casts, one composed of calcium oxalate crystals, one of uric acid (× 300).

cylindric form, or shreds of mucus covered with granules, closely resemble granular casts. Application of gentle heat or appropriate chemicals will serve to differentiate them. When urine contains both mucus and granules, large numbers of these "pseudocasts," all lying in the same direction, can be produced by slightly moving the cover-glass from side to side. It is possible—as in urate

Infarcts of infants—for urates to be molded into cylindric bodies within the renal tubules.

(4) **Hairs** and **fibers** of wool, cotton, etc. These could be mistaken for casts only by beginners. One can easily become familiar with their appearance by suspending them in water and examining with the microscope (see Fig. 78).

(5) **Hyphæ of molds** are not infrequently mistaken for hyaline casts. Their higher degree of refraction, their jointed or branching structure, and the accompanying spores will differentiate them (see Fig. 79).

2. Epithelial Cells.—A few cells from various parts of the urinary tract occur in every urine. A marked increase indicates some pathologic condition at the site of their origin. It is sometimes, but by no means always, possible to locate their source from their form. One should, however, be extremely cautious about making any definite statement as to the origin of any individual cell. Most cells are much altered from their original shape. Any epithelial cell may be so granular from degenerative changes that the nucleus is obscured. Most of them contain fat-globules. They are usually divided into three groups:

(1) **Small, round** or **polyhedral cells** are about the size of pus-corpuscles, or a little larger, with a single round nucleus. Such cells may come from the deeper layers of any part of the urinary tract. They are uncommon in normal urine. When they are polygonal in shape, rather dark in color, very granular, and contain a comparatively large nucleus (Fig. 68), they probably come from the renal tubules, but their origin in the kidney is not proved unless they are found embedded

in casts. In chronic passive congestion of the kidney and in renal infarction some of these cells may contain yellow granules of altered blood-pigment. They are analogous to the "heart-failure cells" of the sputum (see p. 70). Renal cells are abundant in parenchymatous nephritis, especially the acute form. They are nearly always fatty—most markedly so in chronic

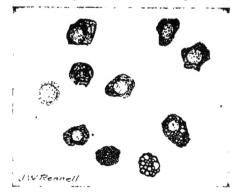

FIG. 68.—Renal epithelial cells from nephritic urine. The four cells below show different grades of fatty degeneration (× 475).

parenchymatous nephritis, where their substance is sometimes wholly replaced by fat-droplets ("compound granule cells") (see Figs. 63, 68, 80 and 81).

(2) **Irregular cells** are considerably larger than the preceding. They are round, pear shaped, or spindle shaped, or may have tail-like processes, and are hence named large round, pyriform, spindle, or caudate cells respectively. Each contains a round or oval, distinct nucleus. Their usual source is the deeper layers of the urinary tract, especially of the bladder. Caudate forms apparently come most commonly from the pelvis of the kidney (see Figs. 69, 70, *b* and 82).

(3) **Squamous or pavement cells** are large flat cells, each with a small, distinct round or oval nucleus (Fig. 70, *a*). They are derived from the superficial layers

FIG. 69.—Caudate epithelial cells from pelvis of kidney (Jakob).

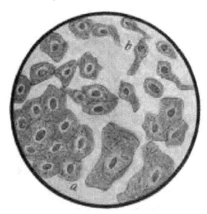

FIG. 70.—Epithelial cells from urethra and bladder: *a*, Squamous cells from superficial layers; *b*, irregular cells from deeper layers (Jakob).

of the ureters, bladder, urethra, or vagina, and when desquamation is active, appear in stratified masses.

Squamous cells from the bladder are generally rounded, while those from the vagina are larger, thinner, and more angular. Great numbers of these vaginal cells, together with pus-corpuscles, may be present when leukorrhea exists (Fig. 71).

Fig. 71.—Squamous epithelial cells, pus-corpuscles and bacteria in urine; vaginal contamination (× 300).

3. Pus-corpuscles.—A very few leukocytes are present in normal urine.. They are more abundant when mucus is present. An excess of leukocytes, mainly of the polymorphonuclear neutrophilic variety, with albumin, constitutes *pyuria*—pus in the urine. At times numerous mononuclear cells (lymphocytes) are seen.

When at all abundant, pus forms a white sediment resembling amorphous phosphates macroscopically. Under the microscope the corpuscles appear as very granular cells, about twice the diameter of a red blood-corpuscle (Figs. 72 and 83). The granules are partly the normal neutrophilic granules, partly granular products of degeneration. In freshly voided urine many exhibit ameboid motion, assuming irregular outlines. Each contains one irregular nucleus or several small, rounded nuclei. The nuclei are obscured or entirely hidden by the granules, but may be brought clearly into view by running a little acetic acid under the cover-glass. This enables one to easily distinguish pus-corpuscles from

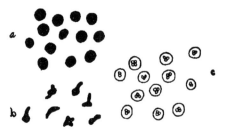

FIG. 72.—Pus-corpuscles: *a*, As ordinarily seen; *b*, ameboid corpuscles; *c*, showing the action of acetic acid (Ogden).

small round epithelial cells, which resemble them in size, but have a single, rather large, round nucleus. In decomposing urine pus is often converted into a gelatinous mass which gives the urine a ropy consistence.

Pyuria indicates suppuration in some part of the urinary tract—urethritis, cystitis, pyelitis, etc.—or may be due to contamination from the vagina, in which case many vaginal epithelial cells will also be present. Of these conditions chronic cystitis usually gives by far the greatest amount of pus. In general, the source of the pus can be determined only by the accompanying structures (epithelia, casts) or by the clinical signs. A considerable amount of pus, appearing suddenly, usually originates from a ruptured abscess.

A fairly accurate idea of the quantity of pus from day to day may be had by shaking the urine thoroughly and

counting the number of corpuscles per cubic millimeter
upon the blood-counting slide. A drop of the urine
is placed directly upon the slide. Dilution is seldom
necessary. The urine must not be alkaline or the corpuscles will adhere in clumps.

Pus always adds a certain amount of albumin to the
urine, and it is often desirable to know whether the
abumin present in a given specimen is due solely to
pus. It has been estimated that 80,000 to 100,000 pus-
corpuscles per cubic millimeter add about 0.1 per cent.
of albumin. If albumin is present in much greater proportion than this, the excess is probably derived from the
kidney.

4. Red Blood-corpuscles.—Urine which contains
blood is always albuminous. Very small amounts do not
alter its macroscopic appearance. Larger amounts alter
it considerably. Blood from the kidneys is generally
intimately mixed with the urine and gives it a hazy
reddish or brown, "smoky" color. When from the
lower urinary tract, it is not so intimately mixed and
settles more quickly to the bottom, the color is brighter,
and small clots are often present. A further clue to
the site of the bleeding may sometimes be gained by
having the patient void his urine in three separate
portions. If the blood be chiefly in the first portion,
the bleeding point is probably in the urethra; if in the
last, it is probably in the bladder. If the blood is uniformly mixed in all three portions, it probably comes from
the kidney or ureter. Microscopically the presence of
tube-casts or of considerable numbers of epithelial
cells of the renal type would be suggestive, while the

presence of blood-casts would of course point definitely to hemorrhage into the kidney tubules.

Red blood-corpuscles are not usually difficult to recognize with the microscope. When very fresh, they have a normal appearance, being yellowish disks of uniform size. When they have been in the urine any considerable time, their hemoglobin may be dissolved out, and they then appear as faint colorless circles or "shadow cells," and are more difficult to see (Fig. 73;

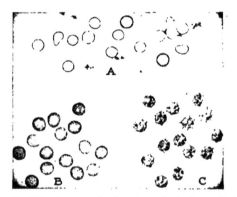

FIG. 73.—Red blood-corpuscles in urine: A, shadow cells from a case of nephritis; B, fresh red corpuscles; C, crenated corpuscles in a urine of high specific gravity (× 475).

see also Figs. 64, 65 and 80). They are apt to be swollen in dilute and crenated in concentrated urines. The microscopic findings may be corroborated by chemic tests for hemoglobin, although the microscope may show a few red corpuscles when the chemic tests are negative.

When not due to contamination from menstrual discharge, blood in the urine, or *hematuria*, is always pathologic, although not always of serious import. A few red blood-corpuscles may be found after strenuous exercise. Blood comes from the *kidney tubules* in severe hyperemia, in acute nephritis and acute exacerbations of chronic nephritis, and in renal tuberculosis and malignant disease. Renal hematuria may also be a manifestation of the "hemorrhagic diseases" and an "idiopathic hematuria," probably of nervous origin, has been observed. The urine of healthy infants frequently contains red blood-corpuscles for weeks at a time. This has been attributed to slight toxic injury to the kidneys. Blood comes from the *pelvis of the kidney* in renal calculus (see Fig. 82), and is then usually intermittent, small in amount, and accompanied by a little pus and perhaps crystals of the substance forming the stone. Considerable hemorrhages from the *bladder* may occur in vesical calculus, tuberculosis, and new growths. Small amounts of blood generally accompany acute cystitis. In Africa the presence of *Schistosomum hæmatobium* in the veins of the bladder is a common cause of hemorrhage (Egyptian hematuria).

5. Spermatozoa are generally present in the urine of men after nocturnal emissions, after epileptic convulsions, and in spermatorrhea. They may be found in the urine of both sexes following coitus. They are easily recognized from their characteristic structure (Fig. 74). The 4-mm. objective should be used, with subdued light and careful focusing.

6. Bacteria.—Normal urine is free from bacteria in the bladder, but becomes contaminated in passing through the urethra. Various non-pathogenic bacteria are always present in decomposing urine. In suppurations of the urinary tract pus-producing organisms may be found. In many infectious diseases the specific bacteria may be eliminated in the urine without producing any local lesion. Typhoid bacilli have been known to persist for months and even years after the attack.

Bacteria produce a cloudiness which will not clear upon filtration. They are easily seen with the 4-mm.

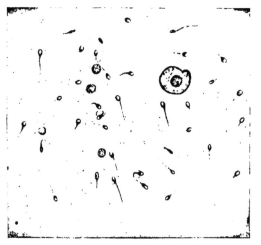

FIG. 74.—Spermatozoa in urine (Ogden).

objective in the routine microscopic examination. Ordinarily, no attempt is made to identify any but the tubercle bacillus and the gonococcus. Others must be studied by cultural methods, the urine being carefully obtained by catheter and received directly into a sterile bottle or test-tube.

Tubercle bacilli are nearly always present in the urine when tuberculosis exists in any part of the urinary tract and are often present in general miliary tuberculosis, but may be difficult to find, especially when the urine contains little or no pus.

Detection of Tubercle Bacilli in Urine.—In order to avoid the smegma bacillus the urine should be obtained aseptically by catheter after careful cleansing of the parts, or by having the patient void urine in three portions, only the last being used for the examination.

1. Centrifugalize thoroughly, or, better, treat by the following method, recommended by Brown:

(a) Acidify 100 c.c. of the urine with 30 per cent. acetic acid.

(b) Add 2 c.c. of 5 per cent. tannic acid solution.

(c) Place in ice chest for twenty-four hours.

(d) Centrifugalize, pipet off the supernatant fluid and re-dissolve the sediment with dilute acetic acid.

(e) Centrifugalize thoroughly once more.

2. Make thin smears of the sediment, adding a little egg-albumen if necessary to make the smear adhere to the glass; dry, preferably in the incubator for three hours, and fix in the usual way. ′

3. Stain thoroughly with carbol-fuchsin in the usual way (see p. 77).

4. Wash in water, and then in 5 to 20 per cent. nitric acid, until only a faint pink color remains.

5. Wash in water.

6. Soak in alcohol fifteen minutes or longer. This de-colorizes the smegma bacillus (see p. 82), which is often present in the urine, and might easily be mistaken for the tubercle bacillus. Some strains of the smegma bacillus are very resistant to alcohol. It is therefore best to avoid it altogether by examining only catheterized specimens, in which case this step may be omitted.

7. Wash in water.

PLATE IV

Tubercle bacilli in urinary sediment; × 800 (Ogden).

8. Apply Löffler's methylene blue solution for one-half minute.

9. Rinse in water, dry between filter-papers, and examine with the oil-immersion objective.

A careful search of several smears may be necessary to find the bacilli. They usually lie in clusters (see Plate IV). Failure to find them in suspicious cases should be followed by inoculation of guinea-pigs; this is the court of last appeal, and must also be sometimes resorted to in order to exclude the smegma bacillus.

In gonorrhea, **gonococci** are sometimes found within pus-cells in the sediment, but more commonly in the "gonorrheal threads" or "floaters." In themselves, these threads are by no means diagnostic of gonorrhea. They are most common in the morning or after massage of the prostate. Detection of the gonococcus is described later (see p. 518). Its recognition in isolated pus-cells in the urine is difficult since these are usually much shrunken. The smears should be thin and quickly dried.

7. Animal parasites are rare in the urine. Hooklets and scolices of *Tænia echinococcus* (Fig. 75) and larvæ of filariæ have been met. In Africa the ova, and even adults, of *Schistosomum hæmatobium* are common, accompanying "Egyptian hematuria." *Trichomonas intestinalis* is a not uncommon contamination. This and other protozoa may be mistaken for spermatozoa by the inexperienced.

A worm which is especially interesting is *Anguillula aceti*, the "vinegar eel." This is generally present in the sediment of table vinegar, and may reach the urine through use of vinegar in vaginal douches, or through contamination of the bottle in which the urine is contained. It has been mistaken for *Strongyloides intes-*

FIG. 75.—1, Scolex of *Tænia echinococcus,* showing crown of hooklets; 2, scolex and detached hooklets (obj. 4 mm.) (Boston).

tinalis and for the larval filaria. It somewhat resembles the former in both adult and embryo stages. The

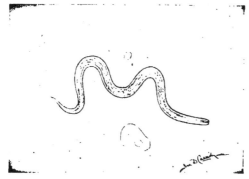

FIG. 76.—Embryo of "vinegar eel" in urine, from contamination; length, 340 μ; width, 15 μ. An epithelial cell from bladder and three leukocytes are also shown.

young embryos have about the same length as the larvæ of *Filaria bancrofti,* but are nearly twice as broad, and the Intestinal canal is comparatively easily seen (compare Figs. 76 and 188).

For fuller descriptions of these parasites the reader is referred to Chapter VI.

C. EXTRANEOUS STRUCTURES

The laboratory worker must familiarize himself with the microscopic appearance of the more common of the

FIG. 77.—Yeasts and calcium oxalate crystals in a urine which had been preserved for two weeks with boric acid (× 450).

numerous structures which may be present from accidental contamination (Fig. 78).

Yeast-cells are smooth, colorless, highly refractive, spheric or ovoid cells. They sometimes reach the size of a leukocyte, but are generally smaller (Fig. 77). They are often mistaken by the inexperienced for red blood-corpuscles and more rarely for fat-droplets, or the spheric crystals of calcium oxalate, but are distinguished by the facts that they are usually ovoid and not of uniform size; that they tend to adhere in short chains; that small buds may often be seen adhering to the larger cells; and that they do not give the hemoglobin test, are not stained by osmic acid or Sudan III, but are colored brown by Lugol's solution, and are insoluble in acids

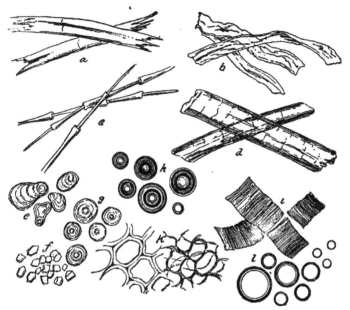

FIG. 78.—Extraneous matters found in urine: *a*, Flax-fibers; *b*, cotton-fibers; *c*, feathers; *d*, hairs; *e*, potato-starch granules; *f*, rice-starch granules; *g*, wheat-starch granules; *h*, air-bubbles; *i*, muscular tissue; *k*, vegetable tissue; *l*, oil-globules.

and alkalies. Yeast-cells multiply rapidly in diabetic urine, and may reach the bladder and multiply there.

Mold fungi (Fig. 79) are characterized by refractive, jointed, or branched rods (hyphæ), often arranged in a network, and by highly refractive spheric or ovoid spores. They are common in urine which has stood exposed to the air. Not infrequently a spore with a short hypha growing from it is reported as a spermatozoön.

Fibers of wool, cotton, linen, or silk, often colored, derived from towels, the clothing of the patient, or the dust in the air, are present in almost every urine. **Fat-droplets** are most frequently derived from unclean bottles or oiled catheters. **Starch-granules** may reach

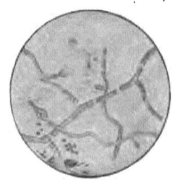

FIG. 79.—Aspergillus from urine (Boston).

the urine from towels, the clothing, or dusting-powders. They are recognized by their concentric striations and their blue color with iodin solution. **Lycopodium granules** (see Fig. 9) may also reach the urine from dusting-powders. They might be mistaken for the ova of parasites. **Bubbles of air** (see Fig. 78, *h*) are often confusing to beginners, but are easily recognized after once being seen.

Scratches and **flaws** in the glass of slide or cover are often most assiduously studied by beginners, and are not infrequently reported as rare crystals, tube-casts, or even worms. **Dirt** upon the top of the cover (especially

16

when this is taken directly from the original box without
cleaning) is likewise a common source of confusion.
It often takes the form of crystals which, because they
are more prominent than the structures in the urine
beneath the cover, receive the student's whole attention. Fibers of **muscle** (Figs. 78, *i*, and 154) and other
particles which are evidently of fecal origin are usually
the result of contamination, but may rarely be present
in catheterized specimens. They then indicate recto-
vesical fistula.

V. THE URINE IN DISEASE

In this section the characteristics of the urine in those
diseases which produce distinctive urinary changes will
be briefly reviewed.

1. Renal Hyperemia.—*Active hyperemia* is usually
an early stage of acute nephritis, but may occur inde-
pendently as a result of temporary irritation. The urine
is generally decreased in quantity, highly colored, and
strongly acid. Albumin is always present—usually in
traces only, but sometimes in considerable amount for a
day or two. The sediment contains a few hyaline and
finely granular casts and an occasional red blood-cell.
In very severe hyperemia the urine approaches that of
acute nephritis.

Passive hyperemia occurs most commonly in diseases
of the heart and in pregnancy. The quantity of urine
is somewhat low and the color high, except in pregnancy. Albumin is present in small amount only. As
the liver is usually deranged in these cases, ·small or
moderate amounts of urobilin may be found. The
sediment contains a very few hyaline or finely granular

casts. In pregnancy the amount of albumin should be carefully watched, as any considerable quantity, and

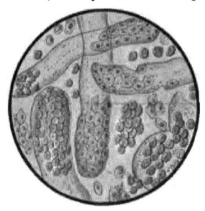

FIG. 80.—Sediment from acute hemorrhagic nephritis: Red blood-corpuscles; leukocytes; renal cells not fattily degenerated; epithelial and blood casts (Jakob).

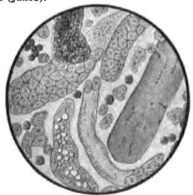

FIG. 81.—Sediment from chronic parenchymatous nephritis: Hyaline (with cells attached), waxy, brown granular, fatty, and epithelial casts; fattily degenerated renal cells, and a few white and red blood-corpuscles (Jakob).

especially a rapid increase, strongly suggests approaching eclampsia.

2. Nephritis.—The various degenerative and inflammatory conditions grouped under the name of nephritis have certain features in common. The urine in all cases contains albumin and tube-casts, and in all well-marked cases shows a decrease of normal solids, especially of urea and the chlorids. In chronic nephritis, especially of the interstitial type, there may be remissions during which the urine is practically normal. The degree of functional derangement is probably best ascertained by the phenolsulphonephthalein test (see p. 112). The characteristics of the different forms are well shown in the table on page 245.

3. Renal Tuberculosis.—The urine is pale, usually cloudy. The quantity may not be affected, but is apt to be increased. In early cases the reaction is faintly acid and there are traces of albumin and a few renal cells. In advanced cases the urine is alkaline, has an offensive odor, and is irritating to the bladder. Albumin in varying amounts is always present. Pus is nearly always present, though frequently not abundant. It is generally intimately mixed with the urine, and does not settle so quickly as the pus of cystitis. Casts, though present, are rarely abundant, and are obscured by the pus. Small amounts of blood are common. Tubercle bacilli are nearly always present, although animal inoculation may be necessary to detect them.

4. Renal Calculus.—The urine is usually somewhat concentrated, with high color and strongly acid reaction. Small amounts of albumin and a few casts may be present as a result of kidney irritation. Blood is frequently present, especially in the daytime and after severe exercise. Crystals of the substance composing the cal-

THE URINE IN NEPHRITIS

	Physical.	Chemic.	Microscopic.
Acute nephritis.	Quantity diminished, often very greatly. Color dark; may be red or smoky. Specific gravity, 1.020 to 1.030.	Urea and chlorids low. Much albumin: up to 1.5 per cent. Reaction acid.	Sediment abundant, red or brown. Many casts, chiefly granular, and epithelial varieties. Red blood-cells abundant. Numerous renal epithelial cells and leukocytes.
Chronic parenchymatous nephritis. (Large white kidney.)	Quantity usually diminished. Color variable, often pale and hazy. Specific gravity, 1.010 to 1.020.	Urea and chlorids low. Largest amounts of albumin: up to 3 per cent. Reaction acid.	Sediment rather abundant. Many casts of all varieties: fatty casts and casts of degenerated epithelium most characteristic. Blood present in traces: abundant only in acute exacerbations. Numerous fattily degenerated renal epithelial cells, often free globules of fat, and a few leukocytes.
Chronic interstitial nephritis. (Contracted kidney.)	Quantity markedly increased, especially at night. Color pale, clear. Specific gravity, 1.005 to 1.015.	Urea and chlorids low in well-marked cases. Albumin present in traces (often overlooked), increasing in late stages. Reaction acid.	Sediment very slight. Few narrow hyaline and finely granular casts. No blood except in acute exacerbations. Very few renal cells. Uric acid and calcium-oxalate crystals common.
Amyloid degeneration of kidney.	Quantity moderately increased. Color pale, clear. Specific gravity, 1.012 to 1.018.	Slight decrease of urea and chlorids. Variable amounts of albumin and globulin.	Sediment slight. Moderate number of hyaline, finely granular, and sometimes waxy casts.

culus—uric acid, calcium oxalate, cystin—may often be found. The presence of a calculus generally produces pyelitis, and variable amounts of pus then appear, the urine remaining acid in reaction.

5. Pyelitis.—In pyelitis the urine is slightly acid, and contains a small or moderate amount of pus, together with many spindle and caudate epithelial cells. Pus-casts may appear if the process extends up into the kid-

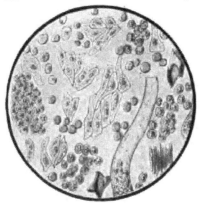

FIG. 82.—Sediment from calculous pyelitis: Numerous pus-corpuscles, red blood-corpuscles, and caudate and irregular epithelial cells; a combination of hyaline and pus casts, and a few uric-acid crystals (Jacob).

ney tubules (Fig. 82). Albumin is always present, and its amount, in proportion to the amount of pus, is decidedly greater than is found in cystitis. This fact is of much value in differential diagnosis. Even when pus is scanty, albumin is rarely under 0.15 per cent., which is the maximum amount found in cystitis with abundant pus.

6. Cystitis.—In *acute* and *subacute* cases the urine is acid and contains a variable amount of pus, with many epithelial cells from the bladder—chiefly large round, pyriform, and rounded squamous cells. Red blood-corpuscles are often numerous.

In *chronic* cases the urine is generally alkaline. It is pale and cloudy from the presence of pus, which is abundant and settles readily into a viscid sediment. The sediment usually contains abundant amorphous phosphates and crystals of triple phosphate and ammonium

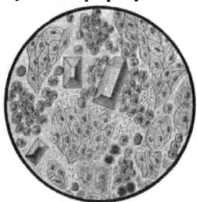

Fig. 83.—Sediment from cystitis (chronic): Numerous pus-corpuscles, epithelial cells from the bladder, and bacteria; a few red blood-corpuscles and triple phosphate and ammonium urate crystals (Jakob).

urate. Vesical epithelium is common. Numerous bacteria are always present (Fig. 83).

7. Vesical Calculus, Tumors, and Tuberculosis. —These conditions produce a chronic cystitis, with its characteristic urine. Blood, however, is more frequently present and more abundant than in ordinary cystitis. With neoplasms, especially, considerable hemorrhages are apt to occur. Particles of the tumor are sometimes passed with the urine. No diagnosis can be made from the presence of isolated tumor cells. In tuberculosis tubercle bacilli can generally be detected.

8. Diabetes Insipidus.—Characteristic of this disease is the continued excretion of very large quantities of pale, watery urine, containing neither albumin nor sugar. The specific gravity varies between 1.001 and 1.005. The daily output of solids, especially urea, is increased.

9. Diabetes Mellitus.—The quantity of urine is very large. The color is generally pale, while the specific gravity is nearly always high—1.030 to 1.050, very rarely below 1.020. Sometimes in mild or early cases the urine varies little from the normal in quantity, color, and specific gravity. The persistent presence of glucose is the essential feature of the disease. The amount of glucose may be small, but is often very great, sometimes exceeding 8 per cent., while the total elimination may exceed 500 gm. in twenty-four hours. It may be absent temporarily. Acetone, indicating acidosis, is generally present in advanced cases. Diacetic and oxybutyric acids may be present, and usually warrant an unfavorable prognosis. Accompanying the acidosis there is a corresponding increase in amount of ammonia which may be taken as an index of the degree of acidosis.

CHAPTER III

THE BLOOD

Preliminary Considerations.—The blood consists of a fluid of complicated and variable composition, the plasma, in which are suspended great numbers of microscopic structures: viz., red corpuscles, white corpuscles, blood-platelets, and blood-dust.

Red corpuscles, or **erythrocytes,** appear as biconcave disks, red when viewed by reflected light or in thick layer, and straw colored when viewed by transmitted light or in thin layer. They give the blood its red color. They are cells which have been highly differentiated for the purpose of carrying oxygen from the lungs to the tissues. This is accomplished by means of an iron-bearing protein, hemoglobin, which they contain. In the lungs hemoglobin forms a loose combination with oxygen, which it readily gives up when it reaches the tissues. Normal erythrocytes do not contain nuclei. They are formed from preëxisting nucleated cells in the bone-marrow.

If a small drop of blood be taken upon a clean slide and covered with a clean cover-glass as in diagnosis of malaria (see p. 355) the red corpuscles in the thicker portions of the preparation will often show a striking tendency to lie with overlapping edges, like piles of coins which have been tilted over. Formerly much attention was paid to this *"rouleaux formation"* as a point in diagnosis of certain diseases, but it is now little regarded. Also, in such preparations of fresh blood, many of the red corpuscles are seen to be globular in shape and covered with knob- or spine-like processes (see Fig. 73). This is called *"crenation"* and has little or no clinical significance. It is favored by concentration of the fluid due to evaporation at the edge of the cover. Crenated corpuscles are often seen in concentrated urine and other body-fluids and should always be recognized.

White corpuscles, or **leukocytes,** are less highly differentiated cells. There are several varieties. They all contain nuclei, and most of them contain granules which vary in size and staining properties. They are formed chiefly in the bone-marrow and lymphoid tissues. Their function is not fully understood. It appears to be concerned chiefly with the protection of the body against harmful agencies, in part through phagocytosis, in part through production of antitoxic substances and of ferments which play an important rôle in pathology.

Blood-platelets, or *blood-plaques*, are colorless or slightly bluish, spheric or ovoid bodies, usually about one-third or one-half the diameter of an erythrocyte, sometimes even as large as an erythrocyte. They appear to be constricted-off portions of the pseudopodia of certain giant cells of the bone marrow. Their function is not fully known, but is in some way connected with coagulation.

The **blood-dust of Müller** (*"hemoconia"*) consists of fine granules which have vibratory motion. The larger granules resemble micrococci. Little is known of them and they are given no consideration in clinical blood-examinations. It has been suggested that they are granules from disintegrated leukocytes.

The **total amount** of blood as shown by the new method of Keith, Rowntree and Geraghty, averages about one-twelfth of the body-weight. Little attention is paid to this subject in clinical work, but it is clear that fluctuations in volume, which are common in pathologic conditions, must have a marked influence upon the percentage of hemoglobin and the blood-cell counts.

The **reaction** is alkaline to litmus.

The **color** is due to the presence of hemoglobin in the red corpuscles, the difference between the bright red of arterial blood and the purplish red of venous blood depending upon the relative proportions of oxyhemoglobin and reduced hemoglobin. The depth of color depends upon the amount of hemoglobin. In very severe anemias the blood may be so pale as to be designated as "watery." The formation of carbon-monoxid-hemoglobin in coal-gas-poisoning gives the blood a bright cherry-red color; while formation of methemoglobin in poisoning with potassium chlorate and certain other substances gives a chocolate color.

The clear, pale, straw-colored fluid which remains after coagulation (see p. 257) and separation of the clot is called **serum.** In the serum are found the numerous substances which the tissues elaborate for protection against bacterial and other harmful agents. In most cases these substances, or "antibodies," are elaborated only when the harmful agent is present in the body, and they are "specific," that is, they are effective only against the one disease which has called them forth. A test for the presence of the antibody is, therefore, a test for the existence of the particular disease. The various tests based upon these principles have within recent years become a very important part of clinical laboratory work. They are discussed in the chapter upon Serodiagnostic Methods.

Clinical study of the blood may be discussed under the following heads: I. Methods of obtaining blood for examination. II. Coagulation. III. Hemoglobin. IV. Enumeration of erythrocytes. V. Color index. VI. Volume index. VII. Enumeration of leukocytes. VIII. Enumeration of platelets. IX. Study of stained blood. X. Blood parasites. XI. Tests for recognition of blood. XII. Less frequently used methods. XIII. Special blood pathology.

FIG. 84.—Daland's blood-lancet.

I. METHODS OF OBTAINING BLOOD

For most clinical examinations only one drop of blood is required. This may be obtained from the lobe of the ear, the palmar surface of the tip of the finger, or, in the case of infants, the plantar surface of the great toe. In the case of the ear, the edge of the lobe, not the side, should be punctured. In general, the finger will be found most convenient. With nervous children the ear is preferable, as it is less sensitive and its situation prevents their seeing what is being done. An edematous or congested part should be avoided; also a cold, apparently bloodless one. The site should be well rubbed with alcohol to remove dirt and epithelial débris and to increase the amount of blood in the part. After allowing sufficient time for the circulation to equalize, the skin is punctured with a blood-lancet (of which there are several patterns upon the market) or some substitute, as a large Hagedorn needle, aspirating needle, trocar, a spicule of glass, or a pen with one of its nibs broken off. The Hagedorn needle may be recommended as being cheap, easily obtained and fully as efficient as an expensive lancet. As suggested by Bass it may be fixed in the cork of a small vial of alcohol and thus kept immersed in the fluid. Nothing is more unsatisfactory than an ordinary round sewing-needle. The lancet should be cleaned with alcohol before and after using, but need not be sterilized. The puncture is practically painless if properly done with a sharp needle. It is made *with a firm, quick stab*, which, however, must not be so quick nor made from so great a distance that its site and depth are uncertain. The depth may be guarded with the thumb-nail if the lancet is not provided with a guard, but this should not be necessary. The first drop of blood which appears should be wiped away, and the second used for examination. The skin at the site of the puncture must be dry else the blood will not form a rounded drop as it exudes. The blood should not be pressed out, since this dilutes it with serum from the tissues; but moderate pressure some distance above the puncture is allowable.

For serologic, bacteriologic and chemic examinations a larger amount of blood is required. When 10 to 20 drops will suffice they can be obtained from a deep puncture of the lobe of the ear. For this, a spring-lancet

(Fig. 85) is best. Larger amounts are usually drawn from a vein as described below. For some purposes, particularly in children when puncture of a vein is not practicable, the blood can be obtained by means of a "wet cup."

Method of Obtaining Blood from a Vein.—Prepare the skin at the bend of the elbow as for a minor operation or simply rub well with alcohol or paint with tincture of iodin. The iodin is efficient as a germicide but makes it difficult to see the vein.

Bind a rubber or muslin bandage tightly around the upper arm. The cuff of the blood-pressure apparatus answers admirably. Instead of a bandage it will often

FIG. 85.—Spring-lancet.

be sufficient for an assistant or even the patient to grasp the upper arm firmly.

Have the patient open and close the fist a few times and when the veins are sufficiently distended insert a sterile hypodermic needle attached to a sterile syringe into any vein that is prominent. The needle should be large—about 19 to 21 gauge. It should go through the skin about ⅛ inch from the vein with the bevel at its tip upper-most and should enter the vein obliquely from the side in a direction opposite to the blood-current. If the needle is pushed through the skin directly over the vein, the vein is likely to roll to one side thus escaping the needle. Unless too small a needle is used, blood will begin to rise in the syringe as soon as the needle has entered the vein. Suction is rarely or never necessary.

When sufficient blood is obtained the bandage is first

removed and the needle is then withdrawn, this order being followed to avoid a hematoma. It is usually easy to secure 5 to 10 c.c. of blood. The procedure causes the patient surprisingly little inconvenience, seldom more than does an ordinary hypodermic injection. There is rarely any difficulty in entering a vein except in children and in adults when the arm is fat and the veins are small. If desired, one of the veins about the ankle can be used.

Instead of a syringe it is convenient to use a large glass tube which has been drawn out at the ends and one end ground to fit a "slip on" needle. A rubber tube like that

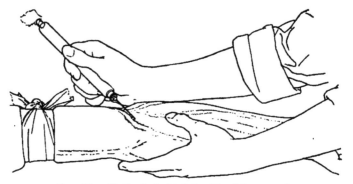

Fig. 86.—Method of obtaining blood from a vein.

used on hemacytometer pipets may be attached to the other end, thus allowing of aspiration if the blood does not enter readily. This little instrument (Fig. 86) can be made by any glass blower at a cost of less than fifty cents, and several of them can be kept on hand in large cotton-plugged test-tubes sterilized ready for use. Other devices for securing blood from a vein are shown in Figs. 87, 88 and 89 which indicate their construction in sufficient detail. They possess the advantage that the blood can be drawn directly into any desired reagent or culture medium (see p. 347).

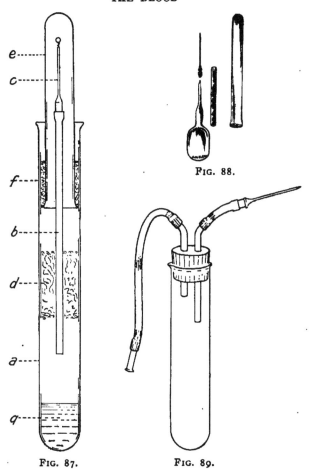

FIG. 88.

FIG. 87.   FIG. 89.

FIG. 87.—McJunkin's device for obtaining blood for a blood culture: *a*, large test tube; *b*, rubber tube; *c*, hypodermic needle; *d*, cotton, which is wrapped around the rubber tube before it is inserted; *e*, small test-tube used as a protecting cap; *f*, cotton; *g*, oxalate solution or culture medium.

FIG. 88.—Keidel's vacuum tube for collecting blood from a vein, consisting of a sealed ampoule, a needle with rubber connection and a glass cap. After the needle has entered the vein the stem of the ampoule is crushed within the rubber connection and blood enters because of the vacuum. Similar tubes containing sterile culture media are upon the market.

FIG. 89.—Device for drawing blood from a vein using a large test-tube, a 50 c.c. centrifuge tube, or a small flask.

II. COAGULATION

Coagulation consists essentially in the transformation of fibrinogen, one of the proteins of the blood-plasma, into fibrin by means of a ferment called thrombin. The presence of calcium salts is necessary. The resulting coagulum is made up of a meshwork of fibrin fibrils with entangled corpuscles and platelets. The clear, straw-colored fluid which is left after separation of the coagulum is called *blood-serum*. Normally, coagulation takes place in two to eight minutes after the blood leaves the vessels. The usual time is about four and one-half minutes. The time is affected by the temperature, the size of the drop, cleanliness of the instruments, and other factors. Clotting is more rapid when blood is squeezed from a puncture than when it flows freely, owing to admixture with tissue juice. For this reason, some prefer to take the blood from a vein. Pathologically, it is delayed in hemophilia, purpura, scurvy, and icterus. Estimation of coagulation time is very important as a preliminary to operation when there is any reason to expect dangerous capillary oozing, as in tonsilectomies and in operations upon jaundiced persons. In treatment, calcium salts, especially the lactate and acetate, are used to hasten coagulation; citric acid, to retard it.

For certain purposes, notably in bacteriologic and opsonic work, it is desirable to prevent coagulation of blood which has been withdrawn. This may be accomplished by receiving it directly into a solution of 1 per cent. sodium citrate (or ammonium oxalate) in normal salt solution or into a tube containing a very little finely powdered potassium oxalate. This

17

precipitates the calcium salts which are necessary to coagulation.

There are many methods of ascertaining the coagulation time and results by the different methods are not comparable because their end-points—the point in the process when coagulation is assumed to have taken place—are not uniform. It is therefore well to adopt a single method for one's routine work. The simplest method is to receive several drops of blood (well rounded drops 4 to 5 mm. in diameter) on a clean slide and to draw a needle through one or another of them at one-minute intervals. When the clot is dragged along by the needle, coagulation has taken place. Duke uses a glass slide to which two glass disks 5 mm. in diameter are cemented. Well-rounded drops of blood are received on the disks and the slide is inverted across the top of a glass or beaker containing water at 40°C. and covered with a towel. Coagulation is judged by the shape of the drop when the slide is held in a vertical position (Fig. 90).

FIG. 90.—Showing difference in shape of blood-drops before and after coagulation (Duke's method).

For more accurate work the method of Russell and Brodie as modified by Boggs is now generally used.

Boggs' Method.—The instrument is shown in Fig. 91. The bottom of the box (A) and the cone (B) are of glass.

The instrument must be absolutely clean. Obtain the blood from a freely flowing puncture. When a large drop has formed, touch the small end of the cone to its surface.

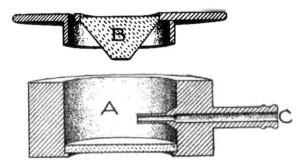

FIG. 91.—Boggs' coagulation instrument: A, chamber with glass bottom; B, glass cone; C, tube through which air is blown.

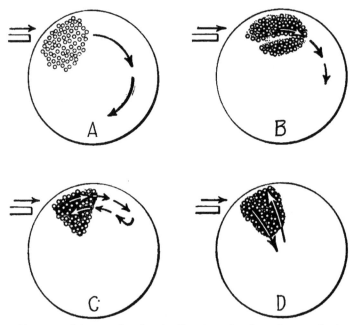

FIG. 92.—Diagram showing the direction taken by red corpuscles in Boggs' method for coagulation time: Radial movement of the corpuscles, D, indicates the end-point (after Boggs).

Quickly invert the cone into the box. Place the instrument on the microscope and blow puffs of air against the drop of blood at intervals by means of a rubber bulb attached to *C*, meanwhile watching the motion of the corpuscles with a low power of the microscope. Coagulation has occurred when the corpuscles move *en masse* in a ·radial direction and spring back to their original position (Fig. 92, *D*). The time is counted from the first appearance of the blood from the puncture to the end-point.

III. HEMOGLOBIN

Hemoglobin is an iron-bearing protein which normally occurs in the circulating blood in two forms: *oxyhemoglobin* chiefly in arterial blood; and *reduced* hemoglobin (more correctly called simply *hemoglobin*) chiefly in venous blood. Through the action of acids, alkalies, oxidizing and reducing substances, heat, and other agencies it is readily converted into a series of derivative compounds which can be distinguished by means of the spectroscope.

Most of these derivative compounds are formed only in blood which has left the vessels; a few, however, may be produced in the circulation.

Methemoglobin is formed in the circulating blood in the rare condition known as "enterogenous cyanosis" and in poisoning with potassium chlorate, nitrites, nitro-benzol, acetanilid, phenacetin, antipyrin, and other substances. Clinically there is marked cyanosis, and in severe cases the blood has a chocolate-brown color when withdrawn. Methemoglobinemia is easily recognized spectroscopically (see p. 370).

Carbon monoxid hemoglobin, formed in carbon monoxid poisoning, gives the blood a brighter red color than is normal. It may, in some cases, be identified with the spectroscope, but the following test is more sensitive:

Receive about ten drops of blood in twenty drops of 10 per cent. sodium hydroxid solution, and mix. Blood containing carbon monoxid remains bright red, while normal blood takes on a dirty brownish-green color.

Normally hemoglobin is confined to the red corpuscles. When it is dissolved out of these cells and appears in the plasma, the condition is known as *hemoglobinemia*. This occurs in a great variety of conditions, among which may be mentioned: severe types of infectious diseases; the hemorrhagic diseases (scurvy, etc.); paroxysmal hemoglobinuria; severe burns and frost bites; and poisoning with potassium chlorate, mushrooms, etc.

To recognize hemoglobinemia, receive a little blood in a small test-tube and allow it to stand in a cool place for twenty-four hours. The serum, which separates after coagulation, will be colored red or pink instead of pale yellow as is normally the case.

The normal amount of hemoglobin is usually given as about 14 Gm. per 100 c.c. of blood. The absolute amount is, however, seldom estimated clinically: it is the relation which the amount present bears to an arbitrarily-fixed normal that is determined. Thus the expression, "50 per cent. hemoglobin," when used clinically, means that the blood contains 50 per cent. of the normal. Theoretically, the normal would be 100 per cent., but with the methods of estimation in general use the blood of healthy adults ranges from 80 to 105 per cent.; these figures may, therefore, be taken as representing normal limits. There are, moreover, marked fluctuations with age and sex which must be

taken into account in any careful case-study. These
are well shown in Fig. 93, which is based upon Williamson's careful spectrophotometric study of the blood of
919 healthy persons in Chicago.

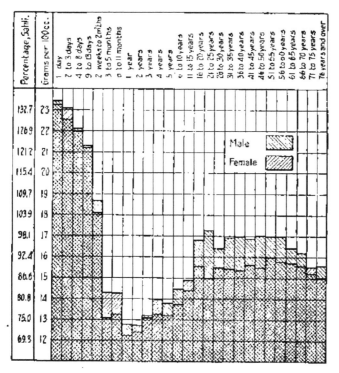

FIG. 93.—Diagram showing average hemoglobin values for both
sexes at different ages. The corresponding percentages on the Sahli
hemoglobinometer are also shown.

The custom of recording hemoglobin in terms of per-
centage· of the normal is grossly inaccurate and leads to
much confusion. From what has just been said it is clear
that no single normal standard can be applied to all ages

and both sexes. The situation is complicated by the fact that no two hemoglobinometers use the same standard. A blood, for example, which reads 100 per cent. on the Dare instrument will read about 80 per cent. on the Sahli. A record therefore means little unless one knows what instrument was used and the age and sex of the patient. This confusion could be avoided if records were made in terms of the actual percentage of hemoglobin, *i.e.*, in grams per 100 c.c. of blood, and this will doubtless soon become customary. The reading on any type of instrument can readily be converted into absolute percentage if one knows what amount of hemoglobin was adopted by the makers as normal. This calculation will be given with the description of the various instruments.

Increase of hemoglobin, or *hyperchromemia,* is uncommon, and is probably more apparent than real. It accompanies an increase in number of erythrocytes, and may be noted in change of residence from a lower to a higher altitude; in poorly compensated heart disease with cyanosis; in concentration of the blood from any cause, as the severe diarrhea of cholera; and in "idiopathic polycythemia."

Decrease of hemoglobin, or *oligochromemia,* is very common and important. It is the distinctive and most striking feature of the anemias (see p. 380). In secondary anemia the hemoglobin loss may be slight or very great. In mild cases a slight decrease of hemoglobin is the only blood change noted. In very severe cases, especially in repeated hemorrhages, malignant disease, and infection by the hookworm and *Dibothriocephalus latus,* hemoglobin may fall to 15 per cent. Hemoglobin is always diminished, and usually very greatly, in chlorosis (average about 40 to 45 per cent.), pernicious anemia (average about 20 to 25 per cent.), and leukemia (usually about 40 to 50 per cent.).

Estimation of hemoglobin is less tedious and usually more helpful than a red corpuscle count. It offers the simplest and most certain means of detecting the existence and degree of anemia, and of judging the

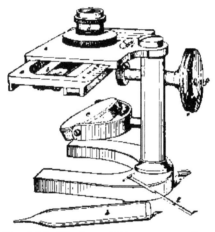

FIG. 94.—Von Fleischl's hemoglobinometer: *a*, Stand; *b*, narrow wedge-shaped piece of colored glass fitted into a frame (*c*), which passes under the chamber; *d*, hollow metal cylinder, divided into two compartments, which holds the blood and water; *e*, plaster-of-Paris plate from which the light is reflected through the chamber; *f*, screw by which the frame containing the graduated colored glass is moved; *g*, capillary tube to collect the blood; *h*, pipet for adding the water; *i*, opening through which may be seen the scale indicating percentage of hemoglobin.

effect of treatment in anemic conditions. Pallor, observed clinically, does not always denote anemia.

There are many methods, but none is entirely satisfactory. Those which are most widely used are here described:

1. **Von Fleischl Method**.—The apparatus consists of a stand somewhat like the base and stage of a microscope (Fig. 94). Under the stage is a movable bar of colored glass, shading from pale pink at one end to deep red at the other. The frame in which this bar is held is marked with a scale of hemoglobin percentages corresponding to the different shades of red. By means of a rack and pinion the color-bar can be moved from end to end beneath a round opening in the center of the stage. A small metal cylinder, which has a glass bottom and which is divided vertically into two equal compartments, can be placed over the opening in the stage so that one of its compartments lies directly over the color-bar. Accompanying the instrument are a number of short capillary tubes in metal handles.

Having punctured the finger-tip or lobe of the ear, as already described, wipe off the first drop of blood, and from the second fill one of the capillary tubes. Hold the tube horizontally, and touch its tip to the drop of blood, which will readily flow into it if it be clean and dry. Avoid getting any blood upon its outer surface. With a medicine-dropper rinse the blood from the tube into one of the compartments of the cylinder, using distilled water, and mix well. Fill both compartments level full with distilled water, and place the cylinder over the opening in the stage, so that the compartment which contains only water lies directly over the bar of colored glass. If there are any clots in the hemoglobin compartment, clean the instrument and begin again.

In a dark room, with the light from a candle reflected up through the cylinder, move the color-bar along with a jerking motion until both compartments have the same depth of color. The number upon the scale corresponding to the portion of the color-bar which is now under the cylinder gives the percentage of hemoglobin. While comparing the two colors, place the instrument so that they will fall upon the right and left halves of the retina, rather than upon the upper and lower halves; and protect the eye from the light with a cylinder of paper or pasteboard. After use, clean the metal cylinder with water, and wash the capillary tube with water, alcohol, and ether, successively. Results with this instrument are accurate to within about 5 per cent.

2. The **Fleischl-Miescher** instrument, a modification of the preceding, is generally considered the most accurate hemoglobinometer available. It is, however, better adapted to laboratory use than to the needs of the clinician. Detailed instructions accompany each instrument. The chief differences from the von Fleischl are: (1) The blood is more accurately measured and diluted, a pipet like that accompanying the hemacytometer being used; (2) 0.1 per cent. solution of sodium carbonate is used instead of water for diluting; (3) the glass bar is more accurately colored; (4) there are two cylindric cells, one four-fifths the depth of the other; and (5) the cell is covered with a glass disk and a metal cap with a slit through which the reading is made. Each Miescher hemoglobinometer is accompanied by a chart showing the actual hemoglobin values in grams for each reading upon that particular instrument.

3. The **Sahli hemoglobinometer** (Fig. 95) is an improved form of the well-known Gowers instrument. It consists of an hermetically sealed comparison tube containing a suspension of acid hematin, a graduated test-tube of the same diameter, and a pipet of 20-cu. mm. capacity. The two tubes are held in a black frame with a white ground-glass back.

Place decinormal hydrochloric acid solution in the graduated tube to the mark 10. Obtain a drop of blood and draw it into the pipet to the 20-cu.mm. mark. Wipe off the tip of the pipet, blow its contents into the hydrochloric acid solution in the tube, and rinse well. The hemoglobin is changed to acid hematin. Place the two tubes in the com-

partments of the frame; let stand one minute; and dilute the
fluid with water drop by drop, mixing after each addition,
until it has exactly the same color as the comparison tube.
The graduation corresponding to the surface of the fluid
then indicates the percentage of hemoglobin. Mixing
may be done by closing the tube with the finger and inverting, but care should be exercised to see that none of the fluid
is removed by adhering to the finger. Slightly waxing the
finger will aid. Decinormal hydrochloric acid solution is
prepared with sufficient accuracy for this purpose by adding

Fig. 95.—Sahli's hemoglobinometer.

12 c.c. of the concentrated acid to 988 c.c. distilled water.
A little chloroform should be added as a preservative.

This method is very satisfactory in practice, and is accurate to within 5 per cent. Unfortunately, not all the
instruments upon the market are well standardized, and the
comparison tube does not keep its color unchanged indefinitely. Usually, however, the apparent fading is due to
the fact that the hematin is in suspension and settles out
when the instrument lies unused for some time. This can
be remedied by inverting the tube a number of times. Most
tubes contain a glass bead to facilitate mixing.

The reading upon the Sahli multiplied by 0.173 gives the hemoglobin value in grams per 100 c.c. of blood. The **Kuttner** modification of this instrument (Fig. 31) uses a standard color tube which corresponds to 15 Gm. of hemoglobin and its readings should be multiplied by 0.15 to find the actual percentage.

4. **Dare's hemoglobinometer** differs from the others in using undiluted blood. The blood is allowed to flow by capillarity into the slit between two small plates of glass (Fig. 96, *W*). It is then placed in the instrument and

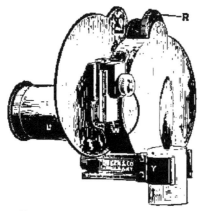

FIG. 96.—Dare's hemoglobinometer.

compared, by looking through the tube, U, with different portions of a circular disk of colored glass which is revolved by means of the wheel, R. The two colors appear side by side, and will show their true values only when viewed in a darkened room by the light of the candle, Y. Usually a dark corner of the office will suffice. The reading is taken at the beveled edge of a slot not shown in the figure, and it must be made quickly, before clotting takes place. This instrument is easy to use and to clean, and is one of the most accurate. One hundred per cent. on the scale corresponds to 13.77 Gm. of hemoglobin por 100 c c of blood. When, therefore, it is desired to express results in actual percentage, the readings must be multiplied by 0.1377.

5. **Tallqvist Method.**—The popular Tallqvist hemoglobinometer consists simply of a book of small sheets of absorbent paper and a carefully printed scale of colors (Fig. 97).

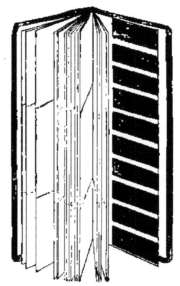

FIG. 97.—Tallqvist's hemoglobin scale.

Take up a large drop of blood with the absorbent paper, and when the humid gloss is leaving, before the air has darkened the hemoglobin, compare the stain with the color scale. The color which it matches gives the percentage of hemoglobin. Except in practised hands, this method is accurate only to within 10 or 20 per cent.

Of the methods given, the physician should select the one which best meets his needs. With any method, practice is essential to accuracy.   The von Fleischl was for many years the standard instrument, but is now little used.   For accurate work the best instruments are the von Fleischl-Miescher and the Dare.   The former is essentially a laboratory instrument.   The Dare is easy to use and to clean, and is probably the best for clinical work.   The Sahli, although less easy to use and probably less accurate, is inexpensive and is very satisfactory, provided a well-standardized color-tube is obtained. The Kuttner modification of the Sahli is an improvement upon the original.   The Tallqvist scale is so inexpensive and so convenient that it should be used by every physician at the bedside and in hurried office work; but it should not supersede the more accurate methods.

IV. ENUMERATION OF ERYTHROCYTES

In health there are about 5,000,000 red corpuscles per cubic millimeter of blood.   The number is generally a little less—about 4,500,000—in women.   Age variations in general follow those already given for hemoglobin.   Hawk finds the normal for athletes in training to be 5,500,000.

Increase of red corpuscles, or *polycythemia,* is unimportant.   There is a decided increase following change of residence from a lower to a higher altitude, reaching a maximum after several days sojourn. The increase, however, is not permanent.   In a few months the erythrocytes return to nearly their original number.   At the University of Colorado (altitude 5400 feet) the average for healthy medical students is about 5,600,000.   Three views have been offered in explanation of this effect of altitude: (a) Concentration of the blood, owing to increased evaporation from the skin; (b) altered distribution of corpuscles, the reserve cells in the splanchnic vessels being thrown into the peripheral circulation; (c) new formation of corpuscles, this giving a compensatory increase of aëration surface. The work of Schneider at Colorado Springs strongly supports the second of these views, although the third must be accepted to explain the moderate permanent increase.

Pathologically, polycythemia is uncommon. It may occur in: (a) Concentration of the blood from severe watery diarrhea; (b) chronic heart disease, especially the congenital variety, with poor compensation and cyanosis; and (c) idiopathic polycythemia, which is considered to be an independent disease, and is characterized by dark red cast of countenance, blood-counts of 7,000,000 to 10,000,000, hemoglobin 120 to 150 per cent., and a normal number of leukocytes.

Decrease of red corpuscles, or *oligocythemia*. Red corpuscles and hemoglobin are commonly decreased together, although usually not to the same extent.

Oligocythemia occurs in all but the mildest symptomatic anemias. The blood-count varies from near the normal in moderate cases down to 1,500,000 in very severe cases. There is always a decrease of red cells in chlorosis, but it is often slight, and is relatively less than the decrease of hemoglobin. Leukemia gives a decided oligocythemia, the average count being about 3,000,000. The greatest loss of red cells occurs in pernicious anemia, where counts below 1,000,000 are not uncommon.

Method of Counting.—The most widely used instrument for counting the corpuscles is the Thoma hemacytometer, although the original Thoma ruling has been largely displaced by more convenient ones. The Bürker type of counting chamber, exemplified by the Bürker and the Levy instruments, is a decided improvement and when supplied with the Neubauer ruling is probably the most satisfactory of all. The Thoma-Metz hemacytometer is convenient for routine working. The hematocrit, which at one time promised much, is not to be recommended for accuracy, since in anemia, where blood-counts are most important, the red cells vary greatly in size and probably also in elasticity. The hematocrit is, however, useful in determining the relative volume of corpuscles and plasma (see Volume Index, p. 285).

Although simple in principle, accurate counting of blood-corpuscles involves a technic which is acquired only after considerable practice. Exact and fairly rapid work is demanded. Before beginning, one should familiarize himself with the instrument and its ruling, and should read the directions carefully, giving especial attention to sources of error. It is likewise an advantage to practice sucking the diluting fluid into the pipet and stopping it at a predetermined height and also to practice adjusting the cover-glass after placing a drop of diluting fluid in the counting chamber. In our class work we have found Emerson's plan satisfactory: after students think they have acquired the technic they are required to count their own red corpuscles at the same hour upon successive days until the difference between successive counts falls below

100,000 cells. Only by rigid adherence to such a plan can a student realize his inaccuracies.

The various hemacytometers are here described in order, but detailed directions are given only for the Thoma instrument. Except for minor variations made necessary by differences in construction, the same directions apply to all the instruments.

The **Thoma-Zeiss instrument** consists of two pipets for diluting the blood and a counting chamber (Fig. 98). The

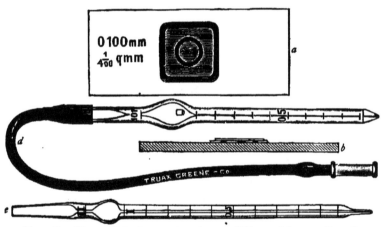

FIG. 98.—Thoma-Zeiss hemacytometer: *a*, Slide used in counting; *b*, sectional view; *d*, red pipet; *e*, white pipet.

rubber tubes which come with the pipets are too short and too flexible and should be replaced. For this purpose nothing is so good as a rubber catheter. The counting chamber is a glass slide with a square platform in the middle. In the center of the platform is a circular opening, in which is set a small circular disk in such a manner that it is surrounded by a "ditch," and that its surface is exactly 0.1 mm. below the surface of the square platform. Upon

18

this disk is ruled a square millimeter, subdivided into 400 small squares. Each fifth row of small squares has double rulings for convenience in counting (Fig. 99). This ruling, known as the Thoma, constitutes the central square millimeter of most of the more recent forms, such as the Neu-

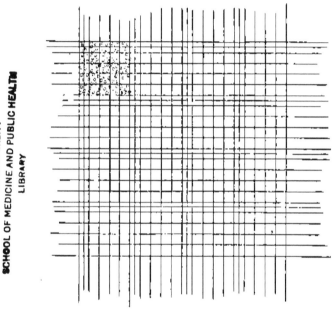

FIG. 99.—Essential portion of Thoma ruling of counting chamber, showing red corpuscles in left upper corner. This ruling constitutes the central portion of most of the other rulings.

bauer and the Türck (see Figs. 106, 107, 108). A thick cover-glass, ground perfectly plane, accompanies the counting chamber. Ordinary cover-glasses are of uneven surface, and should not be used with this instrument. For use with objectives of short working distance, heavy coverglasses can be obtained with a flat-bottomed excavation or "well" in the center. This combines the advantages of a thin cover with the rigidity of a thick one.

It is evident that, when the cover-glass is in place upon the platform, there is a space exactly 0.1 mm. thick between it and the disk; and that, therefore, the square millimeter ruled upon the disk forms the base of a space holding exactly 0.1 cubic millimeter.

Technic.—To count the red corpuscles, use the pipet with 101 engraved above the bulb. It must be clean and dry.

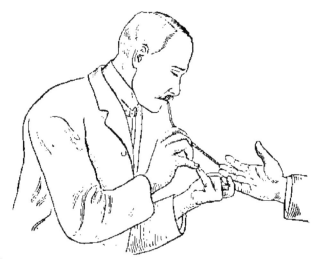

FIG. 100.—Method of drawing blood into the pipet (Boston). The pipet should be held more nearly horizontal than here represented.

Puncture the skin, wipe off the first drop of blood, and fill the pipet from the second, sucking the blood to the mark 0.5 or 1.0, according to the dilution desired. While doing this, hold the pipet horizontally at nearly right angles to the line of vision, so that the exact height of the column may be easily seen. The side of the tip should rest against the skin, but the end must be free. Air-bubbles will enter if the drop is too small or if the tip is not kept immersed. Should the blood go slightly beyond the mark, draw it · back by touching the tip of the pipet. to a moistened hardkerchief. Quickly wipe off the blood adhering to the tip, plunge it into the diluting fluid, and suck the fluid up to the mark 101, slightly rotating the pipet meanwhile. At this stage it is best to hold the pipet nearly vertically in order to avoid inclusion of a large air bubble in the bulb. This dilutes the blood 1:200 or 1:100, according to the amount of blood taken. Except in cases of severe anemia, a dilution of 1:200 is preferable. Close the ends of the pipet with the fingers, and shake vigorously until the blood and diluting fluid are well mixed, keeping the pipet horizontal meanwhile. One minute's shaking is usually sufficient. Some workers mix by rotating the pipet rapidly upon its long axis.

· When it is not convenient to count the corpuscles at once, place a heavy rubber band around the pipet so as to close the ends, inserting a small piece of rubber-cloth or other tough, non-absorbent material, if necessary, to prevent the tip from punching through the rubber. It may be kept thus for twenty-four hours or longer.

When ready to make the count, clean the counting chamber and cover-glass, and place a sheet of paper over them to keep off dust. Mix the fluid thoroughly by shaking; blow 2 or 3 drops from the pipet, wipe off its tip, and then place a small drop (the proper size can be learned only by experience) upon the disk of the counting chamber. Adjust the cover immediately. Hold it by diagonal corners above the drop of fluid so that a third corner touches the slide and rests upon the edge of the platform. Place a finger upon this corner, and, by raising the finger, allow the cover to fall quickly into place. If the cover be properly adjusted, faint concentric lines of the prismatic colors—Newton's bands— can be seen between it and the platform when the slide is viewed obliquely. They indicate that the two surfaces are in close apposition. If they do not appear at once, slight pressure upon the cover may bring them out. Failure to obtain them is usually due to dirty slide or cover—both must be perfectly clean and *free from dust*. The drop placed upon the disk must be of such size that, when the cover is adjusted, it nearly or quite covers the disk, and that none of it runs over into the "ditch." There should be no bubbles upon the ruled area.

The following is an easier method of applying the cover: Place a drop of fluid upon the ruled disk. The size of the drop is of no great consequence, if only it be large enough. Immediately place the cover-glass flat upon one side of the platform with its edge close to the drop of fluid, and hold it firmly down with the two index-fingers, or with the index-finger and middle finger' of the right hand. Now slide it firmly and quickly into place. If the drop of fluid is too large, the excess will be caught on the top of the cover. A moderately thin cover is best.

Allow the corpuscles to settle for a few minutes, and then examine with a low power to see that they are evenly distributed. If they are not *evenly distributed over the whole disk*, the counting chamber must be cleaned and a new drop placed in it.

Probably the most satisfactory objective for counting is the 8-mm. or the 4-mm. with long working distance. To understand the principle of counting, it is necessary to remember that the square millimeter (400 small squares) represents a capacity of 0.1 c.c. Find the number of corpuscles in the square millimeter, multiply by 10 to find the number in 1 cu.mm. of the diluted blood, and finally, by the dilution, to find the number in 1 cu.mm. of undiluted blood. Instead of actually counting all the corpuscles, it is customary to count those in only a limited number of small squares, and from this to calculate the number in the square millimeter. Nearly every worker has his own method of doing this. The essential thing is to adopt a method and adhere to it.

In practice a convenient procedure is as follows: *With a dilution of 1:200, count the cells in 80 small squares, and to the sum add 4 ciphers; with dilution of 1:100, count 40 small squares and add 4 ciphers.* Thus, if with 1:200 dilution, 450 corpuscles were counted in 80 squares, the total count would

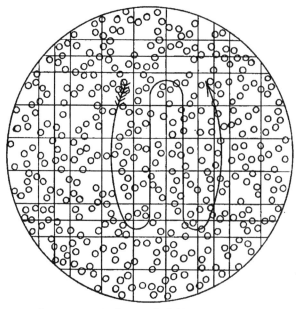

Fig. 101.—Appearance of microscopic field in counting red corpuscles. The arrow indicates the 20 squares to be counted.

be 4,500,000 per cu.mm. This method is sufficiently accurate for all clinical purposes, provided the corpuscles are evenly distributed and 2 drops from the pipet be counted. It is convenient to count a block of 20 small squares, as indicated in Fig. 101, in each corner of the square millimeter. Four columns of 5 squares each are counted. The double rulings show when the bottom of a column has been reached and also indicate the fourth column. In the writer's opinion it is easier to count in vertical than horizontal rows. If distribution be even, the difference between the number of cells in any two such blocks should not exceed twenty. Instead of four blocks of 20 squares, five blocks of 16 squares may be counted, one block in each corner of the ruled area and one block in the center.

In order to avoid confusion in counting cells which lie upon the border-lines, the following rule is generally adopted: *Corpuscles which touch the upper and left sides should be counted as if within the squares, those touching the lower and right sides, as outside; or vice versâ.*

Diluting Fluids.—The most widely used are Hayem's and Toisson's. Both of these have high specific gravities, so that, when well mixed, the corpuscles do not separate quickly. Toisson's fluid is perhaps the better for beginners, because it is colored and can easily be seen as it is drawn into the pipet. It stains the nuclei of leukocytes blue, but this is no real advantage. It must be filtered frequently because of the ready growth of fungi in it. Hayem's fluid is to be preferred for routine work. For convenience in filling pipets the fluids should be kept in small wide-mouth bottles.

Hayem's Fluid		Toisson's Fluid	
Mercuric chlorid	0.5	Sodium chlorid	1.0
Sodium sulphate	5.0	Sodium sulphate	8.0
Sodium chlorid	1.0	Glycerin	30.0
Distilled water	200.0	Distilled water	160.0
		Methyl-violet, 5 B to give a strong purple color.	

Sources of Error.—The most common sources of error in making a blood-count are:

(*a*) Inaccurate dilution, usually from faulty technic,

occasionally from inaccurately graduated pipets. Only an instrument of standard make can be relied upon.

(b) Too slow manipulation, allowing a little of the blood to coagulate and remain in the capillary portion of the pipet.

(c) Inaccuracy in depth of counting chamber usually due to imperfect application of the cover-glass, but sometimes to faulty manufacture or to softening of the cement by alcohol or heat. The slide should not be cleaned with alcohol nor left to lie in the warm sunshine.

(d) Uneven distribution of the corpuscles. This results when the blood has partially coagulated, when it is not thoroughly mixed with the diluting fluid, or when the cover-glass is not applied soon enough after the drop is placed upon the disk.

(e) The presence of yeasts, which may be mistaken for corpuscles, in the diluting fluid.

Cleaning the Instrument.—The instrument should be cleaned immediately after using, and the counting chamber and cover must be cleaned again just before use.

Transfer the rubber tube to the small end of the pipet and draw through it, successively, water, alcohol, ether, and air. This can be done with the mouth, but it is much better to use a rubber bulb or suction filter pump. When the mouth is used, the moisture of the breath will condense upon the interior of the pipet unless the fluids be shaken, and not blown, out. If blood has coagulated in the pipet—which happens when the work is done too slowly—dislodge the clot with a horsehair, never with a wire, and clean with strong sulphuric acid, or let the pipet stand over night in a test-tube of the acid. Even if the pipet does not become clogged, it should be occasionally cleaned in this way. When the etched graduations on the pipets become dim, they can be renewed by rubbing with a wax pencil.

Wash the counting-chamber and the cover with water and

dry them with clean soft linen. Alcohol may be used to clean the latter, but never the former, although a handkerchief *slightly* moistened with alcohol may be used to wipe off the surface of the ruled disk and the platform.

Bürker's hemacytometer (Fig. 102).—This modification of the Thoma-Zeiss instrument allows of greater accuracy.

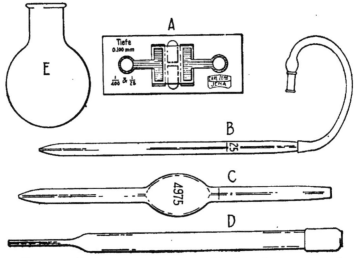

FIG. 102.—Bürker's hemacytometer with pipets for counting red-corpuscles: *A*, counting slide with cover-glass in place; *B*, pipet for measuring blood; *C*, pipet for measuring diluting fluid; *D*, pipet for transferring diluted blood to slide; *E*, mixing flask. In place of the flask and pipets here shown the instrument is now generally supplied with the regular Thoma mixing pipets.

Originally it consisted of a counting slide with cover-glass, three pipets—for (*a*) measuring blood, (*b*) measuring diluting fluid, and (*c*) transferring the diluted blood to the slide—and one or more small flasks for mixing blood and diluting fluid. As usually sold at present it consists simply of the Bürker counting slide and the regular Thoma mixing pipets.

The floor-piece of the counting slide, instead of being circular, as in the Thoma-Zeiss instrument, consists of a plate of glass 5 mm. wide and 25 mm. long, which extends across the slide. This is divided across the middle by a deep groove 1.5 mm. wide, and upon each portion is a ruled area. On each side of the floor-piece and separated from it by a ditch is a glass platform 0.1 mm. higher than the ruled areas. When the cover-glass is adjusted upon the platform, the ends of the floor-piece project beyond it. There are two types of the counting slide: one with cover-glass clamps, as shown in Fig. 102; one lacking them. The clamps are by no means necessary but are a decided advantage.

When the count is to be made, the cover-glass is carefully adjusted so as to show Newton's bands, and is clamped in place if the counting slide is supplied with clamps. A drop of the diluted blood is then placed on each of the projecting ends of the floor-piece. The fluid will run under the cover by capillary attraction. Care must be exercised to use just enough fluid to fill the space between the cover and the floor-piece. The slide is now placed on the microscope with the diaphragm wide open and viewed obliquely with the unaided eye. If the film of corpuscles as seen in this way is not uniform, the slide must be cleaned and filled again. The count is made in the usual way. As originally supplied the instrument had a special ruling, but the Neubauer ruling is now generally used.

Since this counting chamber has two independent rulings it can be filled for both the red and white counts at the same time.

Levy Counting Chamber (Fig. 103).—In this new American-made counting slide the Bürker principle is utilized but the construction is somewhat different and apparently more substantial. The slide has a matte surface which makes it impossible to bring out Newton's bands but which has certain compensating advantages. The makers state that it will soon be obtainable with cover-glass clamps which will be a decided improvement. The ordinary Thoma mixing pipets are supplied with it.

Thoma-Metz Hemacytometer (Fig. 104).—This new instrument introduces certain conveniences into the routine counting of both red cells and leukocytes. Its special

FIG. 103.—Levy's counting chamber, Bürker type.

feature is that the ruling is engraved upon a disk in the ocular instead of upon the counting slide. This disk is ruled with a large circle and with a square, which in turn is subdivided into four smaller squares.

For the red count the squares are used. The four small squares have each the same value as the small squares of the

FIG. 104.—Thoma-Metz hemacytometer and diagram showing ruling in ocular.

Thoma ruling ($\frac{1}{400}$ sq. mm.), and the count may be conducted as already described.

The circle is used for counting leukocytes. Its area corresponds to 0.1 sq. mm. when the correct magnification is used. The leukocytes are counted as in the author's circle method described on page 298.

A decided advantage of this instrument is that the ruled lines are always sharp and clear. The eye-lens of the ocular can be focused to suit different eyes. The chief disadvantage and a source of inaccuracy lies in the fact that the values of the ruled areas vary according to magnification. The makers say that values are correct as above given when the Leitz No. 6 objective (4-mm. focus) is used with tube length of 170 mm. Slight variations with other objectives can be compensated by altering the tube length. For accurate evaluation a square is ruled on the counting slide, and the tube length should be so adjusted that this square exactly coincides with the large square in the ocular.

V. COLOR INDEX

This is an expression which indicates the amount of hemoglobin in each red corpuscle compared with the normal amount. For example, a color index of 1.0 indicates that each corpuscle contains the normal amount of hemoglobin; of 0.5, that each contains one-half the normal.

The color index is most significant in chlorosis and pernicious anemia. In the former it is usually much decreased; in the latter, generally much increased. In symptomatic anemia it is moderately diminished.

To obtain the color index, divide the percentage of hemoglobin by the percentage of corpuscles. The percentage of corpuscles is found by multiplying the first two figures of the red corpuscle count by 2. This simple method holds good for all counts of 1,000,000 or more. Thus, a count of 2,500,000 is 50 per cent. of the normal. If, then, the hemoglobin has been estimated at 40 per cent., divide 40 (the percentage of hemoglobin) by 50 the percentage of corpuscles). This gives $\frac{4}{5}$, or 0.8, as the color index.

From what has already been said regarding the variations in hemoglobin-instruments, and of the impossibility of fixing a normal standard for either red cells or hemoglobin which is applicable to all ages and in all localities, it would appear that color-index calculations, as above described, have little value. Certainly only marked variations can be considered in diagnosis unless one takes into account the age and sex of the patient, the locality, and the hemoglobin-instrument used. It has been suggested that the index be worked out upon a basis of grams of hemoglobin per 1,000,000 red cells, which would make it of real value, but this is not yet in vogue.

VI. VOLUME INDEX

The term "volume index" was introduced by Capps to express the average size of the red cells of an individual compared with their normal size. It is the quotient obtained by dividing the *volume* of red corpuscles (expressed in percentage of the normal) by the *number* of red corpuscles, also expressed in percentage of the normal.

The volume index more or less closely parallels the color index, and variations have much the same significance. The following are averages of the examinations reported by Larrabee in the "Journal of Medical Research:"

	Red corpuscles per cubic millimeter	Hemoglobin per cent. by Sahli instrument	Color index	Volume index
Normal males	5,267,250	103.0	0.98	1.007
Normal females	4,968,667	106.0	1.06	1.001
Primary pernicious anemia	1,712,166	50.0	1.47	1.270
Secondary anemia	3,737,160	61.0	0.81	0.790
Chlorosis	3,205,000	34.5	0.55	0.695

Method.—The red cells are counted and the percentage of red cells calculated as for the color index.

The volume percentage is obtained with the hematocrit as follows: Fill the hematocrit tubes (Fig. 105) with blood, and before coagulation takes place insert them in the frame and centrifugalize for three minutes at about 8000 to 10,000 revolutions a minute. The red cells collect at the bottom and, normally, make up one-half of the total column of blood. Multiply the height of the layer of red cells (as indicated by the graduations upon the side of the tube) by 2 to obtain the volume percentage. When the examination cannot be made immediately after the blood is obtained, the method of Larrabee is available. This consists in mix-

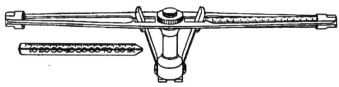

FIG. 105.—Daland hematocrit for use with the centrifuge.

ing a trace of sodium oxalate with a few drops of blood to prevent coagulation, drawing this mixture into a heavy-walled tube of about 2-mm. caliber, closing the ends with a rubber band, and waiting until sedimentation is complete—usually about three days. The height of the column is then measured with a millimeter scale and the percentage relation to the normal calculated.

After the volume of the red cells and the red corpuscle count are thus expressed in percentages, divide the former by the latter to find the volume index. Example: Suppose the volume percentage is 80 (the reds reaching to mark 40 on hematocrit tube) and that the red count is 50 per cent. of the normal (2,500,000 per cubic millimeter), then $\frac{80}{50}$, or 1.6, is the volume index.

VII. ENUMERATION OF LEUKOCYTES

The normal number of leukocytes varies from 5000 to 10,000 per cubic millimeter of blood. The number is larger in robust individuals than in poorly nourished ones, and, if disease be excluded, may be taken as a rough index of the individual's nutrition. Since it is well to have a definite standard, 7500 is generally adopted as the normal for the adult. With children the number is somewhat greater, averaging about 10,000 to 12,000 in infants and somewhat below 10,000 in older children.

DECREASE IN NUMBER OF LEUKOCYTES

Decrease in number of leukocytes, or *leukopenia*, is not important. It is common in persons who are poorly nourished, although not actually sick. The infectious diseases in which leukocytosis is absent (see p. 291) often cause a slight decrease of leukocytes. Chlorosis may produce leukopenia, as also pernicious anemia, which usually gives it in contrast to the secondary anemias, which are frequently accompanied by leukocytosis. Leukocyte counts are, therefore, of some aid in the differential diagnosis of these conditions.

INCREASE IN NUMBER OF LEUKOCYTES

Increase in number of leukocytes is common and of great importance. It may be considered under two heads:

A. Increase of leukocytes due to chemotaxis and stimulation of the blood-making organs, or *leukocytosis*. The increase affects one or more of the normal varieties.

B. Increase of leukocytes due to *leukemia*. Normal

varieties are increased, but the characteristic feature is the appearance of great numbers of abnormal cells.

The former may be classed as a *transient*, the latter, as a *permanent*, increase.

A. LEUKOCYTOSIS

This term is variously used. By some it is applied to any increase in number of leukocytes; by others it is restricted to increase of the polymorphonuclear neutrophilic variety. As has been indicated, it is here taken to mean a transient increase in number of leukocytes, that is, one caused by chemotaxis and stimulation of the blood-producing structures, in contrast to the permanent increase caused by leukemia.

By *chemotaxis* is meant that property of certain agents by which they attract or repel living cells—positive chemotaxis and negative chemotaxis respectively. An excellent illustration is the accumulation of leukocytes at the site of inflammation, owing to the positively chemotactic influence of bacteria and their products. A great many agents possess the power of attracting leukocytes into the general circulation. Among these are many bacteria and certain organic and inorganic poisons.

Chemotaxis alone will not explain the continuance of leukocytosis for more than a short time. It is probable that substances which are positively chemotactic also stimulate the blood-producing organs to increased formation of leukocytes; and in at least one form of leukocytosis such stimulation apparently plays the chief part.

As will be seen later, there are several varieties of leukocytes in normal blood, and most chemotactic agents attract only one variety, and either repel or do not influence the others. It practically never happens that all are increased in the same proportion. The most satisfactory classification of leukocytoses is, therefore, based upon the type of leukocyte chiefly affected.

Theoretically, there should be a subdivision for each variety of leukocyte, e.g., polymorphonuclear leukocytosis, lymphocytic leukocytosis, eosinophilic leukocytosis, large mononuclear leukocytosis, etc. Practically, however, only two of these, polymorphonuclear leukocytosis and lymphocytic leukocytosis, need be considered under the head of Leukocytosis. Increase in number of the other leukocytes will be considered when the individual cells are described (see pp. 326–339). They are present in the blood in such small numbers normally that even a marked increase scarcely affects the total leukocyte count; and, besides, substances which attract them into the circulation frequently repel the polymorphonuclears, so that the total number of leukocytes may actually be decreased.

The polymorphonuclear neutrophiles are capable of active ameboid motion, and are by far the most numerous of the leukocytes. Lymphocytes are about one-third as numerous and have little independent motion. As one would, therefore, expect, marked differences exist between the two types of leukocytosis: polynuclear leukocytosis is more or less acute, coming on quickly and often reaching high degree; whereas lymphocytic leukocytosis is more chronic, comes on more slowly, and is never so marked.

1. Polymorphonuclear Neutrophilic Leukocytosis.—Polymorphonuclear leukocytosis may be either

19

physiologic or pathologic. A count of 20,000 would be considered a marked leukocytosis; of 30,000, high; above 50,000, extremely high.

(1) **Physiologic Polymorphonuclear Leukocytosis.**—This is never very marked, the count seldom exceeding 12,000 per cubic millimeter. It may occur: (*a*) In the new-born; (*b*) in pregnancy; (*c*) during digestion; and (*d*) after cold baths. There is moderate leukocytosis in the moribund state: this is commonly classed as physiologic, but is probably due mainly to terminal infection.

The increase in these conditions is not limited to the polymorphonuclears. Lymphocytes are likewise increased in varying degrees, most markedly in the new-born.

In view of the leukocytosis of digestion, which usually increases the leukocytes by about 30 per cent., the hour at which a leukocyte count is made should always be recorded. Digestive leukocytosis is most marked three to five hours after a hearty meal rich in protein, especially when such a meal follows a long fast. It is absent in pregnancy and when leukocytosis from any other cause exists. It is usually absent in cancer of the stomach, a fact which may be of some help in the diagnosis of this condition, but repeated examinations and careful technic are essential.

(2) **Pathologic Polymorphonuclear Leukocytosis.**—In general, the response of the leukocytes to chemotaxis is a conservative process. It has been compared to the gathering of soldiers to destroy an invader. This is accomplished partly by means of phagocytosis—actual ingestion of the enemy—and partly by means of chemic substances which the leukocytes produce.

In those diseases in which leukocytosis is the rule the degree of leukocytosis depends upon two factors: the *severity of the infection* and the *resistance of the individual*. A well-marked leukocytosis usually indicates good resistance. A mild degree means that the body is not reacting well, or else that the infection is too slight to call forth much resistance. Leukocytosis may be absent altogether when the infection is extremely mild, or when it is so severe as to overwhelm the organism before it can react. When leukocytosis is marked, a sudden fall in the count may be the first warning of a fatal issue. These facts are especially true of pneumonia, diphtheria, and abdominal inflammations, in which conditions the degree of leukocytosis is of considerable prognostic value.

The classification here given follows Cabot:

(a) *Infectious and Inflammatory.*—The majority of infectious diseases produce leukocytosis. The most notable exceptions are influenza, measles, German measles, tuberculosis, except when invading the meninges or when complicated by mixed infection, and typhoid fever, in which leukocytosis indicates an inflammatory complication.

All inflammatory and suppurative diseases cause leukocytosis, except when slight or well walled off. Appendicitis has been studied with especial care in this connection, and the conclusions now generally accepted probably hold good for most acute intra-abdominal inflammations. A marked leukocytosis (20,000 or more) nearly always indicates abscess, peritonitis, or gangrene, even though the clinical signs be slight. Absence of or mild leukocytosis indicates a mild process, or else an overwhelmingly severe one; and operation may safely be postponed unless the abdominal signs are very marked. On the other hand, no matter how low the count, an increasing leukocytosis—counts being made hourly —indicates a spreading process and demands operation, regardless of other symptoms.

Leukocyte counts alone are often disappointing, but are of much more value when considered in connection with a differential count of polymorphonuclears (see p. 331).

(b) *Malignant Disease.*—Leukocytosis occurs in about one-half of the cases of malignant disease. In many instances it is probably independent of any secondary infection, since it occurs in both ulcerative and non-ulcerative cases. It seems to be more common in sarcoma than in carcinoma. Very large counts are rarely noted.

(c) *Posthemorrhagic.*—Moderate leukocytosis follows hemorrhage and disappears in a few days. In cases of ruptured tubal pregnancy with hemorrhage into the peritoneal cavity the count usually reaches 18,000 to 30,000.

(d) *Toxic.*—This is a rather obscure class, which includes gout, chronic nephritis, acute yellow atrophy of the liver, ptomain-poisoning, prolonged chloroform narcosis, and quinin-poisoning. Leukocytosis may or may not occur in these conditions, and is not important.

(e) *Drugs.*—This also is an unimportant class. Most tonics and stomachics and many other drugs produce a slight leukocytosis. A moderate leukocytosis may also occur as a result of prolonged ether anesthesia.

2. Lymphocytic Leukocytosis.—This is characterized by an increase in the total leukocyte count, accompanied by an increase in the percentage of lymphocytes.

The word "lymphocytosis" is often used in the same sense. It is better, however, to use the latter as referring to any increase in the absolute number of lymphocytes, without regard to the total count, since an absolute increase in number of lymphocytes is frequently accompanied by a normal or subnormal leukocyte count, owing to loss of polymorphonuclears.

Lymphocytic leukocytosis is probably due more to stimulation of blood-making organs than to chemotaxis. It is less common, and is rarely so marked as a polymorphonuclear leukocytosis. When marked, the blood cannot be distinguished from that of lymphatic leukemia.

A marked lymphocytic leukocytosis occurs in pertussis. It is said to appear early in the catarrhal stage and to reach its maximum at the height of the paroxysmal stage, after which it gradually subsides. In 30 well-marked cases studied by Schneider the average leukocyte count was 19,000 in the first week, rising to about 27,000 in the third. His lowest counts in the first week were 12,600 and in the third 16,800. Leukocyte counts would therefore seem to have great value in diagnosis, but in our experience they have often been disappointing since in many mild cases the count does not rise above what may be regarded as a high normal for children before the characteristic whoop begins. There is moderate lymphocytic leukocytosis in other diseases of childhood, as rickets, scurvy, and especially hereditary syphilis, where the blood picture may approach that of pertussis. It must be borne in mind in this connection that lymphocytes are normally more abundant in the blood of children than in that of adults.

Slight lymphocytic leukocytosis occurs in many other pathologic conditions, but is of little significance.

B. LEUKEMIA

This is an idiopathic disease of the blood-making organs, which is accompanied by an enormous increase in number of leukocytes. The leukocyte count sometimes reaches 1,000,000 per cubic millimeter, and leukemia is always to be suspected when it exceeds 50,000. Lower counts do not, however, exclude it. The subject is more fully discussed later (see p. 387).

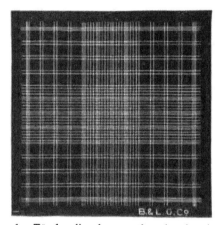

FIG. 106.—Türck ruling for counting chamber (× 15).

METHOD OF COUNTING LEUKOCYTES

The leukocytes may be counted with any one of the hemacytometers already described (see pp. 273–284). Numerous modifications of the original ruling have been introduced, notably the Türck, the Zappert-Ewing, and the Neubauer (Figs. 106, 107, 108), which give a ruled area of 9 sq. mm., the center having the Thoma ruling. Of those the Neubauer may be especially commended. Some of them were originally devised

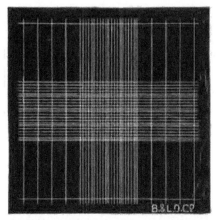

FIG. 107.—Zappert-Ewing ruling for counting chamber (× 15).

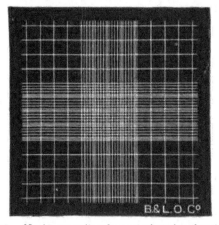

FIG. 108.—Neubauer ruling for counting chamber (× 15).

for counting the leukocytes in the same dilution with the red corpuscles. The two kinds of cell are easily distinguished, especially when Toisson's diluting fluid is used. The red cells are counted in the central portion in the usual manner, after which all the leukocytes in the whole area of 9 sq. mm. are counted; and the number in a cubic millimeter of undiluted blood is then calculated. Bass' new ruling (Fig. 109) covers 4 sq. mm. and is used in a similar manner. With the older Thoma ruling the reds and the leukocytes may be counted in the same preparation by adjusting the

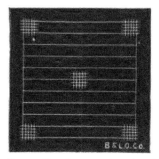

FIG. 109.—Bass ruling for counting chamber (× 15).

microscopic field to a definite size, and counting a sufficient number of fields, as described later.

Although less convenient, it is more accurate to count the leukocytes separately, with less dilution of the blood, as follows:

Technic.—A larger drop of blood is required than for counting the erythrocytes, and more care in filling the pipet, since the bore is considerably larger than that of the "red" pipet. Boggs has suggested a device (Fig. 110) which enables one to draw in the blood more slowly and hence more accurately. He cuts the rubber tube and inserts a Wright "throttle." This consists of a section of glass tubing within which a capillary tube drawn out to a fine thread is cemented with sealing wax. After sealing in place the tip is broken off with forceps, so that upon gentle suction it will just allow air to pass.

Use the pipet with 11 engraved above the bulb.[1] Suck the blood to the mark 0.5 or 1.0, and the diluting fluid to the mark 11. This gives a dilution of 1:20 or 1:10, respectively.

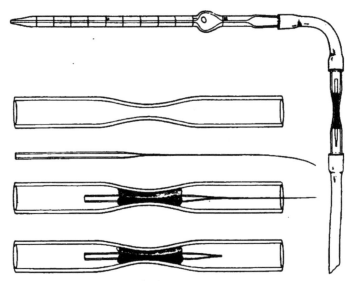

FIG. 110.—Boggs' "throttle control" for blood-counting pipet, and enlarged diagram showing construction of the throttle.

The dilution of 1:20 is easier to make. Mix well by shaking in all directions except in the long axis of the pipet; blow out 2 or 3 drops, place a drop in the counting chamber, and adjust the cover as already described (see pp. 276, 277).

Examine with a low power to see that the cells are evenly distributed. Count with the 16-mm. objective and a high

[1] In some cases of leukemia with very high count it may be necessary to use the "red" pipet with dilution of 1:100

eye-piece, or with the long-focus 4 mm. and a low eye-piece. An 8-mm. objective will be found very satisfactory for this purpose. As one gains experience one will rely more upon the lower powers.

With the Thoma ruling count all the leukocytes in the square millimeter, multiply by 10 to find the number in 1 cu. mm. of diluted blood, and by the dilution to find the number per cubic millimeter of undiluted blood. In every case at least 200 leukocytes must be counted as a basis for calculation, and it is much better to count 500. This will necessitate examination of several drops from the pipet. With the rulings which cover 9 sq. mm. a sufficient number can usually be counted in one drop, but the opportunity for error is very much greater when only one drop is examined.

In routine work the author's modification of the "circle" method is very satisfactory. It requires a 4-mm. objective, and is, therefore, especially desirable for beginners, who are usually unable accurately to identify leukocytes with a lower power. The student is frequently confused by particles of dirt, remains of red cells, and yeast cells which are prone to grow in the diluting fluid. Draw out the sliding tube of the microscope until the field of vision is such as shown in Fig. 111. One side of the field is tangent to one of the ruled lines, A, while the opposite side just cuts the corners, B and C, of the seventh squares in the rows above and below the diameter line. When once adjusted, a scratch is made upon the draw-tube, so that for future counts the tube has only to be drawn out to the mark. The area of this microscopic field is 0.1 sq. mm. With a dilution of 1:20, count the leukocytes in 20 such fields upon different parts of the disk without regard to the ruled lines, and to their sum add two ciphers. With dilution of 1:10, count 10 such fields, and add two ciphers. Thus, with 1:10 dilution, if 150 leukocytes were counted in 10 fields, the leukocyte count would be

15,000 per cubic millimeter. To compensate for possible unevenness of distribution, it is best to count a row of fields horizontally and a row vertically across the disk. This method is applicable to any degree of dilution of the blood, and is simple to remember: *one always counts a number of*

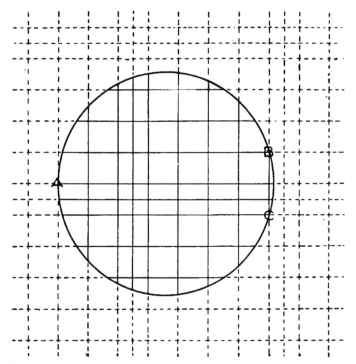

FIG. 111.—Size of field required in counting leukocytes as described in the text.

fields equal to the number of times the blood has been diluted, and adds two ciphers. Evidence of the convenience of using a circle of this size is afforded by its adoption in the new Thoma-Metz instrument.

It is sometimes impossible to obtain the proper size of

field with the objectives and eye-pieces at hand. In such case place a cardboard or stiff paper disk with a circular opening upon the diaphragm of the eye-piece, and adjust the size of the field by drawing out the tube. The circular opening can be cut with a sharp cork-borer. .

Diluting Fluids.—The diluting fluid should dissolve the red corpuscles so that they will not obscure the leukocytes. The simplest fluid is a 1 per cent. solution of acetic acid. More satisfactory is the following: glacial acetic acid, 1 c.c.; 1 per cent. aqueous solution of gentian-violet, 1 c.c.; distilled water, 100 c.c. These solutions must be filtered frequently to remove yeasts and molds.

VIII. ENUMERATION OF BLOOD-PLATELETS

The average normal number of platelets is variously given as 200,000 to 700,000 per cubic millimeter of blood. Many of the counts were obtained by workers who used inaccurate methods. Using their own reliable method, Wright and Kinnicutt found the normal average to range from 263,000 to 336,000. Physiologic variations are marked; thus, the number increases as one ascends to a higher altitude, and is higher in winter than in summer. There are unexplained variations from day to day; hence a single abnormal count should not be taken to indicate a pathologic condition.

Pathologic variations are often very great. Owing to lack of knowledge as to the function of the platelets and to the earlier imperfect methods of counting, the clinical significance of these variations is uncertain. The following facts seem, however, to be established:

(a) In acute infectious diseases the number is subnormal or normal. If the fever ends by crisis, the crisis is accompanied by a rapid and striking increase.

(*b*) In secondary anemia platelets are generally increased, although sometimes decreased. In pernicious anemia they are always greatly diminished, and an increase should exclude the diagnosis of this disease.

(*c*) They are decreased in chronic lymphatic leukemia, and greatly increased in the myelogenous form.

(*d*) In purpura hæmorrhagica the number is enormously diminished.

(*e*) The platelets are somewhat increased in tuberculosis.

Blood-platelets are difficult to count, owing to the rapidity with which they disintegrate, and to their great tendency to adhere to any foreign body and to each other.

Method of Wright and Kinnicutt.—This method is simple, appears to be accurate, and certainly yields uniform results.

The platelets are counted with the hemacytometer already described, using a dilution of 1:100. The diluting fluid consists of 2 parts of an aqueous solution of brilliant cresyl blue (1:300) and 3 parts of an aqueous solution of potassium cyanid (1:1400). These two solutions must be kept in separate bottles and mixed and filtered immediately before using. The cresyl blue solution is permanent but molds have a tendency to grow in it. The cyanid solution deteriorates after about ten days. Rapid work is necessary in order to prevent clumping of the platelets. After the blood is placed in the counting-chamber it is allowed to stand for ten minutes or longer in order that the platelets may settle. The count is made with a high dry objective and a high ocular. The platelets appear as rounded, lilac-colored bodies; the reds are decolorized, appearing only as shadows.

The leukocytes are stained and may be counted at the same time.

Ottenberg and Rosenthal have recently suggested 3 per cent. sodium citrate as a diluting fluid to be used in the same manner as that of Wright and Kinnicutt. This may be colored by adding 0.2 per cent. of brilliant cresyl blue or methyl violet, but the fluid must then be made up freshly each day since it deteriorates rapidly.

IX. STUDY OF STAINED BLOOD
A. MAKING AND STAINING BLOOD-FILMS

1. Spreading the Film.—Thin, even films are essential to accurate and pleasant work. They more than compensate for the time spent in learning to make them. There are certain requisites for success with any me-- thod: (a) The slides and covers must be perfectly clean: thorough washing with soap and water, rubbing with alcohol and drying on a clean handkerchief will usually suffice; (b) the drop of blood must not be too large; (c) the work must be done quickly, before coagulation begins.

The blood is obtained from the finger-tip or the lobe of the ear, as for a blood count; only a very small drop is required, usually about the size of a large pin-head. The size of the drop largely determines the thickness of the film. The proper thickness will depend upon the purpose for which the film is made. For the structure of blood cells and the malarial parasite it should be so thin that, throughout the greater part of the film, the red corpuscles lie in a single layer, close together but not overlapping. In our class work we insist that all films meet this requirement. For routine differential counting of leukocytes a film in which the red cells are piled up somewhat is best because the number of leukocytes in a given area is thus greatly increased and the tedium of counting is corresponding lessened. The film must not, upon the other hand, be so thick that identification of the various leukocytes becomes difficult.

Nearly all ordinary slides are curved. In order that they may lie firmly upon the microscope stage without rocking, the blood film should be spread upon the convex side, which is recognized by laying the slide flat upon the table and twirling it rapidly by snapping the

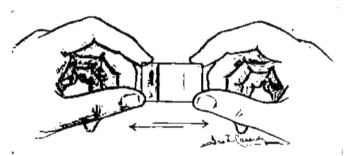

FIG. 112.—Spreading the film: two cover-glass method. It is better to place the top cover diagonally and to grasp it by opposite covers.

end with a finger. The side upon which it twirls the better is the convex side.

Ehrlich's Two Cover-glass Method.—This method is very widely used, but considerable practice is required to get good results. Touch a cover-glass to the top of a small drop of blood, and place it, blood side down, upon another cover-glass. If the drop be not too large, and the covers be perfectly clean, the blood will spread out in a very thin layer. Just as it stops spreading, before it begins to coagulate, pull the covers quickly but firmly apart on a plane parallel to their surfaces (Fig. 112). It is best to handle the covers

with forceps, since the moisture of the fingers may produce
artifacts. The forceps must have a firm grasp.

This method is especially to be recommended for very
accurate differential counts, since all the leukocytes in
the drop will be found on the two covers and thus the
possible error due to unequal distribution can be excluded.
One of the covers is usually much better spread than the
other.

Two-slide Method.—Take a small drop of blood upon a
clean slide about ¾ inch from the end, using care that the
slide does not touch the skin. Place the end of a second

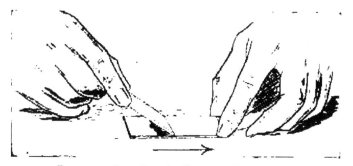

FIG. 113.—Spreading the film: two-slide method.

slide against the surface of the first at an angle of 30 to 40
degrees, and draw it up against the drop of blood, which
will immediately run across the end, filling the angle between the two slides. Now push the "spreader slide"
back along the other in the manner indicated in Fig. 113.
The blood will follow. The thickness of the smear can be
regulated by changing the angle.

It is very easy to make large, thin, even films by this
method.

Cigarette-paper Method.—This gives excellent results in
the hands of the inexperienced if directions are carefully followed, but its only advantage over the two-slide method is

that it may be used with covers as well as with slides. A very thin paper, such as the "Zig-zag" brand, is best. Ordinary cigarette paper and thin tissue-paper will answer, but do not give nearly so good results.

Cut the paper into strips about ¾ inch wide, *across the ribs*. Pick up one of the strips by the gummed edge, and touch its opposite end to the drop of blood. Quickly place the end which has the blood against a slide or a large cover-

FIG. 114.—Spreading the film: cigarette-paper method applied to cover-glasses.

glass held in a forceps. The blood will spread along the edge of the paper. Now draw the paper evenly across the slide or cover. A thin film of blood will be left behind (Fig. 114).

The films may be allowed to dry in the air, or may be dried by gently warming high above a flame (where one can comfortably hold the hand). Such films will keep for years, but for some stains they must not be more than a few weeks old. They must be kept away from flies—a fly can work havoc with a film in a few minutes.

20

When slides are used the label can be written with a
soft lead pencil directly on the blood-film, as was suggested by von Ezdorf.·

2. Fixing the Film.—In general, films must be
"fixed" before they are stained. Fixation may be accomplished by chemicals or by heat. *Those stains
which are dissolved in methyl alcohol combine fixation with
the staining process.*

Chemic Fixation.—Soak the film one to five minutes in
pure methyl alcohol or absolute ethyl alcohol, or one-half
hour or longer in equal parts of absolute alcohol and ether.
One minute in saturated solution of mercuric chlorid or

Fig. 115.—Kowarsky's plate for fixing blood (Klopstock and
Kowarsky).

in 1 per cent. formalin in alcohol is preferred by some,
especially for the carbol-thionin stain. Chemic fixation
may precede hematoxylin-eosin and other simple stains.

Heat Fixation.—This may precede any of the methods
which do not combine fixation with the staining process; it
is almost imperative with Ehrlich's triple stain. The best
method is to place the film in an oven, raise the temperature
to 150°C., and allow to cool slowly. Without an oven,
the proper degree of fixation is difficult to attain. Kowarsky has devised a small plate of two layers of copper
(Fig. 115), upon which the films are placed together with a
crystal of urea. The plate is heated over a flame until the
urea melts, and is then set aside to cool. Some prefer to

use slides and to place the crystal of urea directly upon the slide. This is said to give the proper degree of fixation, but in the writer's experience the films have always been underheated. He obtains better results by use of tartaric acid crystals (melting-point, 168°–170°C.). The plate, upon which have been placed the cover-glasses, film side down, and a crystal of the acid, is heated over a low flame until the crystal has completely melted. It should be held sufficiently high above the flame that the heating will require five to seven minutes. The covers are then removed. Freshly made films of normal blood should be allowed to remain upon the plate for a minute or two after heating has ceased. Fresh films require more heat than old ones, and normal blood more than the blood of pernicious anemia and leukemia.

Blood-films can be satisfactorily fixed for most purposes by covering with absolute alcohol, quickly dashing off the excess, and igniting the remainder.

3. Staining the Film.—The anilin dyes, which are extensively used in blood work, are of two general classes: basic dyes, of which methylene-blue is the type; and acid dyes, of which eosin is the type. Nuclei and certain other structures in the blood are stained by the basic dyes, and are hence called *basophilic*. Certain structures take up only acid dyes, and are called *acidophilic*, *oxyphilic*, or *eosinophilic*. Certain other structures are stained by combinations of the two, and are called *neutrophilic*. Recognition of these staining properties marked the beginning of modern hematology.

(1) **Hematoxylin and Eosin.**—This method is most useful in studying eosinophilic cells and the structure of nuclei, hematoxylin being in fact one of our best nuclear stains. It may therefore be recommended for the

Arneth count (see p. 334).   Red corpuscles are pink or
red, all nuclei blue, eosinophilic granules bright red;
neutrophilic granules and platelets are not stained.
Neither polychromatophilia nor basophilic granular
degeneration of the red cells are demonstrated.

1. Fix by heat or chemicals.

2. Stain with any standard hematoxylin solution until
nuclei are well colored, usually three to five minutes.

3. Wash well with water.

4. Apply a weak aqueous or alcoholic solution of eosin
(about 0.5 per cent.) for a minute or two.

5. Wash well in water, dry and examine. If the eosin
stains too deeply, longer washing in water will usually remove some of the excess.

The procedure may be simplified by mixing the hematoxylin and eosin.   Such a mixture was much used before
modern staining methods were introduced. Almost any
of the standard hematoxylin solutions may be employed;
to this is added enough of a saturated aqueous solution of
eosin to color the reds properly while the hematoxylin is
staining the nuclei.   The combined stain keeps well.   The
fixed smear is immersed in the staining fluid for the required time, usually five to fifteen minutes, and is then
rinsed, dried, and mounted.

(2) **Ehrlich's Triple Stain.**—This was the standard
blood-stain for many years, but is now little used.   It
is probably the best for neutrophilic granules.   It is
difficult to make, and should be purchased ready prepared from a reliable dealer.   Nuclei are stained pale
blue or greenish blue; erythrocytes, orange; neutrophilic granules, violet; and eosinophilic granules, copper
red.   Basophilic granules and blood-platelets are not
stained.

Success in staining depends largely upon proper fixation. The film must be carefully fixed by heat: underheating causes the erythrocytes to stain red; overheating, pale yellow. Immersion in pure acetone for five minutes has been recommended as a satisfactory substitute for heat fixation. The staining fluid is applied for five to fifteen minutes, and the preparation is rinsed quickly in water, dried, and mounted. Subsequent application of Löffler's methylene-blue for one-half to one second will bring out the basophilic granules and improve the nuclear staining, but there is considerable danger of overstaining. The fluid should not be filtered regardless of any precipitate that may form.

(3) **Polychrome Methylene-blue-eosin Stains.**—These stains, outgrowths of the original time-consuming Romanowsky method, have largely displaced other blood-stains for clinical purposes. They may be recommended for all routine work. They stain differentially every normal and abnormal structure in the blood. Most of them are dissolved in methyl alcohol and combine the fixing with the staining process. Numerous methods of preparing and applying these stains have been devised, among the best known being Giemsa's, Wright's, Hastings' and Leishman's.

Wright's Stain.—This is one of the best and is the most widely used in this country. The practitioner will find it convenient to purchase the stain ready prepared, but, since much of the solution offered for sale is unsatisfactory, it is best to purchase the powder and dissolve it in methyl alcohol as needed. Most microscopic supply-houses carry it in stock. Wright's most recent directions for its preparation and use are as follows:

Preparation.—To a 0.5 per cent. aqueous solution of sodium bicarbonate add methylene-blue (B. X. or "medicinally pure") in the proportion of 1 Gm. of the dye to each 100 c.c. of the solution. Heat the mixture in a steam sterilizer at 100°C. for one full hour, counting the time after the sterilizer has become thoroughly heated. The mixture is to be contained in a flask, or flasks, of such size and shape that it forms a layer not more than 6 cm. deep. After heating, allow the mixture to cool, placing the flask in cold water, if desired, and then filter it to remove the precipitate which has formed in it. It should, when cold, have a deep purple-red color when viewed in a thin layer by transmitted yellowish artificial light. It does not show this color while it is warm.

To each 100 c.c. of the filtered mixture add 500 c.c. of a 0.1 per cent. aqueous solution of "yellowish water-soluble" eosin and mix thoroughly. Collect the abundant precipitate, which immediately appears, on a filter. When the precipitate is dry, dissolve it in methylic alcohol (Merck's "reagent") in the proportion of 0.1 Gm. to 60 c.c. of the alcohol. In order to facilitate solution, the precipitate is to be rubbed up with the alcohol in a porcelain dish or mortar with a spatula or pestle. This alcoholic solution of the precipitate is the staining fluid. We frequently find that freshly made solutions stain the red cells blue; such solutions usually work properly after a few months.

Application.—1. Without previous fixation cover the film with a noted quantity of the staining fluid by means of a medicine-dropper. There must be plenty of stain in order to avoid too great evaporation and consequent precipita- ... lides are used, the stain may be confined to ... by two heavy wax pencil marks.

2. After one minute add to the staining fluid on the film the same quantity of distilled water by means of a medicine-dropper. This may be done by counting drops. The drops of water are about twice as large as the drops of stain, but this rarely does any harm and is often an advantage. The quantity of the fluid on the preparation must not be so large that some of it runs off. Allow the mixture to remain for three to six minutes, according to the intensity of the staining desired. A longer period of staining may produce a precipitate. Eosinophilic granules are best brought out by a short period of staining.

3. Wash the preparation in water for thirty seconds or until the thinner portions of the film become yellow or pink in color. The preparation should be flooded with water while the stain is still upon it. If the stain is poured off before rinsing, the scum tends to settle upon the blood-film, where it clings in spite of subsequent washing.

4. Dry and mount in balsam.

The stain is more conveniently applied upon cover-glasses than upon slides. Films much more than a month old do not stain well, red cells and most other structures taking a slate blue color. In some localities ordinary tap-water will answer both for diluting the stain and for washing the film; in others, distilled water must be used. The difficulty here is probably that the tap-water is acid in reaction. This causes the nuclei to stain too palely. Other causes of pale nuclei are addition of too much or too little water and the development of formic acid from the methyl alcohol of the staining fluid.

When properly applied, Wright's stain gives the following picture (see Plates I, V, VII): erythrocytes, yellow or pink; nuclei, various shades of bluish purple;

neutrophilic granules, reddish lilac, sometimes pink; eosinophilic granules, bright red; basophilic granules of leukocytes and degenerated red corpuscles, very dark bluish purple; blood-platelets, dark lilac; bacteria, blue. The cytoplasm of lymphocytes is generally robin's-egg blue; that of the large mononuclears may have a faint bluish tinge.   Malarial parasites stain characteristically: the cytoplasm, sky-blue; the chromatin, reddish purple. These colors are not invariable: two films stained from the same bottle sometimes differ greatly. *In general a preparation is satisfactory when both nuclei and neutrophilic granules are distinct, regardless of their color, and when the film is free from precipitated dye.* In addition, it is desirable, but not essential, that the red corpuscles show a clear pink or yellowish-pink; they should not be blue.   The colors are prone to fade if the preparation is mounted in a poor quality of balsam or exposed much to the light.

It is well known that pathologic bloods will sometimes not stain well with fluids which are satisfactory for normal blood.   Peebles and Harlow have shown that the various polychrome methylene-blue-eosin stains can be modified to suit any blood by adding a trace of alkali or acid.   The alkali used is a weak solution of "potassium hydrate by alcohol" in methyl alcohol; the acid, glacial acetic in methyl alcohol.   The alkali solution also serves to "correct" old fluids which, by reason of development of formic acid in the methyl alcohol, do not stain sufficiently with the blue.

Other Polychrome Methylene-blue-eosin Stains.—While Wright's stain suffices for most clinical work and is to be recommended if only one blood stain is to be used, certain others demand brief mention.

1. **Giemsa's Stain.**—This widely used stain is probably the best modification of the Romanowsky stain for blood parasites and other protozoa, and is also very satisfactory as a routine blood stain. It consists of:

Azur II–eosin..................... 3 Gm.
Azur II........................ 0.8 Gm.
Glycerin (Merck, C. P.).......... 250 Gm.
Methyl alcohol (Kahlbaum I or
    Merck's reagent)............. 250 Gm.

The solution is expensive to make and is best purchased ready prepared. Blood films are fixed in methyl alcohol and are then immersed for twenty minutes or longer in a freshly prepared mixture of 1 c.c. of stain and 10 c.c. distilled water. In order to prevent precipitates falling upon them, the slides or covers should be placed upon edge in the stain.

The use of this stain for *Treponema pallidum* is described later (p. 550).

2. **Pappenheim's Panoptic Method.**—In order to combine the advantages of the several stains, Pappenheim recommended the following procedure: Stain for one minute with the May-Grünwald stain; add an equal quantity of water; after one minute pour off the fluid and stain fifteen minutes with the diluted Giemsa solution. The May-Grünwald stain is the same as Jenner's. Wright's stain, diluted with an equal quantity of water, may be substituted for the Giemsa solution but the time of staining should then not exceed five minutes.

It is difficult to see that slides stained in this way offer any advantages over good Wright or Giemsa preparations.

(4) **Jenner's Stain.**—Jenner's eosinate of methylene-blue, dissolved in methyl alcohol, brings out leukocytic

granules well, and is, therefore, especially useful for differential counting. It stains nuclei poorly and is much inferior to Wright's stain for the malarial parasite since it does not give the so-called "Romanowsky staining."

It may be purchased in solution, in the form of tablets, or as a powder, 0.5 Gm. of which is to be dissolved in 100 c.c. neutral absolute methyl alcohol. The unfixed blood-film is covered with the staining solution and after three to five minutes is rinsed with water, dried in the air, and mounted.

(5) **Carbol-thionin** is especially useful for the study of basophilic granular degeneration of the red cells. The method is described on page 639. Nuclei, malarial parasites, and basophilic granules are brought out sharply. Polychromatophilia is also evident. Fixation may be by alcohol-formalin (see p. 306) or saturated solution of mercuric chlorid.

(6) **Pappenheim's pyronin-methyl green** (see p. 642) can be used as a blood-stain and is very satisfactory for study of the red cells and of the lymphocytes and for demonstration of Doehle's inclusion bodies (see p. 335). All nuclei are blue to reddish purple; basophilic granules, cytoplasm of lymphocytes, and inclusion bodies, red. Polychromatophilia is well demonstrated, the affected cells taking more or less of the red color. Heat fixation is probably best.

B. STUDY OF STAINED FILMS

It has been said with much truth that an intelligent study of the stained film, together with an estimation of hemoglobin, will yield 90 per cent. of all the diagnostic

Information obtainable from a blood-examination. The stained films furnish the best means of studying the morphology of the blood and blood parasites, and, to the experienced, they give a fair idea of the amount of hemoglobin and the number of red and white corpuscles. An oil-immersion objective is required.

FIG. 116.—Red corpuscles of normal blood. Wright's stain (× 750).

1. **Erythrocytes.**—Normally, the red corpuscles are acidophilic. The colors which they take with different stains have been described. When not crowded together, they appear as circular, homogeneous disks, of nearly uniform size, averaging 7.8 μ in diameter (see Fig. 116). The center of each is somewhat paler than the periphery. Red cells are apt to be crenated when the film has dried too slowly.

Pathologically, red corpuscles vary in hemoglobin content, size and shape, staining properties, and structure.

(1) **Hemoglobin Content.**—The depth of staining furnishes a rough guide to the amount of hemoglobin in the corpuscles, *i.e.*, to the color index. When hemoglobin is diminished the central pale area becomes larger and paler. This condition is known as *achromia.* Usually the periphery retains a fairly deep color, so that the cells become mere rings, the so-called "pessary forms." These are most common in chlorosis. In

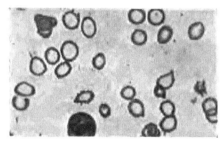

FIG. 117.—Red blood-corpuscles showing deficient hemoglobin (achromia). From a well marked case of chlorosis. Wright's stain (× 750).

pernicious anemia, upon the other hand, as a result of the high color index many of the red corpuscles may stain deeply and lack the pale center entirely.

(2) **Variations in Size and Shape** (See Plate V, Fig. 1). —The cells may be abnormally small (called *microcytes*, 5 μ or less in diameter); abnormally large (*macrocytes*, 10 to 12 μ); or extremely large (*megalocytes*, 12 to 25 μ). Abnormal variation in size is called *anisocytosis.*

Variation in shape is often very marked. Oval, pyriform, caudate, saddle-shaped, and club-shaped corpus-

Abnormal red corpuscles. All drawn from actual specimens and all stained with Wright's stain except where noted. × 1000 (1 mm. = 1 micron).

Fig. 1.—Variations in size, shape, and hemoglobin-content; from cases of pernicious anemia and chlorosis.

Fig. 2.—Polychromatophilia and basophilic granular degeneration; from cases of lead-poisoning and pernicious anemia.

Fig. 3.—Normoblasts, reticulated red cells, and one microblast. The top row represents stages in the development of the normoblast. The two reticulated red cells are stained with brilliant cresyl blue.

Fig. 4.—Megaloblasts from cases of pernicious anemia. Two show polychromatophilia and fairly typical nuclei, two have condensed nuclei, and one of these has basophilic cytoplasmic granules.

Fig. 5.—Nuclear particles or "Howell-Jolly bodies." One cell also shows basophilic granular degeneration.

Fig. 6.—Mitotic figures, two from myelogenous leukemia, one, with polychromatophilic cytoplasm, from von Jaksch's anemia. The last was stained with Leishman's stain.

Fig. 7.—Cabot's ring bodies, from a case of von Jaksch's anemia. Two cells also contain nuclear particles and one shows basophilic granular degeneration. Leishman's stain.

PLATE V

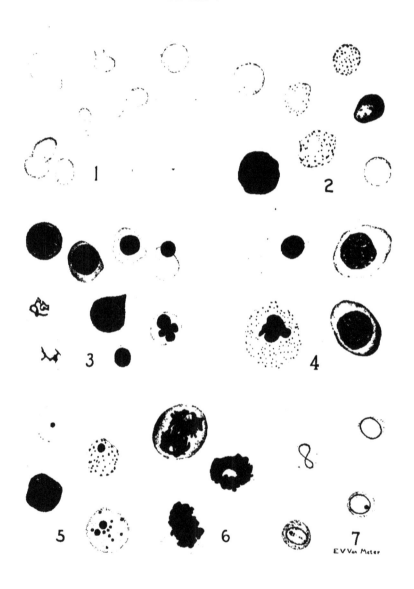

1

2

3

4

5

6

7

E V Van Meter

cles are common (Fig. 118). They are called *poikilo-cytes*, and their presence is spoken of as poikilocytosis.

Red corpuscles which vary from the normal in size and shape are present in most symptomatic anemias, and in the severer grades are often very numerous. Irregularities are particularly conspicuous in leukemia and pernicious anemia, where, in some instances, a normal erythrocyte is the exception. In pernicious anemia there is a decided tendency to large size and oval forms, and megalocytes are rarely found in any other condition.

FIG. 118.—Red corpuscles showing variations in size and shape, from a case of pernicious anemia (× 750).

(3) **Variations in Staining Properties** (See Plate V, Fig. 2).—These include polychromatophilia, basophilic granular degeneration, and malarial stippling. With exception of polychromatophilia they are probably degenerative changes.

(a) *Polychromatophilia.*—Some of the corpuscles partially lose their normal affinity for acid stains and take the basic stain to greater or less degree. Wright's stain gives such cells a faint bluish tinge when the condition is mild, and a rather deep blue when severe. Sometimes

only part of a cell is affected. A few polychromatophilic corpuscles can be found in marked symptomatic
anemias. They occur most abundantly in malaria,
leukemia, and pernicious anemia.

Polychromatophilia has been variously interpreted.
It is thought by many to be evidence of youth in a cell,
and hence to indicate an attempt at blood regeneration.
There are probably several forms referable to different
causes.

(*b*) *Basophilic Granular Degeneration* (*Degeneration of
Grawitz, Basophilic Stippling*).—This is characterized

FIG. 119.—Red blood-corpuscle showing basophilic granular degeneration with large granules. Wright's stain (× 1000).

by the presence, within the corpuscle, of irregular basophilic granules which vary in size from scarcely visible
points to granules nearly as large as those of basophilic
leukocytes (Fig. 119). The number present in a red cell
commonly varies in inverse ratio to their size. They
stain deep blue with carbol-thionin or Wright's stain;
not at all with Ehrlich's triple stain. The cell containing them may stain normally in other respects,
or it may exhibit polychromatophilia. Polychromatophilic cells generally contain the smaller granules, which
may be so fine that the cell appears dusted with them.

Numerous cells showing this degeneration are commonly found in chronic lead-poisoning, of which they were at one time thought to be pathognomonic. They can probably be found in every case with clinical symptoms and in some severe cases are present in nearly every microscopic field. Except in this disease, the degeneration indicates a serious blood condition. It occurs in well-marked cases of pernicious anemia and leukemia, and, much less commonly, in very severe symptomatic anemias.

FIG. 120.—Normoblasts from cases of secondary anemia and leukemia (× 1000).

(c) *Malarial Stippling.*—This term has been applied to the finely granular appearance often seen in red corpuscles which harbor tertian malarial parasites (see Frontispiece, Plates VI and VII). It was formerly classed with the degeneration just described, but is undoubtedly distinct. Not all stains will show it. With Wright's stain it can be brought out by staining longer and washing less than for the ordinary bloodstain. The minute granules, "Schüffner's granules," stain reddish purple. They are sometimes so numerous as almost to hide the parasite.

(4) **Variations in Structure.**—The most important is the presence of a nucleus (see Frontispiece, Plate V, and Figs. 3–7). Nucleated red corpuscles, or *erythroblasts*, are classed according to their size: *microblasts*, 5 μ or less in diameter; *normoblasts*, 5 to 10 μ; and *megaloblasts*, above 10 μ.

Microblasts and **normoblasts** contain one, rarely two, small, round, sharply defined nuclei. As a rule they are the most deeply stained nuclei to be seen in the blood-film, being approached in this respect only by

Fig. 121.—Normoblasts with irregular and fragmented nuclei. Wright's stain (× 1000).

the smaller lymphocytes. The nuclei of the younger normoblasts are relatively large and have their chromatin arranged in a more or less reticular manner with rather clean-cut open spaces. Mitoses are not uncommon in leukemia and pernicious anemia. The older nuclei are smaller and more dense, some being entirely homogeneous and very deeply stained (pyknotic nuclei). These last are apt to be located eccentrically, and sometimes appear as if in process of extrusion from the cell. These characteristics are shown in Fig. 120. As a result of degenerative changes the nuclei may be

Irregular in shape, clover-leaf forms being common; or they may be completely broken up into fragments—the so-called nuclear particles or Howell-Jolly bodies—of which all but one or two may have disappeared from the cell. These nuclear particles are smooth, round, deeply stained bodies, not likely to be mistaken for granules of basophilic degeneration (Fig. 122).

The **megaloblast** is probably a distinct cell, not merely a larger size of the normoblast. In the typical

Fig. 122.—Nuclear particles or Howell-Jolly bodies in red corpuscles. From a case of pernicious anemia. Wright's stain (× 1000).

megaloblast the nucleus is characteristic. This is large, oval, and rather palely staining and it has a more delicate chromatin network with larger and more numerous openings than has the nucleus of the normoblast (see Plates V, VIII, and Fig. 123). Sometimes it appears as if made up of coarse granules. Evidences of age and degeneration (condensation of nucleus, pyknosis, karyorrhexis, etc.), are common.

The recognition of megaloblasts is important, but is not always easy unless the nucleus is typical. Some workers base the distinction from normoblasts upon size of nucleus, requiring this to be larger than a normal red

21

corpuscle if the cell is to be regarded as a megaloblast. Others consider only the size of the cell, regarding as a megaloblast any nucleated red cell over 11 μ in diameter. Neither of these rules, nor the two together, will serve in every case. The exceptions will include, on the one hand, certain old megaloblasts with small condensed nuclei, and upon the other, the very young normo-blast whose diameter may exceed 12 μ and whose nuclei may be larger than a normal red corpuscle. At times one finds cells which must be classed as intermediates.

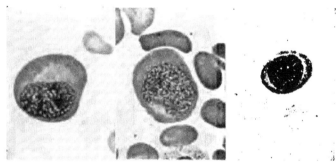

FIG. 123.—Megaloblasts showing typical nuclei; from cases of perni-cious anemia. Wright's stain ($\times$ 1000).

Young nucleated red cells, especially megaloblasts are prone to exhibit polychromatophilia. In some cells the cytoplasm is so blue and shows so little of its characteristic smooth texture that it is diffi-cult to recognize the cell as an erythrocyte except by the character of the nucleus. Such cells might easily be mistaken for lymphocytes or for Türck's irritation leukocytes.

Significance of Nucleated Red Corpuscles.— Normally, erythroblasts are present only in the blood

uf the letus and of very young infants. In the healthy
adult they are confined to the bone-marrow and they
appear in the circulating blood only in disease, where
their presence denotes an excessive demand made
upon the blood-forming organs to regenerate lost
or destroyed red corpuscles. In response to this
demand immature and imperfectly formed cells are
thrown into the circulation. Their number, therefore,
is usually regarded as an indication of the extent to
which the bone-marrow reacts rather than of the severity of the disease. Normoblasts occur in severe symptomatic anemia, leukemia, and pernicious anemia.
They are most abundant in myelogenous leukemia.
While always present in pernicious anemia, they are
often difficult to find. Microblasts have much the same
significance as normoblasts, but are less common.
Nuclear particles, or Howell-Jolly bodies, are most common in pernicious anemia and have been noted in
greatest numbers after splenectomy.

The presence of megaloblasts indicates a change in
the type of blood regeneration. This is seen most
characteristically in pernicious anemia and the finding
of megaloblasts is therefore extremely important in the
diagnosis of this disease. They are always present,
although often in so small numbers as to require a long
search; and they almost invariably exceed the normoblasts in number—a ratio which is practically unknown
in other diseases in which they have been found, such
as myelogenous leukemia, malignant growths in the
bone-marrow, etc.

Cabot's ring bodies are ring- or figure-of-8-shaped
structures (Fig. 124) which have been observed in cer-

tain of the red cells in pernicious anemia, lead-poisoning, and lymphatic leukemia. They stain red or reddish purple with Wright's stain and have been thought to be the remains of a nuclear membrane.

2. The Leukocytes.—An estimation of the number or percentage of each variety of leukocyte in the blood is called a differential count.. It probably yields more helpful information than any other single procedure in blood examinations.

The **differential count** is best made upon a film stained with Jenner's, Wright's, or a similar stain.

FIG. 124.—Cabot's ring bodies in red blood-corpuscles from a case of von Jaksch's anemia of infancy. The cell at the right contains a ring, a nuclear particle, and basophilic granules. Leishman's stain (× 1000).

Wright's stain is probably most widely used but differentiates the leukocytes somewhat less satisfactorily than Jenner's or Ehrlich's. The blood-film need not be quite so thin as is required for study of the red cells, but it must be thin enough to enable one to identify the leukocytes without difficulty. One should first glance over the preparation to find what the general tinting of the cells may be. Two films stained side by side will often show marked differences in the color reactions of the cells.

To make the differential count go carefully over the film with an oil-immersion lens, using a mechanical stage if available. Experienced workers often use the lower powers (even the 16-mm., as recommended by Simon) in routine work; but the film must then be mounted, or wet with water or oil since these lenses cannot be used satisfactorily upon dry, unmounted films. Classify each leukocyte seen, and calculate what percentage each variety is of the whole number classified. For accuracy, 500 to 1000 leukocytes must be classified; for approximate results, 300 are sufficient, but it is imperative to count cells in all parts of the smear, since the different varieties of leukocytes may be unevenly distributed. Track of the count may be kept by placing a mark for each leukocyte in its appropriate column, ruled upon paper. Some workers divide a slide-box into compartments with slides, one for each variety of leukocyte, and drop a coffee-bean into the appropriate compartment when a cell is classified. When a convenient number of coffee-beans is used (any multiple of 100), the percentage calculation is simple.

The actual number of each variety in a cubic millimeter of blood is easily calculated from these percentages and the total leukocyte count, and should form part of the record if this is to be complete. An increase in actual number is an *absolute increase;* an increase in percentage only, a *relative increase.* It is evident that an absolute increase of any variety may be accompanied by a relative decrease.

One should make it a rule, when making a differential count, always to attempt to estimate the total leukocyte count from the appearance of the stained film with the low-power objective. After some practice, this can be done with a considerable degree of accuracy.

The number of nucleated red corpuscles seen while making the count is generally included in the record.

The usual classification of leukocytes is based upon their size, their nuclei, and the staining properties of the granules which many of them contain. It is not altogether satisfactory, but is probably the best which our present knowledge permits. The leukocytes of normal blood fall into two groups. Those in the first group are mononuclear and non-granular. Those in the second group are polymorphonuclear and contain cytoplasmic granules which are distinguished by their size and staining reactions. In its structure the chief abnormal leukocyte, the myelocyte, combines the two groups, being mononuclear like the first and granular like the second.

The leukocytic percentages given here as normal may be taken as representing about the average for this country. Recent studies indicate that variations among healthy individuals may be greater than has been supposed and that climatic factors, altitude, etc., may exert a decided influence. One should therefore hesitate to base diagnostic conclusions upon slight variations in the differential count unless one has previously determined the normal for the individual.

(1) **Normal Varieties.**—(a) **Lymphocytes** are small mononuclear cells without specific granules (see Frontispiece and Plate IX). They are about the size of a red corpuscle or slightly larger (6–10 μ), and consist of a single, sharply defined, deeply staining nucleus, surrounded by a narrow rim of protoplasm. The nucleus

Is generally **round, but is sometimes** indented at one side. Wright's stain gives the nucleus a deep purple color and the cytoplasm a pale robin's-egg blue in typical cells. Larger lymphocytes are frequently found, especially in the blood of children, and are difficult to distinguish from the large mononuclear leukocytes. It is believed that the larger forms are young lymphocytes, which become smaller as they grow older. Some workers record the large and small lymphocytes separately. There is no clear line of distinction, but if it seems desirable to separate them, the terms "immature" and "mature" may appropriately be used. In the cytoplasm of a certain percentage of lymphocytes the Romanowsky stains show a variable number of reddish-purple (azurophilic) granules.

Lymphocytes are formed in the lymphoid tissues, including that of the bone-marrow. They constitute about 25 to 33 per cent. of all leukocytes, or 1200 to 3300 per cubic millimeter of blood. They are more abundant in the blood of children, averaging about 60 per cent. in the first year of life and decreasing to about 36 per cent. in the tenth, the immature cells being especially abundant.

The percentage of lymphocytes is usually moderately increased in those conditions which give leukopenia, especially chlorosis, pernicious anemia, and many debilitated conditions. There is a decided absolute and relative increase at the expense of the polymorphonuclears at high altitudes although the extent of this is somewhat uncertain. A marked increase, accompanied by an increase in the total leukocyte count, is seen in pertussis (Fig. 125) and lymphatic

leukemia. In the former lymphocytes average about
60 per cent. In the latter they sometimes exceed 98
per cent. Exophthalmic goiter commonly gives a
marked relative lymphocytosis, while simple goiter
does not affect the lymphocytes. In pulmonary tuberculosis a high percentage of lymphocytes or, especially, a progressive increase is a favorable prognostic

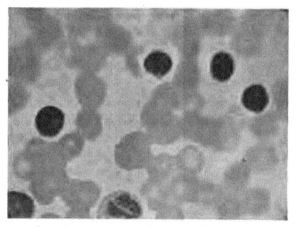

FIG. 125.—Lymphocytosis, case of pertusis (× 1000) (courtesy of
Dr. W. P. Harlow).

sign, while a progressive decline should be looked upon
with apprehension.

There is at present a tendency toward greater con-
servatism in ascribing diagnostic significance to lymphocytosis of moderate degree, *i.e.*, of less than 40 per
cent., unless the normal for the individual has been previously established. Lymphocyte percentages as low
as 15 or as high as 45 are occasionally met with in
apparently healthy individuals.

(b) **Large Mononuclear and Transitional Leuko-cytes** (See Frontispiece).—These cells are two or three times the diameter of the normal red corpuscle.

The large mononuclear contains a single round or oval nucleus, often located eccentrically. The zone of protoplasm surrounding the nucleus is relatively wide. With Wright's stain the nucleus is less deeply colored than that of the lymphocyte, while the cytoplasm is very pale blue or colorless, and sometimes contains a few reddish granules. The size of the cell, the width of the zone of cytoplasm, and the depth of color of the nucleus are the points to be considered in distinguishing between large mononuclears and lymphocytes. When large forms of the lymphocyte are present the distinction is often difficult or impossible. It is then advisable to count the two cells together as lymphocytes. Some workers arbitrarily adopt the size of the polymorpho-nuclear neutrophile as the dividing line between the two cells.

Transitional leukocytes are simply large mononuclear leukocytes whose nuclei are lobulated, deeply indented or horseshoe shape. There is no good reason for plac-ing them in a separate group as is frequently done. Mallory and others class the two cells together as "endothelial leukocytes" or "endotheliocytes." This is convenient but it seems unwise to introduce new names until the nature and origin of the cells are better understood.

Comparatively little is known regarding the origin of the large mononuclear and transitional leukocytes. Some at least appear to be developed from the endo-thelial cells of the blood- and lymph-vessels by pro-

liferation and desquamation. Altogether they constitute 2 to 5 per cent. of the total number of leukocytes;
100 to 600 per cubic millimeter of blood. Only a few
pathologic conditions raise this figure to any marked
degree. A distinct increase is a feature of the blood
in typhoid fever and may be of some value in differential diagnosis. It is also quite constant in malaria,
where sometimes many of the cells contain engulfed
pigment (see Plate VII). Bunting has found it also
constant early in Hodgkin's disease and regards it as an
extremely important point in diagnosis. Late in the
disease it is still evident but is then overshadowed by an
increase of neutrophiles.

(c) **Polymorphonuclear Neutrophilic Leukocytes** (See
Frontispiece).—There is usually no difficulty in recognizing these cells. Their average diameter (about 12 μ)
is somewhat less than that of the large mononuclears. The nucleus stains rather deeply, and is very
irregular, often assuming shapes comparable to letters of
the alphabet, E, Z, S, etc. (Fig. 126). Frequently there
appear to be several separate nuclei, hence the widely
used name, "polynuclear leukocyte." Upon careful inspection, however, delicate nuclear bands connecting
the parts can usually be seen. The cytoplasm is relatively abundant, and contains great numbers of very
fine neutrophilic granules (see Fig. 130, A). With
Wright's stain the nucleus is bluish purple, and the
granules reddish lilac.

Polymorphonuclear leukocytes are formed in the
bone-marrow from neutrophilic myelocytes. Ordinarily
they constitute 60 to 70 per cent. of all the leukocytes:
3000 to 7000 per cubic millimeter of blood. An occa-

sional normal adult may give a count as low as 40 per
cent. or as high as 80 per cent.   In children the average
runs from about 35 per cent. in the first year to 50 per
cent. in the tenth.   Any marked increase in their number practically always produces an increase in the total
leukocyte count, and has already been discussed under
Polymorphonuclear Leukocytosis (see p. 289).   The

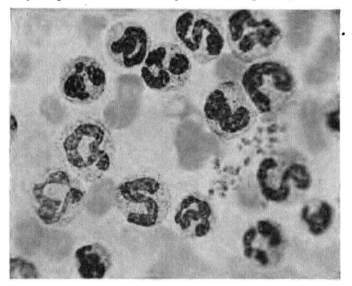

FIG. 126.—Marked   polymorphonuclear   neutrophilic   leukocytosis
(× 1000) (courtesy of Dr. W. P. Harlow).

leukocytes of pus, *pus-corpuscles*, belong almost wholly
to this variety.

A comparison of the percentage of polymorphonuclear
cells with the total leukocyte count yields more information than a consideration of either alone.   In a general
way the percentage represents the severity of the infection or, more correctly, the *degree of toxic absorption;*

while the total count indicates the patient's *power of resistance*. With moderate infection and good resisting powers the leukocyte count and the percentage of polymorphonuclears are increased proportionately. When the polymorphonuclear percentage is increased to a notably greater extent than is the total number of leukocytes, no matter how low the count, either very poor resistance or a very severe infection may be inferred.

Gibson has suggested the use of a chart to express this relationship graphically (Fig. 127). Its arrangement is purely arbitrary, but it will be found very helpful in interpreting counts. An ascending line from left to right indicates an unfavorable prognosis in proportion as the line approaches the vertical. All fatal cases show a rising line. A descending or horizontal line suggests a very favorable prognosis.

It is a matter of observation that in the absence of acute infectious disease or of inflammation directly in the blood stream (*e.g.*, phlebitis, sigmoid sinusitis, septic endocarditis), a polymorphonuclear percentage of 85 or over points very strongly to gangrene or pus formation somewhere in the body. On the other hand, excepting in children, where the percentage is normally low, pus is uncommon with less than 80 per cent. of polymorphonuclears.

Normally, the cytoplasm of leukocytes stains pale yellow with iodin. Under certain pathologic conditions the cytoplasm of many of the polymorphonuclears stains diffusely brown, or contains granules which stain reddish brown with iodin. This is called **iodophilia.** Extracellular iodin-staining granules, which are present normally, are more numerous in iodophilia.

This Iodin reaction occurs in all purulent conditions
except abscesses which are thoroughly walled off and

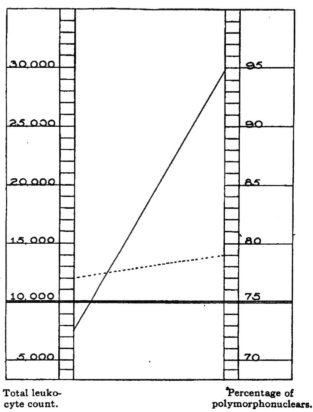

Total leuko-   Percentage of
cyte count.   polymorphonuclears.

FIG. 127.—Gibson chart with blood-count in 2 cases of appendicitis:
Dotted line represents a mild case with prompt recovery; the continu-
ous line, a very virulent streptococcic case with poor resistance, peritonitis, and early death.

purely tuberculous abscesses. It is of some value in
diagnosis between serous effusions and purulent exudates, between catarrhal and suppurative processes in

the appendix and Fallopian tube, etc. Its importance, however, as a diagnostic sign of suppuration has been much exaggerated, since it may occur in any general toxemia, such as pneumonia, influenza, malignant disease, and puerperal sepsis.

To demonstrate iodophilia, place the air-dried films in a stoppered bottle containing a few crystals of iodin until the films become yellow. Mount in syrup of levulose and examine with an immersion objective.

Arneth's Classification of Neutrophiles.—Arneth groups the neutrophilic leukocytes into five classes according to the number of lobes which the nucleus possesses. The forms which fall into each class and the average normal percentages as given by Arneth are indicated in the following list:

Class 1. One round or indented nucleus; 5 per cent.
Class 2. Two nuclear divisions; 35 per cent.
Class 3. Three nuclear divisions; 41 per cent.
Class 4. Four nuclear divisions; 17 per cent.
Class 5. Five or more nuclear divisions; 2 per cent.

This is really a classification of neutrophiles according to their age, the youngest cells being included in Class 1. Among these youngest cells are the myelocytes and metamyelocytes which do not appear in normal blood.

The percentages are fairly constant in the same individual in health, but may show considerable variations in disease, even when the leukocyte count remains unchanged. An increase of the lower classes at the expense of the higher is known as a "shift of the neutrophilic blood picture to the left." The opposite condition is a "shift to the right." In order to simplify comparison many workers in this country use an index number obtained by adding the first, second, and one-half of the third classes. The average normal

"Arneth index" is accordingly about 60. Briggs found variations between 51 and 65 in normal individuals.

The clinical value of an Arneth count is not definitely determined. It appears to have greater usefulness in prognosis than in diagnosis. Most pathologic conditions which produce any change cause a shift to the left, *i.e.*, a high index. Among these are acute infectious diseases, pyogenic infections (appendicitis, etc.), and tuberculosis. In tuberculosis the Arneth count is regarded as having definite prognostic value, the higher the index the more serious being the outlook.

A low index occurs in pernicious anemia. In a series of 23 examinations in 12 cases of pernicious anemia Briggs found an average index of 40.29; lowest, 16.5; highest, 51.25. Eight consecutive cases of severe secondary anemia (malignant disease, syphilis, nephritis, repeated hemorrhages, etc.) gave an average index of 68.23, only one case (a case of syphilis with index of 39) falling below normal limits.

For the Arneth count thin well-stained blood-films are essential. Wright's stain may be used but hematoxylin-eosin is better since it brings out the nuclear structure more clearly. Nuclear parts which are joined by more than a thread should be counted as one.

Döhle's Inclusion Bodies.—In 1911 Döhle called attention to the occurrence of certain peculiar bodies within the cytoplasm of the neutrophiles in cases of scarlet fever (Fig. 128). Their nature has not been definitely determined. The typical "inclusion bodies" are about the size of micrococci or a little larger; some of them are pear-shaped, others appear like short rods or like cocci lying in pairs. Discrete, punctiform granules are sometimes seen but have not the same significance. It now seems well established that typical inclusion bodies have considerable diagnostic value. They are apparently found in many or even the majority of the neutrophilic leukocytes in every case of scarlet fever early in the disease. Upon the other hand, a few may be found in many cases of diphtheria, pneumonia, and some other infectious diseases, but never in German measles and rarely in measles.

The inclusion bodies can be seen in preparations stained with Wright's stain, but long staining with pyronin-methyl-green is preferable. With the latter stain, nuclei are purplish and the bodies bright red.

(d) **Eosinophilic Leukocytes, or "Eosinophiles"** (See Frontispiece).—The structure of these cells is similar to that of the polymorphonuclear neutrophiles, with the

Fig. 128.—Döhle's inclusion bodies in leukocytes. From a case of scarlet fever. Pyronin-methyl-green stain (× 1500) (from a slide prepared by L. W. Hill).

striking difference that, instead of fine neutrophilic granules, their cytoplasm contains coarse round or oval granules having a strong affinity for acid stains. They are easily recognized by the size and color of the granules, which stain bright red with stains containing eosin (see Fig. 130, B). Their cytoplasm has generally a faint sky-blue tinge, and the nucleus stains somewhat less deeply than that of the polymorphonuclear neutrophile.

Eosinophiles are formed in the bone-marrow from eosinophilic myelocytes. Their normal number varies from 50 to 400 per cubic millimeter of blood, or 1 to 4 per cent. of the leukocytes. An increase is called *eosinophilia*, and is better determined by the actual number than by the percentage.

Slight eosinophilia is said to be physiologic during menstruation. Marked eosinophilia is always pathologic. It occurs in a variety of conditions, the most important of which are: infection by animal parasites; bronchial asthma; myelogenous leukemia; scarlet fever; many skin diseases; and tuberculin reactions.

(*a*) Eosinophilia may be a symptom of *infection by any of the worms* and from a diagnostic view-point this is its most important indication. It is fairly constant in trichiniasis, uncinariasis, filariasis, and echinococcus disease, and is usually most marked in the first named condition. In this country an unexplained marked eosinophilia warrants examination of a portion of muscle for *Trichinella spiralis* (see p. 509). The cells usually range between 10 and 30 per cent. of all the leukocytes, but may go much higher.

(*b*) *True bronchial asthma* commonly gives a marked eosinophilia during and following the paroxysms. This is helpful in excluding asthma of other origin. Eosinophiles also appear in the sputum in large numbers.

(*c*) In *myelogenous leukemia* there is almost invariably an absolute increase of eosinophiles, although, owing to the great increase of other leukocytes, the percentage is usually diminished. Dwarf and giant forms are often numerous.

(*d*) *Scarlet fever* is frequently accompanied by eosinophilia, which may help to distinguish it from measles.

(*e*) Eosinophilia has been observed in a large number of *skin diseases*, notably pemphigus, prurigo, psoriasis,

22

and urticaria. It probably depends less upon the variety of the disease than upon its extent.

(*f*) Eosinophilic cells are usually increased to a variable degree in tuberculin reactions and anaphylactic conditions in general.

(*e*) **Basophilic Leukocytes or "Mast-cells"** (See Frontispiece).—In general, these resemble polymorphonuclear neutrophiles except that the nucleus is less irregular (usually merely indented or slightly lobulated) and that

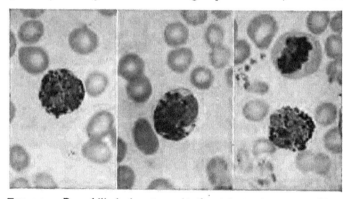

FIG. 129.—Basophilic leukocytes   At the right is also a normoblast undergoing mitosis (X 1000).

the granules are larger and have a strong affinity for basic stains. They are easily recognized (Figs. 129 and 130, C). Sometimes one sees cells from which most of the granules have disappeared, leaving clean-cut openings. With Wright's stain the granules are deep purple, while the nucleus is pale blue and is often nearly or quite hidden by the granules, so that its form is difficult to make out. Basophilic granules are not colored by Ehrlich's stain.

Mast-cells probably originate in the bone-marrow from

basophilic myelocytes. They are least numerous of the leukocytes in normal blood, rarely exceeding 0.5 per

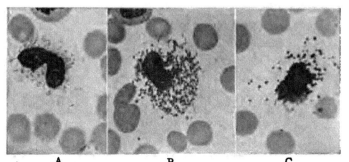

A   B   C

FIG. 130.—Ruptured leukocytes, showing relative size of granules: A, neutrophilic; B, eosinophilic; C, basophilic (× 1000).

cent., or 25 to 50 per cubic millimeter. A notable increase is limited almost exclusively to myelogenous leukemia, where they are sometimes very numerous.

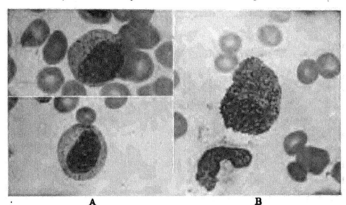

A   B

FIG. 131.—Myelocytes from blood of myelogenous leukemia: A, Neutrophilic; B, eosinophilic (× 1000)

(2) **Abnormal Varieties.**—(a) **Myelocytes** (see Frontispiece and Fig. 131) are large mononuclear cells whose

cytoplasm is filled with granules. Typically, the nucleus
occupies about one-half of the cell, and is round or oval,
or is indented, with its convex side in contact
with the periphery of the cell. It stains rather feebly.
The average diameter of this cell (about 15.75 μ) is
greater than that of any other leukocyte, but there is
much variation in size among individual cells. Myelocytes are named according to the character of their
granules—neutrophilic, eosinophilic, and basophilic
myelocytes. These granules are identical with the corresponding granules in the leukocytes just described.
They are, however, often less distinct and less sharply
differentiated by the various stains than those of the
corresponding polymorphonuclear cells. In some the
granules are few in number, the cells departing but
little from the structure of the parent myeloblast.
Such cells may be called "premyelocytes." In young
neutrophilic myelocytes there is a tendency to relatively large granules which take a purple color with
Wright's stain. Although the occurrence of two kinds
of granules in the same cell is rare, a few basophilic
granules are sometimes seen in young eosinophilic myelocytes. The basophilic myelocyte is usually small;
and its nucleus is commonly so pale and so obscured
by the granules that the cell is not easily distinguished
from the mast-cell.

The small neutrophilic cell with a single small round
deeply staining nucleus which is sometimes encountered
must not be confused with the myelocyte. Such atypic
cells probably result from division of polymorphonuclear neutrophiles.

Myelocytes are the bone-marrow cells from which

the corresponding granular leukocytes are developed.
They in turn are derived from certain non-granular cells
of the bone-marrow, the myeloblasts. Their presence
in the blood in considerable numbers is diagnostic of
myelogenous leukemia. The neutrophilic form is the
least significant. A few of these may be present in very
marked leukocytosis or any severe blood condition, as
pernicious anemia. In the anemia of malignant disease they suggest bone-marrow metastasis. Eosinophilic myelocytes are found only in myelogenous leukemia, where they are often very numerous. The

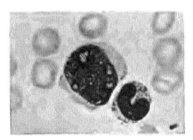

Fig. 132.—A myeloblast and a neutrophilic leukocyte. From a case of
myelogenous leukemia. Wright's stain (× 1000).

basophilic variety is less common, and is confined to
long-standing, severe myelogenous leukemia.

(b) **Myeloblasts.**—These are the parent cells of the
myelocytes, from which they differ chiefly in the absence of cytoplasmic granules. Their round or oval
nuclei are poor in chromatin and contain several rather
indistinct nucleoli (Fig. 132). The cytoplasm, which
is generally not abundant, is markedly basophilic,
staining pure blue with Wright's stain. In some
preparations it is characteristically smooth in texture;
in others it is finely reticular.

Myeloblasts appear in the blood in large numbers in acute myelogenous leukemia and the terminal stages of chronic myelogenous leukemia, when the bone-marrow reverts to the embryonic type. Their number is therefore important in prognosis. They may be indistinguishable morphologically from the large lymphocytes of acute lymphatic leukemia, but can usually be distinguished by the oxydase reaction. In most advanced cases of myelogenous leukemia all stages of transition between the myeloblast and myelocyte may be found.

Indophenol Oxydase Test.—The technic used by Evans is as follows:

1. Fix cover-glass films eight hours in formaldehyd vapor in a closed jar.

2. Stain eight minutes with a saturated aqueous solution of safranin or 2 per cent. solution of pyronin.

3. Wash quickly in water and blot dry.

4. Upon a slide mix 1 drop of a 1 per cent. aqueous solution of dimethylparaphenylendiamin (less than three weeks old) and 1 drop of a 1 per cent. solution of alpha-naphthol in 1 per cent. potassium hydroxid (less than four days old). Upon this place the previously stained cover, film-side down, and examine immediately.

In such preparations the nuclei show a pink or orange color which fades after a few minutes. The cytoplasm of cells containing oxydase—polymorphonuclears, large mononuclears and transitionals, myelocytes, and myeloblasts—gradually becomes a faint diffuse blue, and fine and coarse blue-black granules appear. The reaction endures about ten minutes. Lymphocytes, red corpuscles, and platelets should show no blue.

(c) **Türck's Irritation Leukocytes** —Strictly speaking these ought to be classed with the normal leukocytes since they are often encountered in normal blood, where, however, their number rarely exceeds 1 per cent. of the leukocytes. They are generally included with the large mononuclear leukocytes. In brief they are large, mononuclear, non-granular cells with dense, opaque cytoplasm which stains deep blue with Wright's stain and brown with Ehrlich's and sometimes contains vacuoles (see Plates I and VIII). As a rule they are not difficult to recognize although they might be confused with the large lymphocytes or with very strongly polychromatophilic megaloblasts.

At present Türck's irritation leukocytes have no diagnostic importance. Considerable numbers may appear in the blood in conditions associated with irritation of the bone-marrow, notably primary and secondary anemia, leukemia and malaria.

(d) **Degenerated Forms.**—These are frequently met but have no significance unless present in large numbers. They include (a) vacuolated leukocytes and (b) bare nuclei from ruptured cells. The former are found most frequently in toxemias and leukemia. A few of the latter are present in every blood smear but are especially abundant in leukemia (Fig. 133). They vary from fairly well preserved nuclei to mere strands of palely stained nuclear substance arranged in a coarse network—the so-called "basket-cells" (see Frontispiece).

Occasionally in lymphatic leukemia frayed-out nuclei without cytoplasm exceed the usual lymphocytes in number. In such cases some writers infer involvement of the bone-marrow, holding that the naked nuclei represent very fragile bone-marrow cells which have gone to pieces in the circulation. In many cases at least it seems more likely that such nuclei only represent fragile lymphocytes which have been broken in making the smear.

(*e*) **Atypic Forms.**—Leukocytes which do not fit in with the above classification are not infrequently met, especially in high-grade leukocytosis, pernicious anemia, and leukemia. They are always more abundant in childhood. The nature of many of them is not clear,

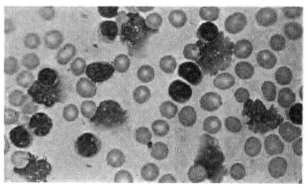

FIG. 133.—Blood in chronic lymphatic leukemia, showing many ruptured lymphocytes (× 750).

and their number is usually so small that they may be classed as "undetermined" in making a differential count.

3. Blood-platelets.—These are not colored by Ehrlich's stain nor by hematoxylin and eosin. With Wright's stain they appear as spheric or ovoid, reddish to violet, granular bodies, 2 to 4 μ in diameter. Occasionally a platelet as large as a red corpuscle is seen. When well stained a delicate hyaline peripheral zone can be distinguished. In ordinary blood-smears they are usually clumped in masses. A single platelet lying upon a red corpuscle may easily be mistaken for a malarial parasite (see Frontispiece and Fig. 134).

Blood-platelets are being much studied at present, but, aside from the facts mentioned under their enumeration (see p. 300), little of clinical value has been learned. They have been variously regarded as very young red corpuscles (the "hematoblasts" of Hayem)

FIG. 134.—A cluster of blood-platelets and two platelets lying upon a red cell and simulating malarial parasites (× 1000).

as disintegration products of leukocytes, as remnants of extruded nuclei of erythrocytes, and as independent nucleated bodies. The most probable explanation of their origin seems to be that of J. H. Wright, who regards them as detached portions of the cytoplasm of certain giant-cells of the bone-marrow and spleen.

X. BLOOD PARASITES
A. BACTERIA

Bacteriologic study of the blood is useful in many conditions, but, in general, the somewhat elaborate technic involved takes it out of reach of the clinician. As applied to the diagnosis of typhoid fever, however, the technic of blood-cultures has been so simplified that it can be carried through by any one who is competent to do the simplest cultural work.

Typhoid bacilli can be detected in the blood in practically every case of typhoid fever in the first week of the disease; in about 80 to 85 per cent. of cases in the second week; and in decreasing percentages in the later weeks. The blood-culture, therefore, offers the most certain means of early diagnosis. It is in a sense complementary to the Widal reaction, the former decreasing and the latter increasing in reliability as the disease progresses. The blood-culture gives best results before the Widal appears, as one would expect from the fact that the Widal test depends upon the presence of antibodies which destroy or, at least, injure the bacilli. The two methods together will establish the diagnosis in practically every case at any stage. Bacilli disappear from the blood in convalescence and reappear in a relapse.

Technic of Blood-cultures in Typhoid Fever.—The blood may be obtained in one of two ways:

(a) With a spring-lancet (see Fig. 85) make a deep puncture in the edge (not the side) of the lobe of the ear, as for a blood-count. Allow the blood to drop directly into a short culture-tube containing the bile medium. By gentle milking, 20 to 40 drops can usually be obtained. This simple method of obtaining blood is especially applicable during the first week of the disease when bacilli are abundant. Contamination with skin cocci is possible, but does not usually interfere when the bile medium is used.

(b) In the later weeks of the disease a larger quantity of blood is needed and must be obtained from a vein as described on page 254.

As special culture-medium, ox-bile is generally used. It favors the growth of the typhoid bacillus and retards the growth of other organisms. A good formula is given on page 569.

As soon as convenient after the blood is added, place the tubes in the incubator. After about twelve hours examine for motile bacilli. If none are found, transfer a few drops to tubes of bouillon or solidified blood-serum and incubate for twelve hours longer. If motile, Gram-negative bacilli are found, they are almost certainly typhoid bacilli. Further study is, however, desirable, especially from a scientific point of view. The only bacilli which might cause confusion are the paratyphoid and colon bacilli, which can be distinguished by gas production in glucose media, indol production, and their effect upon litmus milk (see p. 583). The agglutination test for the identity of the bacillus is not available clinically, since freshly isolated bacilli do not agglutinate well.

Technic for Other Bacteria.—About 10 c.c. of blood are obtained from a vein (see p. 254) under strictly aseptic precautions and immediately distributed among flasks of sterile bouillon. When the pneumococcus or streptococcus is suspected a better medium is nutrient bouillon to which one-fourth its volume of sterile ascitic fluid has been added. Ordinarily not more than 1 c.c. of blood should be added to 50 or 100 c.c. of culture medium. After incubating for twenty-four hours or longer, sub-cultures are made from these flasks upon media appropriate to identify any bacteria which may grow.

When the blood must be obtained at a distance from the laboratory it may be received, by means of one of the devices shown in Figs. 87 and 89, directly into 15 c.c. of a sterile solution consisting of 2 Gm. of ammonium oxalate and 6 Gm. sodium chlorid in 1000 c.c. distilled water. Such a solution will prevent coagulation and will not harm any bacteria that may be present. As soon as convenient the blood is added to appropriate culture media in flasks or tubes, or is mixed with melted agar and poured into Petri plates.

B. ANIMAL PARASITES

Of the animal parasites which have been found in the blood, five are interesting clinically: the spirochete of relapsing fever; trypanosomes; malarial parasites; filarial larvæ; and the larvæ of *Trichinella spiralis*.

1. Spirochæta recurrentis is described on page 460.

2. Trypanosoma gambiense.—Various trypanosomes are common in the blood of fishes, amphibians, birds, and mammals (see Fig. 159). They live in the blood-plasma and do not attack the corpuscles. In some animals they are apparently harmless; in others they are an important cause of disease. They are discussed more fully on page 464.

The trypanosome of human blood, *Trypanosoma gambiense* (Plate VI), is an actively motile, spindle-shaped organism, two or three times the diameter of a red corpuscle in length, with an undulating membrane which terminates at the anterior end in a long flagellum. It can be seen with medium-power objectives in fresh blood, but is best studied with an oil-immersion lens in preparations stained as for the malarial parasite. It will be necessary to search many slides. The concentration method described for the larvæ of *Trichinella* (see p. 363) may be used. Human trypanosomiasis is common in Africa. As a rule, it is a very chronic

PLATE VI

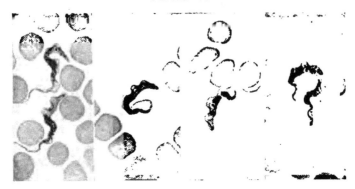

Trypanosoma gambiense.

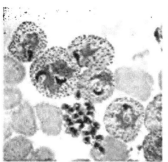

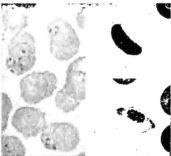

Half-grown tertian malarial parasites in stippled cells and a group of spores from a freshly ruptured segmenter. From a slide of double tertian malarial concentrated by F. M. Johns.

Estivo-autumnal malarial parasites, small ring forms and crescents.

Spirochetes in the blood of a case of relapsing fever originating in Colorado. Reported by Dr. C. N. Meader.

disease. "Sleeping sickness" is a late stage when the organisms have invaded the cerebrospinal fluid. Infection is carried by the tsetse fly, *Glossina palpalis*.

3. The Malarial Parasites.—These protozoa belong to the Sporozoa (see p. 470), order Hemosporidia, the members of which are parasites in the blood of a great variety of vertebrates. Three species, constituting the genus *Plasmodium*, are associated with malarial fever in man: *Plasmodium vivax*, *P. malariæ*, and *P. falciparum*, the parasites respectively of the tertian, quartan, and estivo-autumnal types of malaria. The life histories of the three are so similar that they may well be described together.

(1) **Life Histories.**—There are two cycles of development: one, the *asexual*, in the blood of man; and the other, the *sexual*, in the intestinal tract of a particular genus of mosquito, *Anopheles*.

(a) *Asexual Cycle.*—The young organism enters the blood through the bite of the mosquito. It makes its way into a red corpuscle,[1] where it appears as a small, pale, "hyaline" body. This body exhibits ameboid movement and increases in size. Soon dark-brown granules, derived from the hemoglobin of the corpuscle, make their appearance within it. When it has reached its full size—filling and distending the corpuscle in the case of the tertian parasite, smaller in the others—the

[1] In this section the malarial parasite is described, in accordance with the usual teaching, as living within the parasitized red corpuscle. The recent work of Mary Rowley-Lawson, however, tends to show that the parasite is extra-cellular throughout its whole existence; that it attaches itself to the external surface of the red corpuscle but does not enter it; and that it migrates from corpuscle to corpuscle between paroxysms, destroying each cell before it abandons it.

pigment granules gather at the center or at one side; the organism divides into a number of small hyaline bodies, the spores or *merozoites*; and the red corpuscle bursts, setting spores and pigment free in the blood-plasma. This is called segmentation. It coincides with, and by liberation of toxins causes, the paroxysm of the disease. A considerable number of the spores are destroyed by leukocytes or other agencies; the remainder enter other corpuscles and repeat the cycle. Many of the pigment granules are taken up by leukocytes. In estivo-autumnal fever segmentation occurs in the internal organs and the segmenting and larger pigmented forms are seldom seen in the peripheral blood. This accumulation of parasites in the internal organs explains certain types of pernicious estivo-autumnal malaria, *e.g.*, the comatose type, when the parasites accumulate in the capillaries of the brain.

The asexual cycle of the tertian organism occupies forty-eight hours; of the quartan, seventy-two hours; of the estivo-autumnal, an indefinite time—usually twenty-four to forty-eight hours.

The parasites are thus present in the blood in great groups, all the individuals of which reach maturity and segment at approximately the same time. This explains the regular recurrence of the paroxysms at intervals corresponding to the time occupied by the asexual cycle of the parasite. Not infrequently there is multiple infection, one group reaching maturity while the others are still young; but the presence of two groups which segment upon the same day is extremely rare. Fevers of longer intervals—six, eight, ten days—are probably due to the ability of the body, sometimes of itself, sometimes by aid of quinin, to resist the parasites, so that numbers sufficient to cause a paroxysm do not accumulate in the blood until after several repetitions of the asexual cycle. In estivo-autumnal fever the regular grouping, while usually present at first, is soon lost, thus causing "irregular malaria."

Bass has recently succeeded in cultivating the malarial parasite outside of the body.

(b) *Sexual Cycle.*—Besides the ameboid individuals which pass through the asexual cycle, there are present with them in the blood many individuals with sexual properties. These are called *gametes*. The female is a little the larger. The gametes do not undergo segmentation, but grow to adult size and remain inactive in the blood until taken up by a mosquito. Many of them are apparently extracellular, but stained preparations usually show them to be surrounded by or attached to the remains of a corpuscle. In tertian and quartan malaria they resemble the asexual individuals until a variable time after the blood leaves the body, when the male gamete sends out one or more flagella. In estivo-autumnal malaria the gametes take distinctive ovoid and crescentic forms, and are not difficult to recognize. These sexual forms are very resistant to quinin and often persist in the blood long after the ameboid forms have been destroyed. Under ordinary conditions they are incapable of continuing the disease until they have passed through the cycle in the mosquito, but it seems probable that under certain unusual conditions the female gamete may, without fertilization, undergo further development and sporulate, thus starting a new asexual cycle.

When a malarious person is bitten by a mosquito, the gametes are taken with the blood into its stomach. Here the male sends out one or more flagella. These break off and seek out the females, whom they fertilize much as the sperm fertilizes the ovum. The female soon thereafter becomes encysted in the wall of the intestine. After a time this "oöcyst" ruptures, liberating many minute rods, or *sporozoites*, which have formed

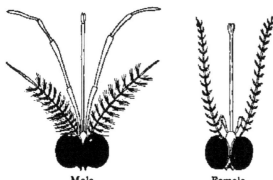

Male.   Female.

FIG. 135.—Head of Culex (after Giles). Showing the straight proboscis, the jointed palpi and, external to these, the hairy antennæ. The male is distinguished from the female by the longer hairs on the antennæ. Note that the palpi of the male are longer than the proboscis, while those of the female are very much shorter (compare with Fig. 136).

within it. These migrate to the salivary glands, and are carried into the blood of the person whom the mosquito bites. Here they enter red corpuscles as young malarial parasites, and the majority pass through the asexual cycle just described.

The sexual cycle can take place only within the body of the female of one genus of mosquito, *Anopheles*. The male does not bite. Absence of this mosquito from certain districts explains the absence of malaria. It is distinguished from our common house mosquito, *Culex*, by the relative lengths of proboscis and palpi (Figs. 135 and 136), which can be seen with a hand-lens, by its attitude when resting, and by its dappled wing (Fig. 137). Anopheles is strictly nocturnal in its habits; it usually flies low, and rarely travels more than a few hundred yards from its breeding-place, although it may

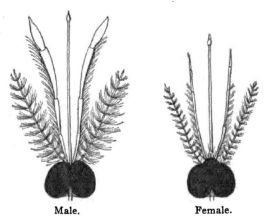

Male. Female.

Fig. 136.—Head of Anopheles (after Giles). The sexes are distinguished by the antennæ as noted under Fig. 135. In this mosquito the palpi of both sexes are nearly the same length as the proboscis.

be carried by winds. These facts explain certain peculiarities in malarial infection; thus, infection occurs practically only at night; it is most common near stagnant water, especially upon the side toward which the prevailing winds blow; and the danger is greater when persons sleep upon or near the ground than in upper stories of buildings. The insects frequently hibernate in warmed houses, and may bite during the winter. A mosquito becomes dangerous in eight to fourteen

23

days after it bites a malarious person, and remains so throughout its life.

(2) **Detection.**—Search for the malarial parasite may be made in either fresh blood or stained films. If pos-

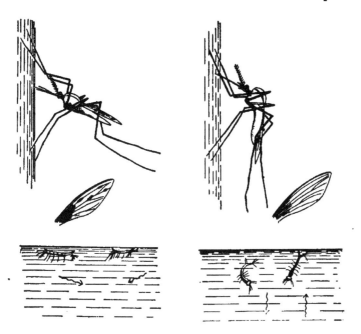

Fig. 137.—Showing, on the left, *Anopheles* in resting position, its dappled wing, and the position of its larvæ in water; on the right, *Culex* in resting position, its plain wing, and the position of its larvæ in water. The arrows indicate the directions taken by the larvæ when the water is disturbed (Abbott).

sible, the blood should be obtained a few hours before the chill—not during it nor within a few hours afterward, since at that time (in single infections) only the very young, unpigmented forms are present, and these are the most difficult to find and recognize. Sometimes many parasites are found in a microscopic field; sometimes, especially in estivo-autumnal infection, owing to accumulation in internal organs, careful search is required to find any, despite very severe symptoms. Quinin causes them rapidly to disappear from the peripheral blood, and few or none may be found after its administration. In the absence of organisms, the presence of pigment granules within leukocytes—especially the large mononuclears—may be taken as presumptive evidence of malaria. Pigmented leukocytes (see Frontispiece and Plate VII) are most numerous after a paroxysm.

(a) *In Fresh Unstained Blood.*—Obtain a small drop of blood from the finger or lobe of the ear. Touch the center of a cover-glass to the top of the drop and quickly place it, blood side down, upon a slide. If the slide and cover be perfectly clean and the drop not too large, the blood will spread out so as to present only one layer of corpuscles. Search with an oil-immersion objective, using very subdued light. The preparation may be kept for many hours if ringed with vaselin or melted paraffin

The young organisms appear as small, round, ring-like or irregular, colorless bodies within red corpuscles. The light spots caused by crenation and other changes in the corpuscles are frequently mistaken for them, but are generally more refractive or have more sharply defined edges. The older forms are larger colorless bodies containing granules of brown pigment. In the case of the tertian parasite, these granules have active vibratory motion, which renders them conspicuous; and as the parasite itself is very pale, one may see only a large pale corpuscle in which fine pigment granules are dancing. Segmenting organisms, when typic, appear as rosets, often compared to daisies, the petals of which represent the segments, while the central brown portion represents the pigment. Tertian segmenting forms are less frequently typic than quartan. Flagellated forms are not seen until ten to twenty minutes after the blood has left the vessels. As Cabot suggests,

VARIETIES OF THE MALARIAL ORGANISM

TERTIAN.	QUARTAN.	ESTIVO-AUTUMNAL.
Asexual cycle, forty-eight hours.	Seventy-two hours.	Usually twenty-four to forty-eight hours.
Substance pale, transparent, comparable to hyaline tube-cast.	Highly refractive, comparable to waxy tube-cast.	Highly refractive.
Outline indistinct.	Distinct.	Distinct.
Ameboid motion active.	Sluggish.	Active.
Mature asexual form large; fills and often distends corpuscle.	Smaller.	Young forms, only, in peripheral blood.
Pigment-granules fine, brown, scattered throughout. Very active dancing motion.	Much coarser, darker in color, peripherally arranged. Motion slight.	Very few, minute, inactive. Distinctly pigmented forms seldom seen.
Segmenting body rarely assumes typical "daisy" form. 15 to 20 segments.	Usually typical "daisy." 6 to 12 segments.	Very rarely seen in peripheral blood.
Gametes resemble asexual forms.	Same as tertian.	Appear in blood as distinctive ovoids and crescents.
Red corpuscles pale and swollen.	Generally darker than normal.	Dark, often bronzed.

one should, while searching, keep a sharp lookout for unusually large or pale corpuscles, and for anything which is brown or black or in motion.

The preceding table contrasts the distinguishing characteristics of the three varieties as seen in fresh blood.

(b) *In Stained Films* (See Frontispiece and Plate VII).—Recognition of the parasite, especially the young forms, is much easier in films stained by Wright's or some similar stain than in fresh blood. The films must be thin and well stained. It is useless to search preparations in which the nuclei of leukocytes are not strongly colored.

In films which are properly stained with Wright's or Giemsa's stain malarial parasites appear as follows:

The **young parasites** are small, round, ring-like or irregular, sky-blue bodies, each with a very small, sharply defined, purplish red chromatin mass. Many structures—deposits of stain, dirt, blood-platelets lying upon red cells (see Fig. 134), etc.—may simulate them, but should not deceive one who looks carefully for both the blue cytoplasm and the purplish red chromatin. A platelet upon a red corpuscle is surrounded by a colorless zone rather than by a distinct blue body and there is no compact chromatin mass. As quartan parasites grow a little older they tend to assume a slender, straight or slightly curved, band-like form which is fairly characteristic of this species. Young estivo-autumnal parasites commonly take the form of small, delicate, blue rings, each with one or two small purplish red chromatin bodies upon its circumference. Their recognition is important because they may be the only form found in a given case. When young tertian and quartan parasites assume this form the ring is usually larger and thicker. Usually it is the dot-like chromatin body which first attracts one's attention to the parasitized cell. In tertian malaria the fact that cells which harbor the parasites are somewhat larger and paler than their fellows is also helpful in attracting one's attention while searching. This may be evident as early as eight hours after the chill. No such enlargement of the red cells is noted in other forms of malaria.

Older tertian and quartan parasites show larger sky-blue bodies with more abundant, paler, and more reticular or granular chromatin, and contain brown granules of pigment, which, however, are less evident than in the living parasite. The chromatin usually lies in a colorless area or "achromatic zone" and is sometimes so pale as to be difficult to see clearly. Not infrequently it appears to lie entirely outside of the cytoplasm. The pigment of the adult tertian parasite is usually fine and scattered uniformly through the cytoplasm. That of the quartan is coarser and more peripherally arranged. The corresponding stage of the estivo-autumnal parasite rarely appears in the blood.

Typical **"segmenters"** present a ring of rounded segments or spores, each with a small, dot-like chromatin mass, but these regular forms are not often seen. With the tertian parasite, especially, the segments much more frequently form an irregular cluster. The pigment is collected near the center or at one side or is scattered among the segments.

Fully grown tertian and quartan **gametes** resemble the fully grown asexual forms in general appearance, but are more compact and less irregular in shape and

PLATE VII

Malarial parasites. Wright's stain. × 1000 (1 mm. = 1 micron).

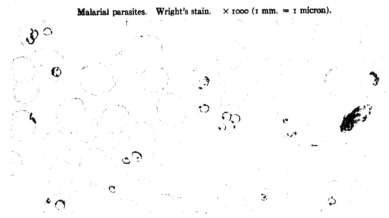

Fig. 1.—Estivo-autumnal malaria, exact reproduction of a portion of a field.

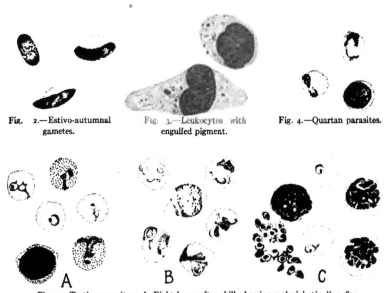

Fig. 2.—Estivo-autumnal gametes.

Fig. 3.—Leukocytes with engulfed pigment.

Fig. 4.—Quartan parasites.

Fig. 5.—Tertian parasites: A, Eight hours after chill, showing malarial stippling, five young parasites, and one gamete; from two slides; B, twenty-four hours after chill, five half-grown parasites, one gamete; C, during chill, one presegmenter, two segmenters, a cluster of freshly liberated merozoites, and two very young parasites; from one slide.

(J. W. Rennell, pinx.)

contain more and larger pigment granules. The female is generally the larger and has more compact chromatin and deeper blue cytoplasm. The crescentic and oval gametes of estivo-autumnal malaria are easily identified. Their length is somewhat greater than the diameter of a red corpuscle. Their chromatin is usually centrally placed, and they contain more or less coarse pigment. The remains of the red cell often form a narrow rim around them or fill the concavity of the crescent.

Concentration Methods for Malarial Parasites.— When parasites are scarce they may sometimes be found, although their structure is not well shown, by the Ross-Ruge thick-smear method. This consists in spreading a very thick layer of blood, drying, placing for a few minutes in a fluid containing 5 per cent. formalin and 1 per cent. acetic acid, which removes the hemoglobin and fixes the smear, rinsing, drying, and finally staining. Carbol-thionin is very useful for this purpose. If Wright's stain be used it is recommended that the preparation be subsequently stained for a half minute with borax-methylene-blue (borax, 5; methylene blue, 2; water, 100). Estivo-autumnal crescents may also be concentrated by the method given for filarial larvæ (p. 363). These older methods are, however, far inferior to the following new method of Bass and Johns, which takes advantage of the fact that parasitized red cells are lighter than the others and rise to the top of the sediment when the blood is centrifugalized at high speed. A centrifuge capable of 2500 revolutions per minute is required.

1. Draw 10 c.c. of blood from a vein (see p. 254) directly into a tube containing 0.2 c.c. of citrate-dextrose

solution. Mix well. The solution is made by dissolving 50 Gm. sodium citrate and 50 Gm. dextrose in 100 c.c. distilled water by the aid of heat.

2. Divide the blood between two centrifuge tubes and centrifugate at 2500 revolutions per minute for the proper length of time, which is determined by the radius of the

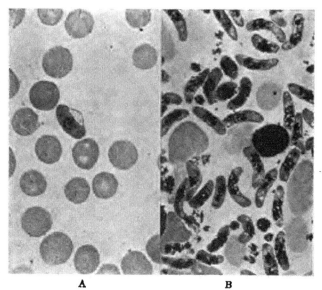

A   B

Fig. 138.—Estivo-autumnal malaria: effect of concentration by Bass and Johns's method. A, direct smear, averaging one crescent in eight fields; B, blood of same patient concentrated. From slides prepared by F. M. Johns. Wright's stain (× 1000).

centrifuge arm and the height of the column of blood in the tube. For a centrifuge whose radius is 18 cm. the proper time is one minute for each centimeter of the blood column. Too long centrifugation will cause the corpuscles to pack so tightly that the subsequent skimming is difficult; too little centrifugation will fail to bring the parasites to the top.

3. The leukocytes and all malarial parasites (except very young forms) will now be concentrated in a layer 1 mm. thick at the top of the sediment. With a capillary pipet (Fig. 228) skim off this layer and place it, together with a like amount of plasma, in a tube about 12 cm. in length and 0.5 cm. in inside diameter. If the column of fluid exceeds 5 cm., two tubes should be used. These tubes are readily made from ordinary glass tubing.

4. Mix thoroughly and centrifugate as before.

5. With a large capillary pipet skim off the top layer of the sediment in these tubes, taking up a column of cells and plasma not exceeding 5 cm. in height.

6. Mix by forcing in and out upon a slide, and then draw the mixture into the pipet away from the tip and seal the tip in a flame. Nick with a file and break off the capillary stem above the blood column.

7. Place this slender tube in the centrifuge and revolve as before. The leukocytes will form a grayish layer upon the surface of the sediment. This and the upper portion of the erythrocyte-layer contains the parasites.

8. Nick with a file and break off the capillary tube at a point 1 to 2 mm. below the bottom of the leukocyte layer.

9. With a capillary pipet whose stem will pass inside the capillary tube remove the small amount of red cells and leukocytes together with a little plasma.

10. Mix well, make smears on slides and stain with Wright's stain in the usual way.

The authors of this method claim that 90 per cent. of the parasites in 10 c.c. of blood can be collected upon one slide. While it is not to be expected that such remarkable concentration can be attained without considerable experience yet the method will yield good results at the first trial if the directions are carefully

ing and coiling movements, and continues for many hours or even days if the preparation be ringed with vaselin and kept in a cool place. If desired, stained smears of the blood may be prepared in the usual way or by the Ross-Ruge method (p. 359). When the micro-filariæ are scarce the following method is efficient:

Receive about 1 c.c. of blood from a puncture of the ear or finger into 5 c.c. of 2 per cent. acetic acid. Mix well and centrifugalize. Spread the sediment, which is not abundant, upon slides and examine in the moist state or after drying, fixing and staining. Hematoxylin is a good stain for the purpose.

The number of micro-filariæ in capillary blood is to be distinctly higher than in that obtained from a vein.

5. Larvæ of Trichinella spiralis.—The worm and its life history are described on page 508. In 1909 Herrick and Janeway demonstrated that diagnosis of trichiniasis can frequently be made by detection of the larvæ in the blood during their migration to the muscles. Of the examinations which have been reported since that time, about one-half have been positive. The earliest time at which the embryos were found was the sixth day after the onset of symptoms; the latest, the twenty-second day.

The approved method is the same as that given above for micro-filariæ except that 5 or 10 c.c. of blood from a vein (see p. 254) and a correspondingly larger quantity of acetic acid solution are required. The larvæ are not difficult to recognize. They are about 125 μ long and 6 μ broad.

XI. TESTS FOR RECOGNITION OF BLOOD

The recognition of red blood-corpuscles microscopically is the surest and simplest means of detecting the presence of blood. In most pathologic material, however, the corpuscles are too much disintegrated for recognition with the microscope, and one has to rely upon a test for hemoglobin or its derivatives. Of such tests, those given in this section are probably the best. Each is reliable within its own sphere, but each has its limitations. The guaiac, benzidin and similar tests are reliable only when *negative*. When, however, proper care is taken to exclude fallacies, they are the most useful and reliable tests for clinical purposes, although they could not be accepted medico-legally. The hemin test is reliable only when *positive*. The spectroscope offers perhaps the most simple and dependable means of identifying blood, but, except under favorable conditions, it is not adapted to the detection of traces. Its particular field lies in distinguishing between the various hemoglobin derivatives.

The only reliable test for human blood as distinguished from that of animals is the precipitin test described on page 611.

1. Guaiac Test.—The technic of this test has been given (see p. 181). It may be applied directly to a suspected fluid, but in order to avoid other substances which might cause the reaction the following procedure is advised: Remove fat if present (*e.g.*, in feces) by shaking with an equal volume of ether and discarding the ether. Add 3 or 4 c.c. of glacial acetic acid to about 10 c.c. of the fat-free fluid; shake thoroughly with an equal volume of ether; decant, and apply the test to the ether. Should the ether not separate well add a little alcohol and mix gently. It should then separate nicely. Jager states that the test is rendered much more sensitive if a few drops of ammonia or sodium hydroxid solution be added to the ether extract. In case of dried stains upon cloth, wood, etc., dissolve the stain in distilled water and test the water, or press a piece of moist blotting-paper against the stain, and touch the paper with drops of the guaiac and the turpentine successively. The test may be applied to microscopic particles by running the reagents under the cover-glass.

The **benzidin test** (see p. 182) is similar to the guaiac test and has the same fallacies, but is distinctly more sensitive.

2. Teichmann's Test.—This depends upon the production of characteristic crystals of *hemin*. It is not sufficiently delicate to detect the minute quantities of blood with which we frequently have to deal in the clinical laboratory, but, when positive, it is absolute proof of the presence of blood. A number of substances —lime, fine sand, iron rust—interfere with production of the crystals; hence negative results are not always conclusive. Dissolve the suspected stain in a few drops of normal salt solution upon a slide. If a liquid is to be tested, evaporate some of it upon a slide and dissolve the residue in a few drops of the salt solution. Let dry, apply a cover-glass, and run glacial acetic acid underneath it. Heat *very gently* until bubbles begin to form, replacing the acid as it evaporates. Allow to cool slowly. When cool, replace the acid with water, and examine for hemin crystals with 16-mm. and 4-mm.

objectives. The crystals are dark-brown rhombic plates, lying singly or in crosses, and easily recognized (Fig. 140). Failure to obtain them may be due to too

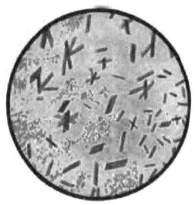

FIG. 140.—Teichmann's hemin crystals (Jakob).

much salt, too great heat, or too rapid cooling. If not obtained at first, let the slide stand in a warm place, as upon a hot-water radiator, for an hour.

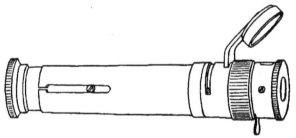

FIG. 141.—Small direct-vision spectroscope with side mirror. About natural size.

3. Spectroscopic Method.—Spectrum analysis de-. pends upon the fact that solutions of many substances, when held so as to intercept the light entering the

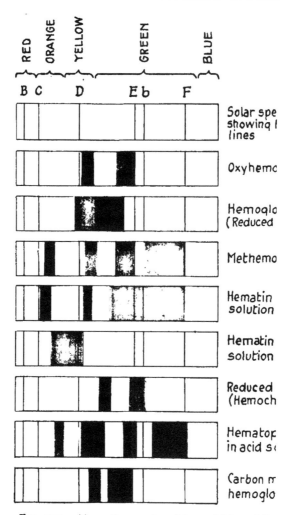

FIG. 142.—Absorption spectra of hemoglobin and its

spectroscope, will absorb certain colors, thus causing dark bands to appear at definite locations in the spectrum. A small direct-vision instrument meets all ordinary requirements and may be recommended as a useful addition to the regular laboratory equipment. The form with a side mirror and reflecting prism (Fig. 141) which gives two spectra side by side is most convenient. Before use, the width of the slit should be so adjusted and the eye-piece so focused that Fraunhofer's lines (Fig. 142, B, C, D, E, b, F) are clearly seen, since it is by means of these lines that the absorption bands are located. The examination is best made by daylight. With artificial light the Fraunhofer lines do not appear. The solution under examination may be held in a test-tube or small beaker. If a test-tube be used, only 1 to 3 c.c. will be required.

The **treatment of the suspected material** will depend upon its condition and the purpose of the examination:

1. When fresh blood is studied for oxyhemoglobin, methemoglobin, etc., a large drop from a skin puncture is received in 1 or 2 c.c. of water in a test-tube and cautiously diluted to the point where the bands become distinct. The optimum dilution is much less for methemoglobin than for oxyhemoglobin.

2. Urine and other fluids suspected to contain blood may be cleared by filtration and examined directly. When this proves unsatisfactory, as is often the case owing to persistent cloudiness or the presence of other pigments which darken the whole spectrum, the blood pigment in 200 to 500 c.c. of the unfiltered fluid should be extracted as follows: Add a little white of egg if the fluid is not already sufficiently albuminous, boil, acidify, centrifugalize, remove

supernatant fluid and treat the sediment as
the following paragraph.

3. Feces, gastric contents and other mat |
be treated with glacial acetic acid and extracte
as described under the guaiac test (p. 364). Bl |
is thus changed to acid hematin, which is tak
acidified ether, giving a clear solution suitable
scopic examination. If the ether does not |
blood-pigment well, a little more acetic aci(
added. In order that the solution may not l :
to show the bands, a less amount of ether th; |
mended for the guaiac test may be used, or the (
may be concentrated by evaporation.

When the result is in doubt the acid hem; :
transformed into the more easily identified hem |
as follows: Render the ethereal extract alkaline '
ammonia, cooling if necessary, mix well, and le(
the fluids separate. The ammonia will co)
hematin. By means of a pipet transfer it to a '
tube and add a few drops of fresh ammoniun
Stokes' reagent. The bands of hemochrom(|
appear at once.

4. Stains of blood dried on clothing, etc.,
dissolved in 1 or 2 c.c. of 10 per cent. caustic so :
heated to a point just short of boiling, cooled, :
with a few drops of ammonium sulphid or Stok|
The solution is then examined for the characte '
of hemochromogen.

5. In very old blood stains the hemoglobin ma;
transformed to the iron-free pigment hemat(
which is very resistent to solution. It will usua
in strong sulphuric acid. It has been advised
few small bits of the dry stain on a slide in a d
centrated sulphuric acid, to apply a cover a
bits of blood between slide and cover. Enou|

24

into solution to admit of spectroscopic examination. Particles of wood, cloth, or other organic material which might blacken the acid should be avoided.

The **characteristic absorption spectra** of the more important hemoglobin derivatives are as follows:

1. **Oxyhemoglobin** is present only in comparatively fresh blood. It gives two dark bands between the lines D and E. In concentrated solution these unite to form a single broad band. Upon addition of a few drops of fresh ammonium sulphid, or, much better, Stokes' reagent,[1] the spectrum changes to that of reduced hemoglobin.

2. **Hemoglobin** (also called *reduced hemoglobin*) gives a single broad band between D and E. By shaking with air it is changed to oxyhemoglobin whose bands in the same dilution are more distinct.

3. **Methemoglobin** occurs in the circulating blood under the conditions which have been described (see p. 260). It may also be found in the urine, in hemorrhagic cyst fluids, etc. In neutral or faintly acid solution, its most characteristic band is situated between the lines C and D. Two less distinct bands lie between D and E and a broad one beyond E; but these are usually not clearly seen. The blood must be diluted cautiously, as it is easy to pass the point where the characteristic band is most distinct. Upon addition of a few drops of fresh ammonium sulphid or Stokes' reagent, methemoglobin is changed to reduced hemoglobin with its single broad band. This will serve to distinguish it from acid hematin.

Methemoglobin can be prepared for purposes of compar-

[1] Stokes' reagent consists of ferrous sulphate, 2 Gm.; tartaric acid 3 Gm.; water, 100 c.c. When needed for use take a few cubic centimeters in a test-tube and add strong ammonia drop by drop until the precipitate which forms at first has entirely dissolved.

ison by diluting 2 drops of blood with 20 d
adding 1 or 2 drops of strong potassium ferricy
and shaking. The solution turns chocolate
may then be diluted until the characteristic ba

4. **Hematin** may be formed through the a
or alkalies as in gastric and intestinal blo
sometimes found in old extravasates, in tl
elsewhere. It is insoluble in water or weak
soluble in acidified ether and weak alkalies.

As seen from Fig. 142, the absorption ba
tin in *acid solution* ("acid hematin") are sir
of methemoglobin. That between C and D
inite; the others may not be clearly seen.
to methemoglobin the addition of ammol
or Stokes' reagent does not produce the sp
duced hemoglobin but rather (if the solut
sufficiently alkalinized to produce alkali herr
hemochromogen.

Hematin in *alkaline solution* ("alkali hema
rather indefinite broad band between C
presence may be confirmed by adding a i
ammonium sulphid or Stokes' reagent.
becomes brighter red in color, and the spectru
the more easily identified one of hemochromo:

5. **Hemochromogen,** also called *reduced a*
gives a narrow, very distinct band between D
not in too dilute solution, a fainter band betw.
This is one of the most definite and charac.
blood-pigment spectra.

6. **Hematoporphyrin** is an iron-free hemop:
tive which may occasionally be present
especially in sulphonal poisoning (see p. 184)
old dried blood. It does not respond to 1
hemin test. It is soluble in strong sulphu:
absorption spectrum is shown in Fig. 142.

For purposes of comparison it can be prepared by adding a drop of blood to 2 or 3 c.c. of concentrated sulphuric acid.

7. Carbon monoxid hemoglobin, which appears in the circulating blood in carbon monoxid poisoning, gives two bands very like those of oxyhemoglobin, but somewhat nearer the violet end of the spectrum. In contrast to oxyhemoglobin, addition of ammonium sulphid or Stokes' reagent leaves these bands unchanged. Owing to the small quantity usually present in poisoning the chemical test is preferable for its detection (see page 261).

XII. LESS FREQUENTLY USED METHODS

In this section brief consideration will be given a number of methods which are not as yet in common use, some because their clinical value has not been proved, others because the technic has not been sufficiently simplified.

1. Chemic Examination.—In routine clinical work chemic study of the blood has been limited to estimations of hemoglobin. The study of other substances has in the past interested the biochemist rather than the clinician. Within the past few years, however, methods have been so simplified and so many facts of clinical value have been gathered that certain chemic examinations are beginning to play an extremely important rôle in clinical medicine. Among the more useful of these are the Lewis and Benedict method for blood sugar; the picric acid method for creatinin; and Folin's new direct Nesslerization methods for urea, nonprotein nitrogen and total nitrogen. All of these are colorimetric methods and detailed directions for some

of them are given in the printed matter
panies the Hellige and the Kuttner colori

2. Vital Staining.—Upon the assi
ordinary staining of dried and fixed blo
the reactions of dead cells and does n
indicate the condition of the living blc
have been made to stain blood cells in th
The information yielded by this so-
staining" is not yet of much value.
chiefly with certain "reticulated" or "
corpuscles which contain a coarse skein
of filaments usually confined to the centr
cell. The filaments stain sharply with
Sometimes discrete granules are also pres
lation is thought to be a characteristic of
red corpuscles. Such cells constitute ab
per cent. of all the red cells in the bloc
adults; an increase may be regarded as a
blood regeneration. They are more abunc
hood. In anemia, particularly in pernic
the percentage is markedly increased.

The following method has proved sa
the writer's laboratory:

1. In a small test-tube (about 10 × 75 mm.
3 drops of the following staining solution whi
freshly mixed:

Saturated solution brilliant cresyl blue in 0.8
cent. salt solution.....................
Saturated solution of neutral potassium oxa

[1] These and other similar methods have recently been mc
the Denison Laboratory colorimeter and will be published so
and A. R. Peebles.

2. Prick the ear and allow one drop of blood to fall into the stain.

3. Mix gently and let stand ten to thirty minutes. A longer time will do no harm.

4. Remove small drops of the fluid, make smears on slides and dry in the air. Examine with an oil-immersion lens. Preparations begin to fade after a day or two.

In such preparations the leukocytes and platelets are colored shades of blue and the red corpuscles, pale greenish yellow. The skein or network in reticulated cells is blue and stands out distinctly (see Plate V).

3. Resistance of the Red Corpuscles.—Many agencies are capable of causing hemoglobin to pass from the red corpuscles into the surrounding medium— a phenomenon which is known as hemolysis and which has a wide interest. Hemolysis sometimes occurs in the circulating blood and may then be due, in part at least, to lowered resistance (or "increased fragility") upon the part of the red cells. The resistance of the red cells can be measured by subjecting them to the action of various agents. For clinical work salt solution is generally used.

Method.—1. Receive 1 or 2 c.c. of blood, preferably from a vein, directly into a graduated centrifuge tube containing about 2 c.c. of citrated salt solution (sodium chloride, 0.9 Gm.; sodium citrate, 0.5 Gm.; water, 100 c.c.). Mix gently.

2. Wash the corpuscles twice with 0.7 per cent. salt solution by centrifugalizing and pipeting off the supernatant fluid, the last time leaving a volume of fluid equal to the volume of corpuscles. Mix gently.

3. Arrange a series of 11 small test-tubes and place in each 1 c.c. of sodium chlorid solution varying in strength

from 0.2 per cent. in the first tube to 0.7 per c
Thus each tube will differ from the next by

4. To each tube add 0.1 c.c. of the
washed corpuscles and mix by inverting
Instead of using washed corpuscles some v
add 1 drop of blood from a skin puncture t

5. Let stand two hours at room tempera
end of that time the corpuscles will have
bottom and hemolysis may be recognized b
the supernatant fluid: faintly pink, if hemo
("initial hemolysis"); red, with little or n
it is complete.

With normal blood, hemolysis usually begi
containing 0.45 per cent. salt solution and i
that containing 0.35 per cent. In chronic
Hill found the figures for initial and compl
to be 0.60 and 0.40 respectively; in obstruc
0.40 and 0.225. In secondary and pernicio
figures vary only slightly from the norm
usually beginning somewhat earlier and endi
later than it does in normal blood.

4. Matching Bloods for Transfusion.

results which sometimes follow transfusion
now known to be due in many instances
or agglutination of the red corpuscles o
donor's or recipient's blood or both. By a
it is possible to ascertain whether the b
individual is suitable in this respect for
into the veins of a given patient. This
"matching bloods," and if possible it shoul
done when transfusion is contemplated
factors to be considered are hemolysis an
tion, but, since hemolysis does not occur wi

tination, it is sufficient in practice to test for agglutination only.   It is an interesting fact that in respect to
the presence or absence of iso-agglutinin every adult
falls into one of four groups which for convenience are
designated I, II, III, and IV.   To explain this grouping
it is necessary to assume the existence of two distinct
agglutinins which may be present alone or in combination or which may both be absent.   A given blood will
agglutinate (and often hemolyze) the red corpuscles of

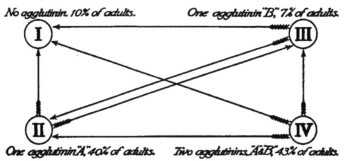

No agglutinin. 10% of adults.   One agglutinin "B," 7% of adults.

One agglutinin "A," 40% of adults.   Two agglutinins "A&B," 43% of adults.

Fig. 143.—Diagram showing the interrelation of the four iso-
agglutination groups of Moss.   The serum of any group will agglutinate the corpuscles of those groups toward which its arrows point.
Thus, serum of an individual belonging to Group IV will agglutinate
red corpuscles belonging to any other group, while the serum of
Group I lacks agglutinating power (after Sanford).

any blood which lacks agglutinin of the same kind and
will have no effect upon blood which contains the same
agglutinin.   There can therefore be neither agglutination nor hemolysis between members of the same
group.   The four groups and their interrelationship is
well shown in Fig. 143.   It is thought that the group
to which an individual belongs is an inherited characteristic which follows Mendel's law, although it does
not become fixed until after the period of infancy.

When transfusion is undertaken, the blo
secured from an individual belonging to th
as the patient. If such a donor can not
may easily happen if the patient belongs
the small groups, I or III, blood belongir
group may be used *provided that the serum
does not agglutinate the corpuscles of the don*

Moss' Method (*Minot's modification*).—1
following from each of the two persons whose
matched:

(*a*) *Red cell suspension.* Puncture finger (
a large drop of blood fall directly into a si
containing 1 c.c. of a 1.5 per cent. solution of s
in 0.9 per cent. salt solution. Mix gently b
few times.

(*b*) *Serum.* Obtain a few drops of blood ii
or Wright capsule (see Figs. 230 and 231).
coagulation has taken place, gently loosen the
wall of the tube. Let stand until serum has si
Separation of serum can be hastened by cer

2. Make thick vaselin rings on two slides.
1 drop each of the patient's serum and the
the donor's corpuscles; in the other mix 1 dro
patient's corpuscles and the donor's serum
may be transferred to the slide by means :
pipet (see Fig. 228) or a platinum loop. Cov
preparations with a cover-glass, and at the
five and ten minutes re-mix corpuscles and se
one edge of the cover.

If hollow-ground slides are at hand hangir
rations are preferable. The corpuscles and s
mixed at intervals by tilting the slide.

3. At intervals examine for agglutination :
cles with a low-power objective. When aggl

place the corpuscles gather into dense irregular clumps or large masses (Fig. 144). These are often so large as to be seen with the unaided eye as fine brick-red granules. Clumping is usually well marked within a few minutes but it is safe to allow half an hour.

The only important source of error is rouleau formation, which may or may not occur and which, although the clumps

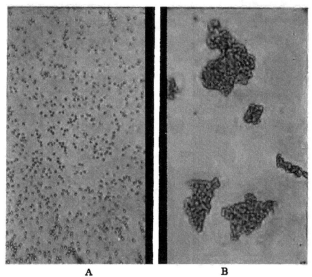

A   B

FIG. 144.—Matching bloods for transfusion. A, corpuscles of a patient with serum of a prospective donor; no agglutination. B, serum of patient with corpuscles of prospective donor; strong agglutination. The blood of the donor is therefore unsuited for use in this case (× 100).

are usually very small, might not be easy to differentiate without close observation with the 4 mm. objective. In the case of rouleau formation the corpuscles can be seen to lie in rows within the groups. Re-mixing of the cells and serum as above directed tends to break up rouleaux and to favor agglutination. If one feels uncertain of one's

interpretation, one should make control sli
patient's serum and corpuscles and the donor
corpuscles. Under these conditions agglutina
occur.

To Determine the Group to which an Indivi(
—As has been indicated above it is sufficient ir
to test the blood of a series of prospective
one is found which matches the patient's blo
pital work, upon the other hand, it will be
more satisfactory to determine in advance ·
of a number of individuals who may be will
as donors upon occasion. When an emer;
it is then only necessary to find the group
patient belongs in order to know at once the
donor, a procedure which does not require
half an hour.

The group to which any individual belo1
ascertained by testing his serum and corpuscle
corpuscles and serum of an individual known
Group II or III, using the simple method des(
Interpretation of results is made clear by Fig.
example, the unknown blood agglutinates Gr
and is not agglutinated by it, then the unknowr
to Group IV. The same end may be accc
testing the corpuscles of the unknown against
Groups II and III. Such sera, if kept sterile
active for months and may be kept on hand :
capsules (Fig. 231) the ends of which are to
the flame.

5. Viscosity.—It is evident that varia
viscosity of the blood must markedly influe
carried by the heart, but viscosity estin
proved of comparatively little value. The
would seem to be in suggesting need for tre:

high viscosity is throwing an excessive burden upon an already weakened heart.

Compared with distilled water, the normal viscosity is about 4.5. It is reduced in primary and secondary anemia (roughly proportional to the grade of anemia), nephritis, cardiac lesions with edema, and usually in leukemia and malaria. It is increased in polycythemia, diabetes mellitus, icterus, and usually in pneumonia. Measurement of viscosity is comparatively simple if one has a suitable instrument. The Hess instrument is one of the best and is accompanied by directions for use.

XIII. SPECIAL BLOOD PATHOLOGY

The more conspicuous characteristics of the blood in various diseases have been mentioned in previous sections. Although the great majority of blood changes are secondary, there are a few blood conditions in which the changes are so prominent, or the etiology so obscure, that they are commonly spoken of as blood diseases. These will receive brief consideration here. They fall into two general groups. In the one group (Anemia) the red cells and hemoglobin are chiefly affected; in the other (Leukemia) changes in the leukocytes constitute the conspicuous feature of the blood-picture.

A. ANEMIA

This is a deficiency of hemoglobin or red corpuscles, or both. It is either primary or secondary. The distinction is based chiefly upon etiology, although each type presents a more or less distinctive blood-picture. Secondary anemia is that which is symptomatic of some

other pathologic condition. Primary an
which progresses without apparent c
such a classification is unsatisfactory fror
standpoint, it has long been current and i
in practice.

1. Secondary Anemia.—The more im
ditions which produce secondary or symp
mia are:

(*a*) *Poor nutrition*, which usually acco
sanitary conditions, poor and insufficient f

(*b*) *Acute infectious diseases*, especially
and typhoid fever. The anemia is more
during convalescence.

(*c*) *Chronic Infectious Diseases.*—Tuberc
ilis, leprosy.

(*d*) *Chronic exhausting diseases*, as he
chronic nephritis, cirrhosis of the liver,
intestinal diseases, especially when asso
atrophy of gastric and duodenal glands. 'I
give an extreme anemia, indistinguishable
cious anemia.

(*e*) *Chronic poisoning*, as from lead, i
phosphorus.

(*f*) *Hemorrhage.*—Either repeated small l
(chronic hemorrhage), as from gastric canc
hemorrhoids, uterine fibroids, etc., or acute
such as may occur in typhoid fever, tul
traumatism.

(*g*) *Malignant Tumors.*—These affect the
through repeated small hemorrhages, pa
toxic products, and partly through inten
nutrition.

(*h*) *Animal Parasites.*—Some cause no appreciable change in the blood; others, like the hookworm and *Dibothriocephalus latus*, may produce a very severe anemia, almost identical with pernicious anemia. Anemia in these cases is probably due both to toxins and to abstraction of blood. In malaria the parasites themselves directly destroy the red cells.

The blood-picture varies with the grade of anemia. Diminution of hemoglobin is the most characteristic feature. In mild cases it is slight, and is the only blood change to be noted. In very severe cases hemoglobin may fall to 15 per cent. or even lower. Red corpuscles are diminished in all but very mild cases, while in the severest cases the red-corpuscle count is sometimes below 1,000,000. The color-index is usually decreased.

Although the number of leukocytes bears no relation to the anemia, leukocytosis is common, being due to the same cause.

Stained films show no changes in very mild cases. In moderate cases variations in size and shape of the red cells and polychromatophilia occur. Very severe cases show the same changes to greater degree, with addition of basophilic degeneration and the presence of normoblasts in small or moderate numbers. Megaloblasts in very small numbers have been encountered in certain severe cases. They are especially abundant and may even predominate over the normoblasts in dibothriocephalus infection and in the anemia of malignant disease when there are metastases in the bone-marrow. Blood-platelets are usually increased.

Posthemorrhagic Anemia.—Within a few hours after an acute hemorrhage the volume of blood is nearly or quite restored by means of fluids from Owing to the fact that some destruction of cles continues for a time, the anemia is r a few days after the hemorrhage. Hemogl cells are diminished according to the amo lost. The color-index is moderately low moderate leukocytosis. Some of the re show polychromatophilia and a few norm be found. In some cases great numbers of appear rather suddenly—a so-called **blood** c mal conditions may be restored within a although the color-index is apt to remain l time thereafter.

2. Primary Anemia.—The commonl varieties of primary anemia are pernici aplastic anemia and chlorosis, but splenic also be mentioned under this head.

(1) Progressive Pernicious Anemia.—It impossible to diagnose this disease from t amination alone. Severe secondary anemi that due to gastro-intestinal cancer, intestir and repeated small hemorrhages, sometin identical picture. Remissions, in which tl proaches the normal, are common. All data must, therefore, be considered, togeth mations of urobilin in the feces (see p. careful analysis of repeated blood examina

The disease is characterized by active d red corpuscles with excessive activity of blastic bone-marrow, and the appearance and abnormal red cells in the circulation.

Hemoglobin and red corpuscles are al

diminished. Several counts in which the red cells were below 150,000 have been recorded. In none of Cabot's 139 cases did the count exceed 2,500,000, the average being about 1,200,000. In more than two-thirds of the cases hemoglobin was reduced to less extent than the red corpuscles; the color-index was, therefore, high. A low color-index probably indicates a mild type of disease. The average hemoglobin value is about 20 to 25 per cent.

The leukocyte count may be normal, but is commonly diminished to about 3000, and is sometimes much lower. The decrease affects chiefly the polymorphonuclear cells, so that the lymphocytes are relatively increased. In some cases a decided absolute increase of lymphocytes occurs. Polymorphonuclear leukocytosis, when present, is due to some complication.

The red corpuscles show marked variation in size and shape (Plate VIII and Fig. 145). There is a decided tendency to large oval forms, and, despite the presence of microcytes, the average size of the corpuscles is generally strikingly increased. Polychromatophilia and basophilic degeneration are common. Nucleated red cells are always present, although *in many instances careful search is required to find them.* In the great majority of cases megaloblasts exceed normoblasts in number. This ratio is practically unknown in other diseases. Blood-platelets are diminished.

As far as the blood is concerned, the chief points to be considered in diagnosis are the high color-index and the presence of megaloblasts.

(2) **Aplastic Anemia.**—The rare and rapidly fatal anemia which has been described under this name was

PLATE VIII

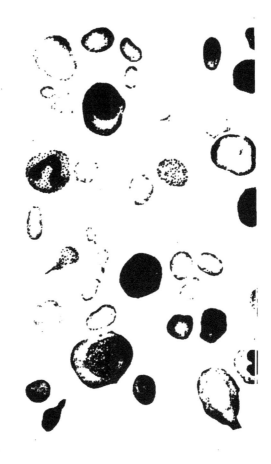

Blood-cells in pernicious anemia. Note variations i
of the red corpuscles; three megaloblasts, one with .
stained nucleus; red corpuscles showing grades o
philia, basophilic granular degeneration, and one nucl
irritation leukocyte, one lymphocyte, and one po
neutrophil. All drawn from actual cells on two slides.
× 800 (1 mm. = 1 micron).

once considered a variety of pernicious
absence of any attempt at blood regener
ing the marked difference in the blood-[
corpuscles and hemoglobin are rapidly c
an extreme degree. The color-index is n

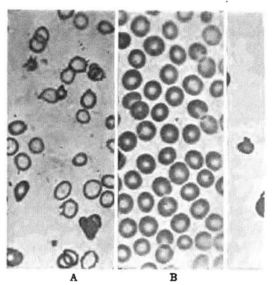

A   B

FIG. 145.—Red blood corpuscles in chlorosis (A)
anemiä (C) contrasted with those of normal blood :
marked case of chlorosis the red corpuscles are pale :.
pernicious anemia they are rich in hemoglobin and sh :
ations in size and shape. The megalocyte in the u|
figure is especially characteristic of pernicious an
stain. × 750.

The leukocyte count is normal or low, '
increase of lymphocytes. Stained smeai :
slight variations in size, shape, and staini: |
of the red cells. There are no megalobl .
or no normoblasts.

25

(3) **Chlorosis.**—This is probably a disease of defective blood formation. It is confined almost exclusively to unmarried girls. The clinical symptoms furnish the most important data for diagnosis. The blood resembles that of secondary anemia in many respects.

The most conspicuous feature is a *marked* decrease of hemoglobin, accompanied by a *slight* decrease in number of red corpuscles. The color-index is thus almost invariably low.

The following figures represent about the average for well-marked cases: hemoglobin, 40 per cent.; red corpuscles, 4,000,000; color-index, 0.5. Much lower figures are frequent; while, upon the other hand, mild cases may show no loss at all in number of red cells.

As in pernicious anemia, the leukocytes are normal or decreased in number, with a relative increase of lymphocytes.

In contrast to pernicious anemia (and in some degree also to secondary anemia), the red cells are of nearly uniform size, are pale (see Fig. 145), and their average diameter is somewhat less than normal. Changes in size, shape, and staining reactions occur only in severe cases. Erythroblasts are rarely present. The number of platelets is generally decreased.

(4) **Splenic Anemia.**—This is an obscure form of anemia associated with great enlargement of the spleen. It is probably a distinct entity, although several types may exist. There is decided decrease of hemoglobin and red corpuscles, with moderate leukopenia and relative lymphocytosis. Osler's 15 cases averaged 47 per cent. hemoglobin and 3,336,357 red cells. Stained films show notable irregularities in size, shape, and staining

properties only in advanced cases. Er: |
uncommon.

B. LEUKEMIA

Except in rare instances, diagnosis i
from the blood alone, usually at the firsı |
stained film. Two types of the disease
distinguished: the *myelogenous* and t
Atypical cases are not uncommon, especia
The disease is characterized by hypeı ı
leukoblastic bone-marrow (myelogenous l :
the lymphoid tissues (lymphatic leukeı
with overflow of many immature leuko
cessive numbers of normal types into t
blood. The more acute the process, the n
are the cells which appear in the blood.

1. Myelogenous Leukemia (Plate
usually a chronic disease, although acu ı
been described.

Hemoglobin and red corpuscles show dec
The red count is usually below 3,500,oc
hemoglobin estimation is difficult becausı
number of leukocytes. The color-index :
low.

Most striking is the immense increase
leukocytes. The count in ordinary cases v
100,000 and 400,000. Counts over 1,000,c
met. During spontaneous remissions, ı
ment with x-ray or benzol, and during
infections the leukocyte count may fal

While these enormous leukocyte count
in no other disease, and approached only

leukemia and extremely high grade leukocytosis, the
diagnosis, particularly during remissions, depends more
upon qualitative than quantitative changes. Although
all varieties are increased, the characteristic and conspicuous cell is the myelocyte. This cell never appears
in normal blood; extremely rarely in leukocytosis; and
never abundantly in lymphatic leukemia. In myelogenous leukemia myelocytes usually constitute more than
20 per cent. of all leukocytes. Da Costa's lowest case
gave 7 per cent. The neutrophilic form is generally
much more abundant than the eosinophilic. Both show
considerable variations in size. Myeloblasts may be
present in small numbers at any stage, and in the terminal stages they may be abundant. An increase in
their number is of grave significance. Very constant in
myelogenous leukemia is a marked absolute, and often
a relative, increase of eosinophiles and basophiles.
Polymorphonuclear neutrophiles and lymphocytes are
absolutely increased, although relatively decreased.

The red cells show the changes characteristic of a
severe secondary anemia, except that nucleated reds are
commonly abundant; in fact, no other disease gives so
many. They are chiefly of the normoblastic type.
Megaloblasts are uncommon. Blood-platelets are generally increased.

In **acute myelogenous leukemia** the myeloblast may
be the predominant cell, and the blood will then resemble that of acute lymphatic leukemia. The myeloblast can be distinguished from the large lymphocyte
by the oxydase reaction (see p. 342) although cases
occur in which the type of blood formation is so embryonic that the oxydase reaction fails. As a matter

PLATE IX

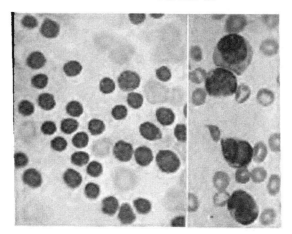

Fig. 1.—Blood in lymphatic leukemia; × 700.   On the le
of the disease; on the right, acute form (courtesy of Dr. \

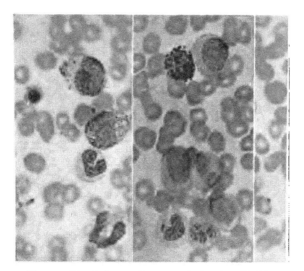

Fig. 2.—Blood in myelogenous leukemia.   Wright's
(photographs by the author).

of fact the test is rarely necessary for c
in most cases of acute myelogenous leuke1
number of myelocytes can be found to pu
right track.

2. Lymphatic Leukemia (Plate :
Form.—There is generally greater loss (
and red corpuscles than in myelogenous le
color-index is usually moderately low.

The leukocyte count is high, but low(
myelogenous type. Counts of 100,000 .
average, but in many cases are much l
low as 15,000. Some cases, on the othe1
high as 1,000,000. This high count is re
wholly to increase of lymphocytes. T
exceed 90 per cent. of the total number a
of the small variety. During remissi(
count may fall below normal, but the
lymphocytes remains high. Myelocytes

The red corpuscles show the changes u
secondary anemia. Erythroblasts are
dant. Blood-platelets are decreased.

Acute Form.—The blood is similar t(
chronic variety. The total leukocyte co
so high, and the large type of lymphocyte
in most cases. The anemia is apt to b(
and the normoblasts more abundant.

3. Anæmia Infantum Pseudoleuk
der this name von Jaksch described a r;
infancy, the proper classification of whicl
There is enlargement of liver and splee
times of lymph-nodes, together with
blood changes: grave anemia with d

degenerated red cells and many erythroblasts of both normoblastic and megaloblastic types; great increase in number of leukocytes (20,000 to 100,000) and great variations in size, shape, and staining of leukocytes, with many atypic forms, and a few myelocytes.

From the work of more recent investigators it appears probable that von Jaksch's anemia is not a distinct disease, and that the reported cases have been atypical forms of leukemia, pernicious anemia, or even secondary anemia with leukocytosis. As is well known, all of these conditions are apt to be atypical in children.

The table on the following page contrasts the distinctive blood-changes in the more common conditions.

DIFFERENTIAL DIAGNOSIS OF BLOOD DISEASES

	Hæmoglobin.	Red Corpuscle Count.	Color-index.	Leukocyte Count.	Red Corpuscles.	Stained Films. Leukocytes.
Secondary anemia	Diminished according to degree of anemia.	Normal in mild cases; diminished in all others.	Normal or slightly diminished.	Not necessarily affected; leukocytosis common.	Variations in size and shape in moderate cases; variations in staining reactions and normoblasts in severe cases.	Normal proportions or increase of polynuclears.
Pernicious anemia.	Diminished	Greatly diminished.	High.	Normal or diminished.	Marked variations in size, shape, and staining reactions. Average size increased. Tendency to large oval forms. Erythroblasts always present; megaloblasts exceed normoblasts.	Lymphocytes relatively, sometimes absolutely, increased.
Chlorosis.	Greatly diminished.	Slightly diminished.	Low.	Normal or diminished.	Nearly uniform size and shape; average size decreased; pale centers. Erythroblasts very rare.	Lymphocytes apt to be relatively increased.
Myelogenous leukemia.	Decidedly diminished.	Decidedly diminished.	Usually slightly diminished.	Extremely high.	Similar to secondary anemia, except normoblasts generally very numerous.	Large numbers of myelocytes (average, 20 per cent.) Absolute increase

CHAPTER IV

THE STOMACH

LABORATORY methods may be applied to the diagnosis of stomach disorders in: I. Examination of the gastric contents removed with the stomach-tube. II. Certain other examinations which give information as to the condition of the stomach.

I. EXAMINATION OF THE GASTRIC CONTENTS

Stomach digestion consists mainly in the action of pepsin upon proteins in the presence of hydrochloric acid and in the curdling of milk by rennin. The fat-splitting ferment, lipase, of the gastric juice has very little activity excepting upon previously emulsified fats such as those of milk and egg-yolk.

Pepsin and rennin are secreted by the gastric glands as zymogens—pepsinogen and renninogen respectively —which are converted into pepsin and rennin by hydrochloric acid. Hydrochloric acid is secreted chiefly by the fundus end of the stomach. It at once combines loosely with the proteins of the food, forming acid-metaprotein, the first step in protein digestion. Hydrochloric acid, which is thus loosely combined with proteins, is called "combined" hydrochloric acid. The acid which is secreted after the proteins present have all been converted into acid-metaprotein remains

as "free" hydrochloric acid, and, together
continues the process of digestion.

At the height of digestion the stomach-c
sist essentially of: (1) Water; (2) free hydro
(3) combined hydrochloric acid; (4) pepsin
(6) mineral salts, chiefly acid phosphates,
importance; (7) particles of undigested ar
gested food; (8) various products of diges
tion. In pathologic conditions there may
in addition, various microscopic structures
organic acids, of which lactic acid is most

A **routine examination** is conveniently can
following order:

1. Give the patient a test-meal upon an em
washing the stomach previously if necessary.

2. At the height of digestion, usually in one
the contents of the stomach with a stomach-

3. Measure and examine macroscopically.

4. Filter. A suction filter is desirable, and
sary when much mucus is present.

5. During filtration, examine microscopica
qualitative tests for—(a) free acids; (b) free
acid; (c) lactic acid.

6. When sufficient filtrate is obtained, mak
estimations of—(a) total acidity; (b) free hyd
(c) combined hydrochloric acid (if necessary)

7. Make whatever additional tests seem de
blood, pepsin, or rennin.

A. OBTAINING THE CONTENTS

Gastric juice is secreted continuously, b
sufficiently large for examination are o

tainable from the fasting stomach. In clinical work, therefore, it is desirable to stimulate secretion with food—which is the natural and most efficient stimulus—before attempting to collect the gastric fluid. Different foods stimulate secretion to different degrees, hence for the sake of uniform results certain standard "test-meals." have been adopted.

1. Test-meals.—It is customary to give the test-meal in the morning, since the stomach is most apt to be empty at that time. If it be suspected that the stomach will not be empty, it should be washed out with water the evening before.

(1) **Ewald's test-breakfast** consists of a roll (or two slices of bread), without butter, and two small cups (300 to 400 c.c.) of water, or weak tea, without cream or sugar. It should be well masticated. The contents of the stomach are to be removed one hour afterward, counting from the beginning, not the end, of the meal. This test-meal has long been used for routine examinations. Its disadvantage is that it introduces, with the bread, a variable amount of lactic acid and numerous yeast-cells. This source of error may be eliminated by substituting a shredded whole-wheat biscuit for the roll. The shredded wheat test-meal is now widely used and is probably the most satisfactory for general purposes.

(2) **Boas' test-breakfast** consists of a tablespoonful of rolled oats in a quart of water, boiled to one pint, with a pinch of salt added. It should be withdrawn in forty-five minutes to one hour. This meal does not contain lactic acid, and is usually given when the detection of lactic acid is important, as in suspected gastric cancer.

The stomach should always be washed wi·
evening previous.

(3) **Riegel's test-meal** consists of 400 c.c
a broiled beefsteak (about 150–200 Gm.), ;
of mashed potato. Since it tends to clog
must be thoroughly masticated.

(4) **Fischer's test-meal** is similar, but pi
erable. It consists of an Ewald breakfasi
¼ pound lean, finely chopped Hambı
broiled, and lightly seasoned. This and l
be removed in three to four hours. The)
what higher acidity values than the Ewald

2. Withdrawal of the Contents.—Th(
ach-tube, with bulb, is the form which is
used. It should be of rather large caliber,
opening in the tip and one or two in the s
tip. When not in use it may be kept iı
borax solution, and should be well washed
both before and after using.

It·is important confidently to assure the
introduction of the tube cannot possibly ha·
that, if he can control the spasm of his th
experience very little choking sensation.
tients are very nervous it is well to spraȷ
with cocain solution.

The tube should be dipped in warm wate
using; the use of glycerin or other lubrican
able. With the patient seated upon a cha
ing protected by towels or a large apron, ;
tilted forward, the tip of the tube, held as
pen, is introduced far back into the phar
then urged to swallow, and the tube is pı

into the esophagus until the ring upon it reaches the incisor teeth, thus indicating that the tip is in the stomach. If, now, the patient cough or strain as if at stool, the contents of the stomach will usually be forced out through the tube. Should it fail, the fluid can generally be pumped out by alternate compression of the tube and the bulb. If unsuccessful at first, the attempts should be repeated with the tube pushed a little further in, or withdrawn a few inches, since the distance to the stomach is not the same in all cases. The tube may become clogged with pieces of food, in which case it must be withdrawn, cleaned, and reintroduced. If, after all efforts, no fluid is obtained, another test-meal should be given and withdrawn after a somewhat shorter period, since, owing to excessive motility, the stomach may empty itself in less than the usual time.

Care must be exercised to prevent saliva running down the outside of the tube and mingling with the gastric juice in the basin.

As the tube is removed, it should be pinched between the fingers so as to save any fluid that may be in it.

The stomach-tube must be used with great care, or not at all, in cases of gastric ulcer, aneurysm, uncompensated heart disease, and marked arteriosclerosis. Except in gastric ulcer, the danger lies in the retching produced, and the tube can safely be used if the patient takes it easily.

The above description applies to the use of the type of stomach tube which for many years has been the standard. The procedure is made much easier both for physician and patient by use of the new Rehfuss stomach tube. This is a modification of Einhorn's

duodenal tube and consists of a slender
with perforated olive-shaped metal tip. W
is to be introduced, the metal tip is pushed v
the patient's pharynx and it is then easil
After it has reached the stomach the hea
to the most dependent portion and as
little of the stomach contents as is desi
drawn off by means of an aspirating sy
tube is especially useful for the fractiona

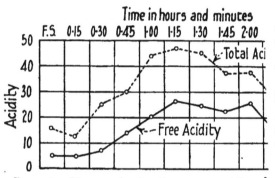

FIG. 146.—Diagram showing the average acidity of
of 24 healthy persons studied by Talbot by the fract
F.S., fasting stomach.

examination described below, as it may be l
for a long time without serious inconvenie
patient.

With the practical appreciation that there is
tion in the time at which the height of digestioɪ
a new method of examination known as the
method" has come into wide use. This is cɛ
follows:

1. Insert a Rehfuss stomach-tube before bː
empty the stomach as far as possible.

2. Remove the tube and give an Ewald test-breakfast, which must be chewed thoroughly.

3. Re-insert the tube and withdraw 5 c.c. or the stomach content at fifteen-minute intervals until the fluid is free from food particles or until the acidity has returned to the same level as was found in the fasting content. The tube is left in place during the whole procedure. Ordinarily it causes very little nausea.

4. Examine each of the 5-c.c. portions and also the fluid from the fasting stomach for total acidity, free hydrochloric acid and lactic acid.

By means of the Rehfuss tube a much larger quantity of gastric juice can often be obtained from the fasting stomach than was formerly believed possible. The quantity is very variable, ranging from 5 c.c. to 150 c.c. or even more, and averaging about 45 c.c. The acidity values are also variable. Averages for the fasting content and each of the fifteen-minute periods are shown in Fig. 146.

B. Physical Examination

Under normal conditions 50 to 100 c.c. of fluid can be obtained one hour after administering Ewald's breakfast. Larger amounts point to motor insufficiency or hypersecretion; less than 20 c.c., to too rapid emptying of the stomach, or else to incomplete removal. Upon standing, it separates into two layers: the lower consisting of particles of food; the upper, of an almost clear, faintly yellow fluid. The extent to which digestion has taken place can be roughly judged from the appearance of the food-particles.

The **reaction** is frankly acid in health and in nearly all pathologic conditions. It may be neutral or slightly alkaline in some cases of gastric cancer and marked

chronic gastritis, or when contaminated by
able amount of saliva.

A small amount of **mucus** is presen
Large amounts, when the gastric contents a
with the tube and not vomited, point to
tritis. Mucus is recognized from its cl
slimy appearance when the fluid is poure
vessel into another. It is more frequen
stomach washings than in the fluid remo
test-meal.

A trace of **bile** is common as a result
straining while the tube is in the stoma
amounts are very rarely found, and general
obstruction in the duodenum. Bile produce
ish or more frequently greenish discolorat
fluid.

Blood is often recognized by simple insp
more frequently requires a chemic test. It is
when very fresh, and dark, resembling coffe
when older. Vomiting of blood, or *hemateme.*
mistaken for pulmonary hemorrhage, or *hemo*
the former the fluid is acid in reaction and us
red or brown in color and clotted, while in l
it is brighter red, frothy, alkaline, and usu;
with a variable amount of mucus. When
is small in amount and bright red the possil
it originates from injury done by the tube n
overlooked.

Particles of food eaten hours or even days
may be found, and indicate deficient motor

Search should always be made for **bits** of t
the gastric mucous membrane or new growth

when examined by a pathologist, will sometimes render the diagnosis clear.

C. Chemic Examination

A routine chemic examination of the gastric contents involves qualitative tests for free acids, free hydrochloric acid, and organic acids, and quantitative estimations of total acidity, free hydrochloric acid, and sometimes combined hydrochloric acid. Other tests are applied when indicated.

1. Qualitative Tests.—(1) **Free Acids.**—The presence or absence of free acids, without reference to the kind, is easily determined by means of Congo-red, although the test is not much used in practice.

Congo-red Test.—Take a few drops of a strong alcoholic solution of Congo-red in a test-tube, dilute with water to a strong red color, and add a few cubic centimeters of filtered gastric juice. The appearance of a *blue color* shows the presence of some free acid (Plate X, B, B'). Since the test is more sensitive to mineral than to organic acids, a marked reaction points to the presence of free hydrochloric acid.

Thick filter-paper soaked in Congo-red solution, dried, and cut into strips may be used, but the test is much less delicate when thus applied.

(2) **Free Hydrochloric Acid.**—In addition to its digestive function, free hydrochloric acid is an efficient antiseptic. It prevents or retards fermentation and lactic-acid formation, and is an important means of protection against the entrance of pathogenic organisms into the body. It is never absent in health.

Dimethylamidoazobenzol Test.—To a little of the filtered gastric juice in a test-tube, or to several drops in a

PLATE X

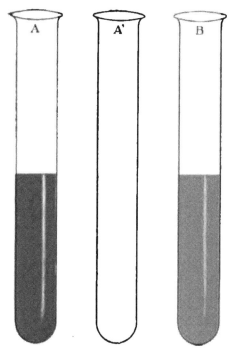

A, Uffelmann's reagent; A', A after the addition
containing lactic acid; B, water to which three drop
solution have been added; B', change induced in B wł
containing free hydrochloric acid is added (Boston).

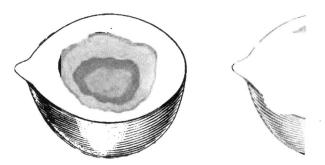

1, Resorcin-test for free hydrochloric acid; 2, Günzburg'
chloric acid (Boston).

porcelain dish, add a drop of o.5 per cent. alc
of dimethylamidoazobenzol. In the presen
drochloric acid there will at once appear a c
varying in intensity with the amount of acid
This test is very delicate; but, unfortunately,
when present in large amounts (above o.5 pe:
similar reaction. The color obtained with o
however, more of an orange red.

Boas' Test.—This test is less delicate than
but is more reliable, since it reacts only to fre
acid. It is probably the best routine test.

In a porcelain dish mix a few drops of the ga
the reagent, and slowly evaporate to dryness
taking care not to scorch. The appearance of a
which fades upon cooling, shows the presence
chloric acid (Plate X, 1).

Boas' reagent consists of 5 Gm. resublimed r
3 Gm. cane-sugar, in 100 c.c. alcohol. The :
well, which, from the practitioner's view-pc
preferable to Günzburg's phloroglucin-va:
(phloroglucin, 2 Gm.; vanillin, 1 Gm.; abs:
30 c.c.). The latter is just as delicate, is :
same way, and gives a sharper reaction (Pl:
is unstable.

(3) **Organic Acids.**—Lactic acid is the n
and is taken as the type of the organic
appear in the stomach-contents. It is :.
bacterial activity. Acetic and butyric ac:
times present. Their formation is close
with that of lactic acid, and they are rare l
When abundant, they may be recognized :
upon heating. Butyric acid gives the o(.
butter.

26

Lactic acid is never present at the height of digestion in health. Although usually present early in digestion, it disappears when free hydrochloric acid begins to appear. Small amounts may be introduced with the food. Pathologically, small amounts may be present whenever there is stagnation of the gastric contents with deficient hydrochloric acid, as in many cases of dilatation of the stomach and chronic gastritis. The presence of notable amounts of lactic acid (more than 0.1 per cent. by Strauss' test) is strongly suggestive of gastric cancer, and is probably the most valuable single symptom of the disease.

As already stated, the Ewald test-breakfast introduces a small amount of lactic acid, but rarely enough to respond to the tests given here. In every case, however, in which its detection is important, the shredded-wheat biscuit or Boas' test-breakfast should be given, the stomach having been thoroughly washed the evening before.

Uffelmann's Test for Lactic Acid.—Thoroughly shake up 5 c.c. of filtered stomach fluid with 50 c.c. of ether for at least ten minutes. Collect the ether and evaporate. Dissolve the residue in 5 c.c. of water and test with Uffelmann's reagent as follows:

In a test-tube mix 3 drops concentrated solution of phenol and 3 drops saturated aqueous solution of ferric chlorid. Add water until the mixture assumes an amethyst-blue color. To this add the solution to be tested. The appearance of a *canary-yellow color* indicates the presence of lactic acid (Plate X, A, A').

Uffelmann's test may be applied directly to the stomach-contents without extracting with ether, but is then neither

sensitive nor reliable, because of the phospha
other interfering substances which may be p

Kelling's Test (*Simon's Modification*).—
more satisfactory than Uffelmann's. To
distilled water add sufficient ferric chlorid sol
faint yellowish tinge. Pour half of this into
tube to serve as a control. To the other add a
of the gastric juice. Lactic acid gives a
distinct yellow color which is readily recognized by comparison with the control.

Strauss' Test for Lactic Acid.—This is
a good test for clinical work, since it gives
a rough idea of the quantity present and
is not sufficiently sensitive to respond to
the traces of lactic acid which some test-
meals introduce. Strauss' instrument
(Fig. 147) is essentially a separatory
funnel with a mark at 5 c.c. and one at
25 c.c. Fill to the 5-c.c. mark with
filtered stomach fluid, and to the 25-c.c.
mark with ether. Shake thoroughly for
ten or fifteen minutes, let stand until the
ether separates, and then, by opening the
stop-cock, allow the gastric juice to run
out. Fill to the 25-c.c. mark with water,
and add 2 drops of a 10 per cent. solution
of ferric chlorid. Shake gently. If 0.1
per cent. or more lactic acid be present,
the water will assume a strong greenish-ye
slight tinge will appear with 0.05 per cent.

(4) **Pepsin and Pepsinogen.**—Pepsino
no digestive power. It is secreted by the
and is transformed into pepsin by the ac
acid. Although pepsin digests proteins

presence of free hydrochloric acid, it has a slight digestive activity in the presence of organic or combined hydrochloric acids.

The amount is not influenced by neuroses or circulatory disturbances. Absence or marked diminution, therefore, indicates organic disease of the stomach. This is an important point in diagnosis between functional and organic conditions. Pepsin is rarely or never absent in the presence of free hydrochloric acid.

Test for Pepsin and Pepsinogen.—With a cork-borer cut small cylinders from the coagulated white of an egg, and cut these into disks of uniform size. The egg should be cooked very slowly, preferably over a water-bath, so that the white may be readily digestible. The disks may be preserved in glycerin, but must be washed in water before using.

Place a disk in each of three test-tubes.

Into tube No. 1 put 10 c.c. distilled water, 5 gr. pepsin, U. S. P., and 3 drops of the official dilute hydrochloric acid.

Into tube No. 2 put 10 c.c. filtered gastric juice.

Into tube No. 3 put 10 c.c. filtered gastric juice and 3 drops dilute hydrochloric acid.

Place the tubes in an incubator or in warm water for three hours or longer. At intervals observe the extent to which the egg-albumen has been digested. This is recognized by the depth to which the disk has become translucent.

Tube No. 1 is used for comparison, and should show the effect of normal gastric juice.

Digestion of the egg in tube No. 2 indicates the presence of both pepsin and free hydrochloric acid.

When digestion fails in tube No. 2 and occurs in No. 3, pepsinogen is present, having been transformed into pepsin by the hydrochloric acid added. Should digestion fail in this tube, both pepsin and pepsinogen are absent.

(5) **Rennin and Renninogen.**—Renn
curdling ferment of the gastric juice. It i
renninogen through the action of hydr
Deficiency of rennin has the same s
deficiency of pepsin, and is more easil

Test for Rennin.—Neutralize 5 c.c. filter
with very dilute sodium hydroxid solution; ;
milk, and place in an incubator or in a ves
about 40°C. Coagulation of the milk in
minutes shows a normal amount of rennin. :
lation denotes a less amount.

(6) **Peptid-splitting Enzyme.**—It has b
in cancer of the stomach there may be pr
logic ferment which is capable of splittin;
amino-acids. No such ferment is pres
the gastric juice being incapable of carr
to the amino-acid stage. Neubauer and
utilized this fact for the diagnosis of ;
by subjecting the dipeptid, glycyl-trypt
action of the gastric fluid and testing fo
of the amino-acid tryptophan. The i
follows:

Place 10 c.c. of the filtered gastric juice a
of glycyl-tryptophan in a test-tube, overlay
prevent bacterial action, and place in an incu
38°C. At the end of twenty-four hours pipet
centimeters and test for tryptophan as follow
a few drops of 3 per cent. acetic acid, add a ve
vapor with a medicine-dropper, and shake.
ance of a rose-red color shows the presence
and hence of the peptid-splitting fermer
quickly disappears if too much bromin is

color appears at first, add more bromin vapor in small quantities. Only when the fluid has become yellow from excess of bromin can the test be considered negative.

Before applying this method, the stomach fluid must be tested for pre-existing tryptophan, blood (see p. 407), and bile (see p. 179). Blood and pancreatic juice each contain peptid-splitting ferments, and pancreatic juice may be assumed to be present if bile is detected.

Glycyl-tryptophan can be purchased in bottles, each containing a little toluol and the correct amount of the di-peptid for one test. The gastric juice is introduced into the bottle to the level of a mark on its side and then incubated. Such an outfit is called a "ferment diagnosticum."

Instead of glycyl-tryptophan, Jacque and Woodyatt and others have used 20 c.c. sterilized filtered 2 per cent. solution of Witte's peptone for each 10 c.c. of stomach fluid. They then estimate amino-acids in 10 c.c. of the mixture before incubating and in 10 c.c. afterward, using the formalin method which is given for ammonia in urine (see p. 147). The difference between the two estimations expresses the degree of peptolysis.

The value of the test is impaired by Warfield's discovery of peptid-splitting ferments in the saliva. Later workers have shown that much, at least, of the peptolytic activity of the saliva is due to ferments of leukocytes and bacteria, which are capable of splitting proteins as well as peptids. The chief source of error, however, appears to be regurgitated trypsin, which may be present in the absence of bile. To exclude these sources of error Friedman and Hamburger propose a control test for proteolytic ferments, using edestin as substrate. If the edestin test is positive, the glycyl-tryptophan test cannot be relied upon. It is performed as follows:

Edestin Test.—The gastric juice is filter
with normal Na_2CO_3 solution, using pher
indicator, and then brought to an alkalinit
Na_2CO_3, in order to inactivate pepsin. Pl
0.1 per cent. solution of edestin[1] in 0.1 per c
each of four test-tubes. To three tubes ad
and 0.5 c.c. of the faintly alkalinized gastric
the fourth tube as a control and adding to it
phenolphthalein solution. Place the four tu
bator at 37°C. At the end of four hours exa
the contents of each of the tubes with 5 per c
When the neutral point is reached all the unc
will be precipitated. The degree of digestion
the amount of turbidity compared with tha
tube. Absence of turbidity denotes comple

(7) **Blood** is present in the vomitus in a
of conditions. When found in the fluid
a test-meal, it commonly points toward
cinoma. Blood can be detected in near
the cases of gastric cancer. The presence
blood and blood from injury done by the
must be excluded.

Test for Blood in Stomach-contents.—Ext
to remove fat. To 10 c.c. of the fat-free
4 c.c. of glacial acetic acid and shake the
oughly with about 5 c.c. of ether. Separate
use half of it for the guaiac or benzidin tes
In the case of a positive reaction the rei
ether-extract may be examined spectros
treating so as to develop the bands of h
(see pp. 369, 371).

[1] Edestin is a protein extracted from hemp
purchased from Eimer and Amend, New York.

When brown particles are present in the fluid, the hemin test should be applied directly to them.

2. Quantitative Tests.—(1) **Total Acidity.**—The acid-reacting substances which contribute to the total acidity are free hydrochloric acid, combined hydrochloric acid, acid salts, mostly phosphates, and, in some pathologic conditions, the organic acids. The total acidity is normally about 50 to 75 *degrees* (see method below), or, when estimated as hydrochloric acid, about 0.2 to 0.3 *per cent*. With Riegel's or Fischer's test-meal the figures are a little higher.

Töpfer's Method for Total Acidity.—In an evaporating dish or small beaker (an "after-dinner" coffee-cup is a very convenient substitute) take 10 c.c. filtered stomach-contents and add 3 or 4 drops of the indicator, a 1 per cent. alcoholic solution of phenolphthalein. When the quantity of stomach fluid is small, 5 c.c. may be used, but results are less accurate than with a larger amount. Add decinormal solution of sodium hydroxid drop by drop from a buret, until the fluid assumes a rose-red color which does not become deeper upon addition of another drop (Plate XI, A, A'). In ordinary titrations the end-point is the appearance of the first permanent pink, but owing to interaction of phosphates it is advised (Wood) to carry the titration of gastric juice a little farther, as here indicated. When this point is reached, all the acid has been neutralized. The end reaction will be sharper if the fluid be saturated with sodium chlorid. A sheet of white paper beneath the beaker facilitates recognition of the color change.

In clinical work the amount of acidity is expressed by the number of cubic centimeters of the decinormal sodium hydroxid solution which would be required to neutralize 100 c.c. of the gastric juice, each cubic centimeter representing

one *degree* of acidity. Hence, multiply the r
centimeters of decinormal solution require
the 10 c.c. of stomach fluid by 10. This give
degrees of acidity. The amount may be exp
of hydrochloric acid, if one remembers that
equivalent to 0.00365 per cent. hydrochlor
one suggests that this is the number of days i
last figure, 5, indicating the number of decir

Example.—Suppose that 7 c.c. of decin
were required to bring about the end reactior
tric juice; then $7 \times 10 = 70$ *degrees* of aci
pressed in terms of hydrochloric acid, $70 \times 0.$
per cent.

Preparation of decinormal solutions is des
books on chemistry. The practitioner will
have them made by a chemist, or to purchase
supply house. Preparation of an approximat
solution is described on page 652.

(2) **Hydrochloric Acid.**—After the Ewa
test-breakfasts the amount of **free hydr**
varies normally between 25 and 50 degrees,
to 0.2 per cent. In disease it may go
higher or may be absent altogether.

When the amount of free hydrochloric a
organic disease of the stomach probably d

Increase of free hydrochloric acid abov
(*hyperchlorhydria*) generally indicates a n
also occurs in most cases of gastric ulcer ar
chronic gastritis.

Decrease of free hydrochloric acid belo
(*hypochlorhydria*) occurs in some neuroses,
tritis, early carcinoma, pellagra, and mo
associated with general systemic depressi

variation in the amount at successive examinations strongly suggests a neurosis. Too low values are often obtained at the first examination, the patient's dread of the introduction of the tube probably inhibiting secretion.

Absence of free hydrochloric acid (*achlorhydria*) occurs in most cases of gastric cancer and far-advanced chronic gastritis, in many cases of pernicious anemia and pellagra, and sometimes in hysteria and pulmonary tuberculosis.

The presence of free hydrochloric acid presupposes a normal amount of **combined hydrochloric acid,** hence the combined need not be estimated when the free acid has been found. When, however, free hydrochloric acid is absent, it is important to know whether any acid is secreted, and an estimation of the combined acid then becomes of great value. The normal average after an Ewald breakfast is about 10 to 15 degrees, the quantity depending upon the amount of protein in the test-meal. Somewhat higher figures are obtained after a Riegel or Fischer test-meal.

Töpfer's Method for Free Hydrochloric Acid.—In a beaker take 10 c.c. filtered stomach fluid and add 4 drops of the indicator, a 0.5 per cent. alcoholic solution of di-methyl-amido-azobenzol. A red color instantly appears if free hydrochloric acid be present. Add decinormal sodium hydroxid solution, drop by drop from a buret, until the last trace of red just disappears, and a canary-yellow color takes its place (Plate XI, C, C'). Read off the number of cubic centimeters of decinormal solution added, and calculate the degrees, or percentage of free hydrochloric acid, as in Töpfer's method for total acidity.

When it is impossible to obtain sufficient fluid for all the

PLATE XI

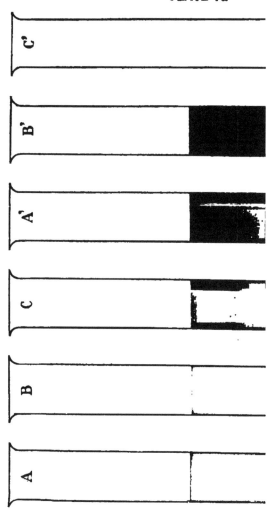

A, Gastric fluid to which a 1 per cent. solution
has been added; B, gastric fluid to which a 1 per cen
has been added; C, gastric fluid to which a 0.5 p
dimethylamido-azobenzol has been added; A', A a
decinormal solution of sodium hydroxid; B', B a
decinormal solution of sodium hydroxid; C', C a
decinormal solution of sodium hydroxid (Boston).

tests, it will be found convenient to estimate t
chloric acid and total acidity in the same por :
is frequently adopted as a routine regar(:
amount of fluid available. After finding th
chloric acid as just described, add 4 drops ph(ı
solution, and continue the titration. The am .
normal solution used in both titrations indic; ı
acidity.

Töpfer's Method for Combined Hydrochlo: :
a beaker take 10 c.c. filtered gastric juice and .
of the indicator, a 1 per cent. aqueous solutic
alizarin sulphonate. Titrate with decinormal ·
droxid until the appearance of a bluish-violet :
does not become deeper upon addition of a ·
(Plate XI, B, B'). It is difficult, without ·
determine when the right color has been :
reddish violet appears first. The shade which
end reaction can be produced by adding 2 or 3 .
indicator to 5 c.c. of 1 per cent. sodium carbon; ı

Calculate the number of cubic centimeters of
solution which would be required for 100 c.c.
fluid. This gives, in degrees, *all the acidit)*
combined hydrochloric acid. The combined l
acid is then found by deducting this amount fr(
acidity, which has been previously determined.

Example.—Suppose that 5 c.c. of decinormal s(
required to produce the purple color in 10 c.c. g
then $5 \times 10 = 50 = $ *all the acidity except coml*
chloric acid. Suppose, now, that the total aci
ready been found to be 70 degrees; then 70
degrees of combined hydrochloric acid; and 20)
0.073 *per cent.*

When free hydrochloric acid is absent, it
more helpful to estimate the **acid deficit** th;

bined hydrochloric acid. The acid deficit shows how far the acid secreted by the stomach falls short of saturating the protein (and bases) of the meal. It represents the amount of hydrochloric acid which must be added to the fluid before a test for free hydrochloric acid can be obtained. It is determined by titrating with $\frac{n}{10}$ hydrochloric acid, using dimethyl-amido-azo-benzol as indicator, until the fluid assumes a red color. The amount of deficit is expressed by the number of cubic centimeters of the decinormal solution required for 100 c.c. of the stomach fluid.

(3) **Organic Acids.**—There is no simple direct quantitative method. After the total acidity has been determined, organic acids may be removed from another portion of the gastric filtrate by shaking thoroughly with an equal volume of neutral ether, allowing the fluids to separate, and repeating this process until the gastric fluid has been extracted with eight or ten times its volume of ether. The total acidity is then determined, and the difference between the two determinations indicates the amount of organic acids.

(4) **Pepsin.**—No direct method is available. The following are sufficient for clinical purposes:

1. **Hammerschlag's Method.**—To the white of an egg add twelve times its volume of 0.4 per cent. hydrochloric acid (dilute hydrochloric acid, U. S. P., 4 c.c.; water, 96 c.c.), mix well, and filter. This gives a 1 per cent. egg-albumen solution. Take 10 c.c. of this solution in each of three tubes or beakers. To A add 5 c.c. gastric juice; to B, 5 c.c. water with 0.5 Gm. pepsin; to C, 5 c.c. water only. Place in an incubator for an hour and then determine the amount of

albumin in each mixture by Esbach's me
shows the amount of albumin in the test
difference between C and B indicates the amc
which would be digested by normal gastı
difference between C and A gives the alb
digested by the fluid under examinatior
shown that the amounts of pepsin in two
portionate to the squares of the product:
Thus, if the amounts of albumin digested in
are to each other as 2 is to 4, the amounts c
each other as 4 is to 16.

Certain sources of error can be eliminated
gastric juice several times before testing.
portant of these are that the law of Schütz h
for comparatively dilute solutions, and tha
of peptic activity inhibit digestion.

2. **Mett's method** is generally preferred to
Put three or four Mett's tubes about 2 cm. lo
beaker with diluted gastric juice (1 c.c. of tl
15 c.c. twentieth-normal hydrochloric acid).
incubator for twenty-four hours, and the:
accurately as possible in millimeters the ·
has been digested, using a millimeter scalı
lens or, better, a low power of the microsco:
piece micrometer. Square the average lı
column (law of Schütz) and multiply by
dilution, 16. The maximum figure obtain:
is 256, representing a digested column of 4 1·

Prepare Mett's tubes as follows:

Beat up slightly the whites of one or two ·
Pour into a wide test-tube and stand in th
capillary glass tubes, 1 to 2 mm. in diametı
tubes are filled, plug their ends with brea:
coagulate the albumin by heating in wate:
boiling. Dip the ends of the tubes in melt:

preserve until needed. Bubbles, if present, will probably disappear in a few days. When wanted for use, cut the tubes into lengths of about 2 cm. Discard any in which the albumin has separated from the wall.

D. MICROSCOPIC EXAMINATION

A drop of unfiltered stomach-contents is placed upon a slide, covered with a cover-glass, and examined with the 16-mm. and 4-mm. objectives. A drop of Lugol's solution allowed to run under the cover will aid in distinguishing the various structures. As a rule, the microscopic examination is of little value.

Under normal conditions little is to be seen except great numbers of starch-granules, with an occasional epithelial cell, yeast-cell, or bacterium. Starch-granules are recognized by their concentric striations and the fact that they stain blue with iodin solutions when undigested, and reddish, due to erythrodextrin, when partially digested.

Pathologically, remnants of food from previous meals, red blood-corpuscles, pus-cells, sarcinæ, and excessive numbers of yeast-cells and bacteria may be encountered (Fig. 148).

Remnants of food from previous meals indicate deficient gastric motility.

Red Blood-corpuscles.—Blood is best recognized by the chemic tests already given. The corpuscles sometimes retain a fairly normal appearance, but are generally so degenerated that only granular pigment is left. When only a few fresh-looking corpuscles are present, they usually come from irritation of the mucous membrane by the tube.

Pus-cells.—Pus is rarely encountered removed after a test-meal. Considerabl pus-corpuscles have been found in some c cancer. The corpuscles are usually part so that only the nuclei are seen. Swall must always be considered.

Sarcinæ.—These are small spheres arra groups, often compared to bales of cotto

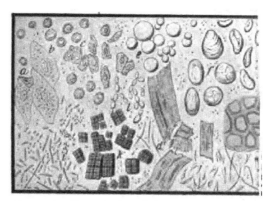

FIG. 148.—General view of the gastric contents: *a,* thelial cells from esophagus and mouth; *b,* leukocytes thelial cells; *d,* muscle-fibers; *e,* fat-droplets and fat-c: granules; *g,* chlorophyl-containing vegetable matte spirals; *i,* bacteria; *k,* sarcinæ; *l,* yeast cells (Jakob).

quently form large clumps and are easil They stain brown with iodin solution. The mentation. Their presence is evidence existence of gastric cancer, in which diseas occur.

Yeast-cells.—As already stated, a fe may be found under normal conditions. of considerable numbers is evidence of r fermentation. Their appearance has been described (see p. 239). They stain brown with iodin solution.

Bacteria.—Numerous bacteria may be encountered, especially in the absence of free hydrochloric acid. The *Boas-Oppler bacillus* is the only one of special significance. It occurs in the majority of cases of cancer, and is rarely found in other conditions. Carcinoma probably furnishes a favorable medium for its growth.

These bacilli (Fig. 149) are large (5 to 10 μ long), non-motile, and usually arranged in clumps or end to

FIG. 149.—Boas-Oppler bacilli from case of gastric cancer (Boston).

end in zig-zag chains. They stain brown with iodin solution, which distinguishes them from *Leptothrix buccalis* (see p. 535), which is not infrequently swallowed, and hence found in stomach fluid. They also stain by Gram's method. They are easily seen with the 4-mm. objective in unstained preparations, but are best recognized with the oil lens, after drying some of the fluid upon a cover-glass, fixing, and staining with a simple bacterial stain or by Gram's method.

A few large non-motile bacilli are frequently seen;

they cannot be called Boas-Oppler
they are numerous and show something
arrangement.

E. THE GASTRIC CONTENTS IN DISI

In the diagnosis of stomach disorders th
must be cautioned against relying too mucl
nations of the stomach-contents. A first
is especially unreliable. Even when repea
tions are made, the laboratory findings w
considered apart from the clinical signs.

The more characteristic findings in cert
are suggested here:

1. Dilatation of the Stomach.—Evi
tention and fermentation are the chief c
of this condition. Hydrochloric acid i
diminished. Pepsin may be normal or sl
ished. Lactic acid may be detected in sn
but is usually absent when the stomach has
before giving the test-meal. Both motil
sorptive power are deficient. The mici
monly shows sarcinæ, bacteria, and grea
yeast-cells. Remnants of food from previo
be detected with the naked eye or micros

2. Gastric Neuroses.—The findings
Successive examinations may show norm
or diminished hydrochloric acid, or even e
of the free acid. Pepsin is usually norma

The presence of more than 100 c.c. of
in the fasting stomach has until lately l
indicate a neurosis characterized by cont
secretion (gastrosuccorrhea), but recer

the fasting contents with the Rehfus tube throw some doubt upon the condition. When the fluid contains food-particles, it is the result of retention, not hypersecretion.

3. Chronic Gastritis.—Free hydrochloric acid may be increased in early cases. It is generally diminished in well-marked cases, and is often absent in advanced cases. Lactic acid is often present in traces, rarely in notable amount. Secretion of pepsin and rennin is always diminished in marked cases. Mucus is frequently present, and is very significant of the disease. Motility and absorption are generally deficient. Small fragments of mucous membrane may be found, and when examined by a pathologist, may occasionally establish the diagnosis.

4. Achylia Gastrica (Atrophic Gastritis).—This condition may be a terminal stage of chronic gastritis. It is sometimes associated with the blood-picture of pernicious anemia. It gives a great decrease, and sometimes entire absence of hydrochloric acid and ferments. The total acidity may be as low as 1 or 2 degrees. Small amounts of lactic acid may be present. Absorption and motility are not greatly affected.

5. Gastric Carcinoma.—As far as the laboratory examination goes, the cardinal signs are absence of free hydrochloric acid and presence of a peptid-splitting ferment, of lactic acid, and of the Boas-Oppler bacillus. These findings are, however, by no means constant.

It is probable that some substance is produced by the cancer which neutralizes the free hydrochloric acid, and thus causes it to disappear earlier than in other organic diseases of the stomach. The peptid-splitting ferment (see p. 405) is also probably a cancer.

The presence of lactic acid is possibly gestive single symptom of gastric cancer, majority of cases its presence in notabl per cent. by Strauss' method) after Boas' stomach having been washed the evenir rants a tentative diagnosis of malignanc

Carcinoma seems to furnish an espec medium for the growth of the Boas-O hence this micro-organism is frequently :

Blood can be detected in the stomac chemic tests in nearly one-half of the cas common when the new growth is situated Blood is present in the stool in nearly ev

Evidences of retention and fermentati in pyloric cancer. Tumor particles a found late in the disease.

6. Gastric Ulcer.—There is excess chloric acid in about one-half of the ca cases the acid is normal or diminished. present. The diagnosis must be based la clinical symptoms, and where ulcer is stroi it is probably unwise to use the stomacl

II. ADDITIONAL EXAMINATIONS WHICH MATION AS TO THE CONDITION OF TI

1. Absorptive Power of the Stoma very unimportant function, only a fe being absorbed in the stomach. It is de organic diseases of the stomach, especiall and carcinoma, but not in neuroses. The practical value.

Give the patient, upon an empty stomach, a 3-gr. capsule of potassium iodid with a glass of water, taking care that none of the drug adheres to the outside of the capsule. At intervals test the saliva for iodids by moistening starch-paper with it and touching with yellow nitric acid. A blue color shows the presence of an iodid, and appears normally in ten to thirty minutes after ingestion of the capsule. A longer time denotes delayed absorption.

Starch-paper is prepared by soaking filter-paper in boiled starch and drying.

2. Motor Power of the Stomach.—This refers to the rapidity with which the stomach passes its contents on into the intestines. It is very important: intestinal digestion can compensate for insufficient or absent stomach digestion only so long as the motor power is good.

Motility is impaired to some extent in chronic gastritis. It is especially deficient in dilatation of the stomach due to atony of the gastric wall or to pyloric obstruction, either benign or malignant. It is increased in most conditions with hyperchlorhydria.

The best evidence of deficient motor power is the detection of food in the stomach at a time when it should be empty, *e.g.*, before breakfast in the morning. A special test-meal containing easily recognized materials (*e.g.*, rice pudding with currants) is sometimes given and removed at the end of six or seven hours. When more than 100 c.c. of fluid are obtained with the tube one hour after an Ewald breakfast, deficient motility may be inferred.

Ewald's salol test is scarcely so reliable as the above. It depends upon the fact that salol is not absorbed until it

reaches the intestines and is decomposed
intestinal juices.

The patient is given 15 gr. of salol with ƚ
and the urine, passed at intervals thereaft
salicyluric acid. A few drops of 10 per cer
solution are added to a small quantity of the
color denotes the presence of salicyluric ac
normally in sixty to seventy-five minutes a
the salol. A longer time indicates impairec

3. To Determine Size and Positior
—After removing the test-meal, while the
place force quick puffs of air into the stc
pression of the bulb. The puffs can be
with a stethoscope over the region of the
nowhere else.

4. Sahli's Desmoid Test of Gastric
Two pills, one containing 0.1 Gm. iodol
0.05 Gm. methylene-blue, are wrapped
made of thin sheets of rubber and tied w
raw catgut, No. ∞. The bags must be c
and tied. Before use they should be pla
in water. If they float or if any of t
blue escapes and colors the water they
the test.

The patient swallows the two bags wi
little water during the noon meal, and the
at intervals thereafter. According to Sa
is digested by gastric juice and not by
intestinal juices. If gastric digestion is
and methylene-blue can be detected in t:
afternoon or evening of the same day.
may occur when digestion is very poor, p:

motility is diminished, but it is then delayed If the reaction does not appear, gastric digestion has not occurred.

Methylene-blue is recognized in the urine by the green or blue color which it imparts. It is sometimes eliminated as a chromogen, in which case a little of the urine must be acidified with acetic acid and boiled to bring out the color.

To detect the iodin, some of the urine is decolorized by gently heating and filtering through animal charcoal. To 10 c.c. are then added 1 c.c. dilute sulphuric acid, and 0.5 c.c. of a 1 per cent. solution of sodium nitrite and 2 c.c. of chloroform. Upon shaking, a rose color will be imparted to the chloroform if iodin be present. Another method of testing for iodin is given on page 194.

CHAPTER V

THE FECES

As commonly practised, an examinatior
limited to a search for intestinal parasite
Much of value can, however, be learni
simple examinations, particularly a care
Anything approaching a complete anal;
other hand, a waste of time for the clinic

The normal stool is a mixture of—
undigested and indigestible remnants of f
granules, particles of meat, vegetable ci
etc.; (c) digested foods, carried out befi
can take place; (d) products of the dige
altered bile-pigments, enzymes, mucus, et
of decomposition, as indol, skatol, fatty a
ous gases; (f) epithelial cells shed from i
intestinal canal; (g) harmless bacteria, wh
present in enormous numbers.

Pathologically, we may find abnorm;
normal constituents, blood, pathogenic b;
parasites and their ova, and biliary
concretions.

The stool to be examined should be pass
vessel, without admixture of urine. Th
should not be delayed more than a few h
the changes caused by decomposition.

odor can be partially overcome with turpentine, 5 per cent. phenol, or a little formalin. When search for amebæ is to be made, the vessel must be warm, and the stool kept warm until examined; naturally, no disinfectant can be used. For other protozoa a saline cathartic may be given and the second stool examined. The first stool is usually too solid, and the later ones too greatly diluted.

I. MACROSCOPIC EXAMINATION

1. Quantity.—The amount varies greatly with diet and other factors. The average is about 100 to 150 Gm. in twenty-four hours. It is much larger upon a vegetable diet.

2. Frequency.—One or two stools in twenty-four hours may be considered normal, yet one in three or four days is not uncommon with healthy persons. The individual habit should be considered in every case.

3. Form and Consistence.—Soft, mushy, or liquid stools follow cathartics and accompany diarrhea. Copious, purely serous discharges without fecal matter are significant of Asiatic cholera, although sometimes observed in other conditions. Hard stools accompany constipation. Rounded scybalous masses are common in habitual constipation, and indicate atony of the muscular coat of the colon. Flattened, ribbon-like stools result from some obstruction in the rectum, generally a tumor or a stricture from a healed ulcer, most commonly syphilitic. When bleeding piles are absent, blood-streaks upon such a stool point to carcinoma.

4. Color.—The normal light or dark-brown color is due chiefly to urobilin, which is formed from bilirubin

by reduction processes in the intest
result of bacterial activity. The stool
yellow, owing partly to their milk diet a
presence of unchanged bilirubin.

Diet and drugs cause marked chang
yellow color; cocoa and chocolate, dar
fruits, reddish or black; iron and bismt
or black; hematoxylin, red, etc.

Pathologically, the color is important
low is generally due to unchanged b
stools are not uncommon, especially in d
hood. They are sometimes met in apr
infants, alternating with normal yellow
little significance unless accompanied
The color is due to biliverdin or, someti
genic bacteria. Putty-colored or "
occur when bile is deficient, either fror
outflow or from deficient secretion. 1
less to absence of bile-pigments than to
Similar stools, which manifestly consis
are common in conditions like tubercu
which interfere with absorption of fa
creatic disease.

Notable amounts of blood produce ta
when the source of the hemorrhage is
upper intestine, and a dark brown to b
source is nearer the rectum. When dia
color may be red, even if the source of t
up. Red streaks of blood upon the out
are due to lesions of rectum or anus.

5. Odor.—Products of decompositic
and skatol, are responsible for the norma

The strength of this odor depends largely upon the amount of meat in the diet and the activity of putrefactive bacteria in the intestine. Upon a vegetable or milk diet the odor is much less. A sour odor due to fatty acids is normal for nursing infants, and is noted in mild diarrheas of older children. In the severe diarrheas of childhood a putrid odor is common. In adults, stools emitting a very foul stench are suggestive of malignant or syphilitic ulceration of the rectum or gangrenous dysentery.

6. Mucus.—Excessive quantities of mucus are easily detected with the naked eye, and signify irritation or inflammation. When the mucus is small in amount and intimately mixed with the stool, the trouble is probably in the small intestine. Larger amounts, not well mixed with fecal matter, indicate inflammation of the large intestine. Stools composed almost wholly of mucus and streaked with blood are the rule in dysentery, ileocolitis, and intussusception.

In the so-called mucous colic or membranous enteritis, shreds and ribbons of altered mucus, sometimes representing complete casts of portions of the bowel, are passed. These may appear as firm, irregularly segmented strands, suggesting tapeworms. The mucus sometimes takes the form of frog-spawn-like masses. In some cases it is passed at variable intervals, with colic; in others, with every stool, with only vague pains and discomfort. It is distinguished from inflammatory mucus by absence of pus-corpuscles. The condition is not uncommon and should be more frequently recognized. It is probably a secretory neurosis, hence the name "membranous enteritis" is inappropriate.

7. Concretions.—Gall-stones sho
for in every case of obscure colicky
Intestinal concretions (enteroliths) are
sand, consisting of sand-like grains, is
rotic conditions, such as mucous coli
tion of considerable amounts of oliv
soap and fat often appear in the feces
taken by the patient for gall-stones, |
the oil has been given for cholelithiasi

Concretions can be found by breal
matter in a sieve (which may be impro
while pouring water over it. It mus
that gall-stones, if soft, may go to pie
Gall-stones are readily identified by 1
faces. When facets are absent the s
tinguished from other concretions by
terol and bile-pigment in them. The
up and as far as possible dissolved in e
ether be slowly evaporated in a watch
cholesterol (Fig. 49) will separate out.
pigments treat the parts of the stone 1
to dissolve in ether with chloroforn
hot alcohol. A yellow color in the c
green in the alcohol show the press
and biliverdin respectively.

8. Animal Parasites.—Segments c
the adults of other parasites are often f
The smaller ones are sought as describe
the material caught by the sieve bei
clear water and examined in a glass
back ground placed some distance bel
should be preceded by a vermicide an

Patients often mistake vegetable tissue for intestinal parasites, and the writer has many times known physicians to make similar mistakes. The most frequent sources of confusion are long fibers from poorly masticated celery or "greens," which suggest round worms; cells from orange, which suggest seat worms; and fibers from banana, which, because of the segmented structure and the presence of oval cells, suggest tapeworms and ova (Fig. 150). Even slight familiarity with the micro-

FIG. 150.—Undigested fiber from center of banana, in feces (× 15). In the lower part of the figure the fiber is shown natural size. The segments are colored reddish-brown when found in the stool. Such fibers are often reported as small tapeworms.

scopic structure of vegetable tissue will prevent the chagrin of such errors.

9. Curds.—The stools of nursing infants frequently contain whitish curd-like masses, due either to imperfect digestion of fat or casein or to excess of these in the diet. When composed of fat, the masses are soluble in ether, and give the Sudan III test. If composed of casein, they will become tough and fibrous-like when placed in formalin (10 per cent.) for twenty-four hours.

II. CHEMIC EXAMINATION

Complicated chemic examinations are of little value to the clinician. Certain tests are, however, important.

1. Blood.—When present in large amount blood produces such changes in the appearance of the stool that it is not likely to be overlooked. Traces of blood (occult hemorrhage) can be detected only by special tests. Recognition of occult hemorrhage has its greatest value in diagnosis of gastric cancer and ulcer. It is constantly present in practically every case of gastric cancer, and is always present, although usually intermittently, in ulcer. Traces of blood also accompany malignant disease of the bowel, the presence of certain intestinal parasites, and other conditions.

Detection of Occult Hemorrhage.—Soften a portion of the stool with water, shake with an equal volume of ether to remove fat, and discard the ether. Treat 10 c.c. of the remaining material with about one-third its volume of glacial acetic acid and extract with about 10 c.c. ether. Should the ether not separate well, add a little alcohol and mix gently. Apply the guaiac or benzidin test to the ether as already described (see p. 181). This will require only a portion of the ether-extract. In case the test is positive, it is a good plan to use the remainder for spectroscopic examination treating it so as to produce the bands of hemochromogen (see pp. 369, 371).

Wagner makes a thick smear of the feces on a glass slide by means of a wooden spatula, allows this to dry, and pour the mixed benzidin reagent on it. The blue color is recognized macroscopically and microscopically.

In every case iron-containing medicines must be stopped, and blood-pigment must be excluded from the food by giving an appropriate diet, *e.g.*, bread, milk, eggs, and fruit. At the beginning of the restricted diet give a gram of powdered charcoal or, better, o.3 Gm. of carmin, in capsules, so as to mark the corresponding stool.

2. Bile.—Normally, unaltered bile-pigment is never present in the feces of adults. In catarrhal conditions of the small intestine bilirubin may be carried through unchanged. It may be demonstrated by the Schmidt test for urobilin which follows, or, if a considerable amount is present, by filtering (after mixing with water if the stool be solid) and testing the filtrate by Gmelin's method, as described under The Urine.

3. Urobilin (Hydrobilirubin).—The urobilin of the urine and the hydrobilirubin which constitutes the principal normal pigment of the feces appear to be identical; and the present tendency is to use the name "urobilin" in both cases. In a general way, the name covers both the pigment, urobilin, and the chromogen, urobilinogen, of which it is an oxidation product, since the two substances have exactly the same significance. For the mode of formation and the significance in the urine the reader is referred to the chapter on the urine. Owing to constipation and other factors, the amount of urobilin in the feces is subject to marked daily variations. The average of a number of successive daily estimations is, however, fairly constant Ordinarily the twenty-four-hour stool gives a dilution value (see method below) of 6000; and 9000 may be taken as the upper normal limit.

Since bilirubin, its mother substance, is a product of blood-pigment, an abnormally large amount of uro-bilin in the feces may be taken as definite evidence of excessive destruction of red blood cells within the circulation; and quantitative estimations are of great value whenever such increased blood destruction is in question. They have been found especially useful in distinguishing the anemias due to excessive hemolysis (*e.g.*, pernicious anemia) from other anemias in which hemolysis is not a prominent factor (carcinoma, hemorrhage); in following the progress of individual cases of pernicious anemia; and in studying the effect of splenectomy performed as a therapeutic measure in this disease. In progressing cases of pernicious anemia the Wilber and Addis method usually gives urobilin dilution values of 20,000 to 30,000 and often much more. Urobilin is nearly or quite absent from the stool in cases of obstruction to the common or hepatic bile-ducts.

Detection.—The chemical tests mentioned on p. 186 may be applied to a watery extract of the stool. Direct spectroscopic examination is impossible owing to the cloudiness of the suspension. The following test is also useful:

Schmidt's Test.—Rub up a small quantity of the fecal matter with saturated mercuric chlorid solution and let stand twenty-four hours. Urobilin will give a red color, which is likewise imparted to such microscopic structures as are stained with urobilin. A green color shows the presence of unchanged bilirubin and is not seen normally.

Quantitative Estimation.—The method of Wilber and Addis is probably most useful clinically. While it does not give the actual quantity of urobilin, it furnishes a rough comparative method which works very well in practice. Because of the instability of urobilin, methods which involve elaborate treatment of the feces are not applicable. Since urobilin and urobilinogen have the same significance and are so readily changed one into the other they are included together in the estimation. Estimations are valueless unless the average of six to ten, made on successive days, is taken.

Method of Wilber and Addis.—1. Collect all the feces for twenty-four hours, keeping them in darkness.

2. Grind the whole quantity with water to a homogeneous paste.

3. Dilute to 1000 c.c. with tap water (or to 500 c.c. or 2000 c.c. if the amount of feces is unusually small or large).

4. Measure off 25 c.c. and add to this 75 c.c. acid alcohol (alcohol 64 c.c., concentrated hydrochloric acid 1 c.c., water 32 c.c.).

5. Place in a mechanical shaker for one-half hour. Constant shaking by hand for a similar period will answer.

6. Add 100 c.c. of saturated alcoholic solution of zinc acetate and filter.

7. To 20 c.c. of the filtrate add 2 c.c. of Ehrlich's reagent (para-dimethylamidobenzaldehyde, 20 Gm.; concentrated hydrochloric acid, 150 c.c.; water, 150 c.c.).

8. Keep in darkness until next day (or at least for six hours) and examine spectroscopically. In the presence of both urobilinogen and urobilin the absorption bands indicated in Fig. 151A, and B, will be seen.

9. Dilute with 60 per cent. alcohol, adding a few cubic centimeters at a time, until first one and then the other band has entirely disappeared when the slit of the spectroscope is wide open but still remains visible when the slit is partly closed. The end-point is fairly definite after one has established his standard upon a series of normal stools. For the sake of uniformity the examination may be made

in a 50-c.c. cylinder graduate, in a dark room by the light of a Mazda electric bulb, with the spectroscope held close to the light.

10. Calculate separately the number of dilutions necessary to cause disappearance of each of the absorption bands and add the two together. The calculation is based not upon the 20 c.c. of filtrate used, but *upon the 2½ c.c. of fecal suspension represented by the filtrate.* The dilution value for the twenty-four-hour stool (1000 c.c. of fecal

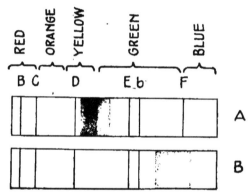

FIG. 151.—Absorption spectra of A, urobilinogen in acid solution with Ehrlich's reagent and B, urobilin in acid solution with zinc acetate.

suspension) is then found by multiplying this figure by 400. When the fecal suspension was made up to 500 c.c. or 2000 c.c. the multiplier would of course be 200 or 800. This final result indicates the number of dilutions which would be necessary if all the urobilin and urobilinogen of the twenty-four-hour stool were concentrated in the 2½ c.c. of fecal suspension examined.

Example.—Suppose that in step 9 the urobilinogen band disappeared when the 20 c.c. of filtrate had been diluted to 25 c.c. and the urobilin band when the volume

28

reached 30 c.c., then the dilution values for the 1½ c.c. of feces would be 10 and 12 respectively and the combined value 10 + 12 = 22. The total dilution value of the twenty-four-hour stool would then be 22 × 400 = 8800.

4. Pancreatic Ferments.—Two of the ferments of the pancreatic juice—amylase and trypsin—are normally present in the feces. Lipase can usually not be detected. In pancreatic disease and in simple obstruction of the pancreatic duct these ferments are diminished or absent. Quantitative estimations therefore furnish a valuable aid in the diagnosis of pancreatic disease, particularly when carried out in conjunction with an estimation of amylase in the urine. Results, although less reliable, have much the same significance as those given by examination of the duodenal contents removed through the duodenal tube—a procedure to which the practitioner will hesitate to resort owing to its technical difficulties and the discomfort to the patient.

Owing to constipation, diet, and other factors there are considerable variations in the amounts of ferments. It is therefore essential that a uniform technic be adopted. The following directions are based upon the method recommended by T. R. Brown for amylase. It is best in every case to estimate both amylase and trypsin, but if the examination is limited to one ferment amylase should be chosen, since the action of trypsin may be simulated by erepsin and the proteolytic activity of bacteria.

Estimation of Pancreatic Ferments in Feces.—1. Upon the evening before the test, limit the patient to a light supper and give a high enema at bed time.

2. At 7:00 next morning, give 750 c.c. (25 ounces) of milk.

3. At 7:30 give ½ ounce of Epsom salts; repeat at 8:00.

4. At 8:30 give a glass of water containing ¼ teaspoonful of sodium bicarbonate.

5. Save all the feces passed up to 2 p.m. in a vessel containing 2 ounces of toluol. Keep in a cool place. If less than 400 c.c. are obtained give an enema of 1 pint of water.

6. Dilute the whole volume of feces to 3000 c.c. with normal salt solution, mix well and centrifugalize a portion for five minutes. Use the supernatant fluid for the following tests:

Estimation of Amylase.—1. Prepare a 1 per cent. solution of soluble starch as follows: To 100 c.c. cold distilled water add 1 Gm. soluble starch (Kahlbaum's recommended) and heat gently with constant stirring until clear.

2. Place 2 c.c. of this solution in each of 13 test-tubes.

3. To these tubes add the supernatant fluid from the centrifugalized feces as follows:

To tube 1 add 1.8 c.c.	To tube 8 add 0.4 c.c.
To tube 2 add 1.6 c.c.	To tube 9 add 0.2 c.c.
To tube 3 add 1.4 c.c.	To tube 10 add 0.1 c.c.
To tube 4 add 1.2 c.c.	To tube 11 add 0.05 c.c.
To tube 5 add 1.0 c.c.	To tube 12 add 0.025 c.c.
To tube 6 add 0.8 c.c.	To tube 13 add none (control)
To tube 7 add 0.6 c.c.	

Bring the quantity in each tube up to 4 c.c. with normal salt solution.

4. Place the tubes in an incubator or water bath at about 38°C.[1] for one-half hour.

---

[1] Variations in reaction and variations in temperature from 37° to 40°C. exert no appreciable effect upon the result.

5. Fill all tubes with tap water and add a drop of weak iodin solution to each. Gram's iodin solution will answer.

6. If amylase be present, the series of tubes will vary from yellow through red-purple to pure blue, depending upon complete or partial digestion of the starch. The tube before the one in which the first definite trace of blue appears is taken as the measure of digestion. Brown found the lowest normal to be the tenth tube, corresponding to 60,000 units.[1]

Test for Trypsin.—The well-known Gross test may be applied as follows:

1. Prepare a 1 : 1000 solution of casein as follows:

Casein (Gruebler's preferred)	0.1 Gm.;
Sodium bicarbonate	0.1 Gm.;
Distilled water	100 c.c.

Boil for one minute, stirring constantly, and cool.

2. Place 5 c.c. of the casein solution in each of 13 test-tubes and add to these tubes the same amounts of the fecal suspension as were used for the amylase test.

3. Place the tubes in the incubator or a water bath at 38°C. for one hour.

4. Test for digestion of casein by adding a few drops of 3 per cent. acetic acid to each tube. Digestion is complete in those tubes in which no white precipitate forms.

III. MICROSCOPIC EXAMINATION

Care must be exercised in selection of portions for examination. A random search will often reveal nothing of interest. A small bit of the stool, or any suspicious-looking particle, is placed upon a slide,

[1] This means the number of cubic centimeters of 1 per cent. starch solution which would be digested by the 3000 c.c. of fecal suspension under the stated conditions of time and temperature.

thinned with water if necessary, and covered with a cover-glass. As emphasized by Bass and Johns the layer should be just thin enough to read news-print through it. A large slide—about 2 by 3 inches—with a correspondingly large cover will be found convenient. Most of the structures which it is desired to see can be found with a 16-mm. objective. Details of structure must be studied with a higher power.

Fig. 152.—Microscopic elements of normal feces: *a,* Muscle-fibers; *b,* connective tissue; *c,* epithelial cells; *d,* white blood-corpuscles; *e,* spiral vessels of plants; *f–h,* vegetable cells; *i,* plant hairs; *k,* triple phosphate crystals; *l,* stone cells. Scattered among these elements are micro-organisms and débris (after v. Jaksch).

The bulk of the stool consists of granular débris. Among the recognizable structures (Fig. 152) met in normal and pathologic conditions are: Remnants of food, epithelial cells, pus-corpuscles, red blood-corpuscles, crystals, bacteria, and ova of animal parasites.

1. Remnants of Food.—These include a great variety of structures which are very confusing to the student. Considerable study of normal feces is necessary for their recognition.

Vegetable fibers are generally recognized from their spiral structure or their pits, dots, or reticulate markings; **vegetable cells,** from their double contour and the chlorophyl bodies which many of them contain. These cells are apt to be mistaken for the ova of parasites. **Vegetable hairs** (Fig. 153) frequently look much like the larvæ of some of the worms. Anything like a careful examination will, however, easily distinguish them, because of the homogeneous and highly refractile wall, the distinct central canal which extends

FIG. 153.—Vegetable hair (down from skin of peach) in feces (× 150). Compare with Fig. 193.

the whole length, and, especially, the absence of motion. **Starch-granules** sometimes retain their original form, but are ordinarily not to be recognized except by their staining reaction. Potato remains appear in colorless translucent masses somewhat like sago grains or flakes of mucus. Starch strikes a blue color with Lugol's solution when undigested; a red color, when slightly digested. **Muscle-fibers** are yellow, and when poorly digested appear as short, transversely striated cylinders with rather squarely broken ends (Fig. 154). Generally, the ends are rounded and the striations faint; or only irregularly round or oval yellow masses which bear little resemblance to normal muscle-tissue are found. If a little eosin solution be run under the cover, muscle-fibers will take up the red color and stand out distinctly.

Fats occur in three modifications: neutral fats, fatty acids, and soaps. *Neutral fats* are present in very small

FIG. 154.—Poorly digested muscle-fiber in feces showing striations (× 200).

amounts or not at all on an ordinary diet. They appear as droplets or yellowish flakes, depending upon the melting point. They stain strongly with Sudan III; and do not stain with dilute carbol-fuchsin as do fatty acids and soaps. *Fatty acids* take the form of flakes like those of neutral fat, or of needle-like crystals which are generally aggregated into thick balls or irregular masses in which the individual crystals are difficult to make out. When treated with Sudan III the amorphous flakes take a lighter orange than do the neutral fats, while the crystals do not stain. *Soaps*—chiefly calcium soap—appear partly as well-defined yellow amorphous flakes or rounded masses sugesting eggs, partly as coarse crystals. They do not stain with Sudan III and do not melt when warmed as do the fatty acids. **Connective tissue** consists of colorless or yellowish threads with poorly defined edges and indefinite longitudinal striations. When treated with 30 per cent. acetic acid the fibers swell up and become clear and homogeneous. **Elastic fibers,** which are often present along with the connective tissue, are more definite in outline and branch and anastomose. They are rendered more distinct by acetic acid.

Excess of any of these structures may result from excessive ingestion or deficient digestion.

2. Epithelial Cells.—A few cells derived from the wall of the alimentary canal are a constant finding. They show all stages of disintegration and are often unrecognizable. A marked excess has its origin in a catarrhal condition of some part of the bowel. Squamous cells come from the anal orifice; otherwise the form of the cells gives no clue to the location of the lesion.

3. Pus.—Amounts of pus sufficient to be recognized with the eye alone indicate rupture of an abscess into the bowel. If well mixed with the stool, the source is high up, but in such cases the pus is apt to be more or less completely digested, and hence unrecognizable. Small amounts, detected only by the microscope, are present in catarrhal and ulcerative conditions of the intestine, the number of pus-cells corresponding to the severity and extent of the process.

4. Blood-corpuscles.—Unaltered red corpuscles are rarely found unless their source is near the anus. Ordinarily, only masses of blood-pigment can be seen. Blood is best recognized by the chemic tests (see p. 429).

5. Bacteria.—In health, bacteria—mostly dead—constitute about one-third of the weight of the dried stool. They are beneficial to the organism, although not actually necessary to its existence. Under certain conditions they may be harmful. It is both difficult and unprofitable to identify them. The great majority belong to the colon bacillus group, and are negative to Gram's method of staining.

In some pathologic conditions the character of the intestinal flora changes, so that Gram-staining bacteria very greatly predominate. As shown by R. Schmidt, of Neusser's clinic in Vienna, this change is most constant and most striking in cancer of the stomach, owing to large numbers of Boas-Oppler bacilli, and is of some value in diagnosis. He believes that a diagnosis of gastric carcinoma should be very unwillingly made with an exclusively "Gram-negative" stool, while a "Gram-positive" stool, due to bacilli (which should also stain brown with Lugol's solution), may be taken as very strong evidence of cancer. A Gram-positive stool due to cocci is suggestive of intestinal ulceration. The technic is the same as when Gram's method is applied to other material (see p. 572), except that the smear is fixed by immersion in methyl-alcohol for five minutes instead of by heat. Pyronin is a good counterstain.

The deep purple Gram-staining bacteria stand out more prominently than the pale-red Gram-negative organisms, and one may be misled into thinking them more numerous even in cases in which they are much in the minority. The number of Boas-Oppler bacilli can be increased by administering a few ounces of sugar of milk the day before the examination. The bacteria can be obtained comparatively free from food remnants by mixing a little of the feces with water, allowing to settle for a short time, and making smears from the supernatant fluid. One must of course be on his guard against *Bacillus bulgaricus* taken with artificial buttermilk.

Owing to the difficulty of excluding swallowed sputum, the presence of the **tubercle bacillus** is less significant in the feces than in other material. It may, however, be taken as evidence of intestinal tuberculosis when clinical signs indicate an intestinal lesion and reasonable care is exercised in regard to the sputum. Success in the search will depend largely upon careful selection of the portion examined. A random search will almost surely fail. Whitish or grayish flakes of mucus or blood-stained or purulent particles should be spread upon slides or covers and stained by the method given upon p. 236. In the case of rectal ulcers, swabs can be made directly from the ulcerated surface. With young children, who swallow all their sputum, an examination of the stool for tubercle bacilli may be the means of diagnosing tuberculosis of the lung.

6. Crystals.—Various crystals may be found, but few have any significance. Slender, needle-like crystals of fatty acids and soaps (see Fig. 49) and triple phosphate crystals (see Fig. 152) are common. Characteristic octahedral crystals of calcium oxalate (see Fig. 51) appear after ingestion of certain vegetables. Charcot-Leyden crystals (see Fig. 18) are not infrequently encountered, and strongly suggest the presence of intestinal parasites. Yellowish or brown, needlelike or rhombic crystals of hematoidin (see Fig. 49) may be seen after hemorrhage into the bowel. The dark color of the feces after administration of bismuth salts is due largely to great numbers of bismuth suboxid crystals. They resemble hemin crystals.

7. Parasites and Ova.—Descriptions will be found in the following chapter. The ova most likely to be encountered are shown in Figs. 178 and 182. The flagellates are usually best found in the second stool after a saline cathartic, the first stool being ordinarily too solid and the later ones too dilute.

To find ova when scarce, they must be concentrated. Stiles advises thoroughly mixing the stool with a quart or more of water, allowing to settle, pouring off the water almost down to the sediment, and repeating the process as long as any matter floats. The final sediment is poured into a conical glass and allowed to settle. Ova will be found in the fine sediment, which can readily be removed with a pipet. The same end may be accomplished more efficiently and more quickly by means of the centrifuge; but one must learn how long his individual centrifuge requires to throw down the ova while the lighter particles still float. These methods are more satisfactory if the larger particles are first removed by passing the fecal suspension through two or three layers of gauze or through a succession of wire screens with mesh apertures ranging from 6 to 100 to the lnch. Such concentration methods are greatly to be preferred to direct microscopic examination of the stool; not only are the ova concentrated, but they are more easily identified than in untreated feces, since bacteria and débris which would otherwise obscure them have been removed. Other and more complicated methods have been devised, but those just given and Pepper's method for hookworm eggs (see p. 505) will probably answer all clinical needs.

IV. FUNCTIONAL TESTS

1. Schmidt's Test Diet.—Much can be learned of the various digestive functions from a microscopic study of the feces, especially when the patient is upon a known diet. For this purpose the standard diet of Schmidt is generally adopted. This consists of:

Morning.............. 0.5 liter milk and 50 Gm. toast.
Forenoon... .... 0.5 liter porridge, made as follows: 40 Gm. oatmeal, 10 Gm. butter, 200 c.c. milk, 300 c.c. water, one egg, and salt to taste.
Midday............... 125 Gm. hashed meat, with 20 Gm. butter, fried so that the interior is quite rare; 250 Gm. potato, made by cooking 190 Gm. potato with 100 c.c. milk and 10 Gm. butter, the whole boiled down to 250 c.c.
Afternoon............. Same as morning.
Evening............... Same as forenoon.

At the beginning of the diet, the stool should be marked off with carmin or charcoal (see p. 447). One should familiarize himself with the feces of normal persons upon this diet. A portion of the stool about the size of a walnut should be rubbed up with water to a consistency of thick soup and examined macroscopically and microscopically. The microscope examination may be facilitated by preparing four slides: one of the diluted feces untreated; one treated with dilute Lugol's solution; one with 30 per cent. acetic acid; one with Sudan III.

Deficiency of starch digestion is recognized by the number of starch-granules which strike a blue color with iodin. With exception of those inclosed in plant cells none are present normally.

The degree of protein digestion is ascertained by the appearance of the muscle-fibers. Striations are clearly visible on any considerable number of the fibers only when digestion is imperfect (see Fig. 154). They are most clearly seen in the acetic-acid preparation. The striations usually disappear after the feces have stood for some time. According to Schmidt, the presence of nuclei in muscle-fibers denotes complete absence of pancreatic function. The presence of connective-tissue shreds indicates deficient gastric digestion, since raw connective tissue is digested only in the stomach. These shreds can be recognized macroscopically by examining in a thin layer against a black background, and microscopically by their fibrous structure and the fact that they swell up and become clear and gelatinous when treated with acetic acid. The only structure likely to cause confusion is elastic tissue and this is rendered more distinct by acetic acid.

Digestion of fats is checked up by the amount of neutral fat, which should not be present in appreciable quantity normally. It is best seen after staining with Sudan III. The amount of fatty acid seen in an acetic acid preparation after it has been heated until bubbles rise is also a good guide if one is familiar with what to expect under normal conditions.

Schmidt's nuclei test for pancreatic insufficiency consists in the administration of a ½-cm. cube of beef or, better, of thymus tied in a little gauze bag with the test-meal. The meat must previously have been hardened in alcohol and well washed in water. When the bag appears in the feces it is opened and its contents examined microscopically by pressing out small bits between a slide and cover. A drop of some nuclear stain may be applied if desired. If the nuclei are for the most part undigested, pancreatic insufficiency may be assumed, since it is probable that nuclei can be digested only by the pancreatic juice. Normally the nuclei are digested, provided the time of passage through the intestine is not less than six hours. Upon the other hand, if the time of passage exceeds thirty hours nuclei may be partially digested in the complete absence of pancreatic juice.

2. Sahli's Glutoid Test.—The Schmidt test diet involves some inconvenience for the patient, and interpretation of results requires much experience upon the part of the physician. A number of other methods of testing the digestive functions have been proposed. The glutoid test of Sahli is one of the most satisfactory. This is similar to his desmoid test of gastric digestion described on page 421. A glutoid capsule containing 0.15 Gm. iodoform is taken with an Ewald breakfast. The capsule is not digested by the stomach fluid, but is readily digested by pancreatic juice. Appearance of iodin in the saliva and urine within four to six hours indicates normal gastric motility, normal intestinal digestion, and normal absorption. Instead of iodoform, 0.5 Gm. salol may be used, salicyluric acid appearing in the urine in about the same time. For tests for iodin and salicyluric acid, see pages 421 and 422.

Glutoid capsules are prepared by soaking gelatin capsules in 10 per cent. formalin. Sahli states that filled capsules can be purchased of A. G. Haussmann, in St. Gall, Switzerland. To make sure that the capsules are soluble one should try a test upon oneself.

3. **Motility.**—Ordinarily, with adults who are upon a mixed diet, fifteen to thirty hours are required for the passage of ingested material through the gastrointestinal tract. With infants the time is about one-third as long. In diarrheal conditions it is usually much shortened, unless the pathologic process is in the colon. In intestinal stasis it may be much prolonged. The time of passage is ascertained by giving 0.5 Gm. of powdered charcoal or 0.3 Gm. of carmine in a capsule with a meal and watching for the resulting discolored feces.

CHAPTER VI

ANIMAL PARASITES

ANIMAL parasites are common in all countries, but are especially abundant in the tropics, where, in some localities, almost every native is host for one or more species. Because of our growing intercourse with these regions the subject is assuming increasing importance in this country. Many parasites, hitherto comparatively unknown here, will probably become fairly common.

Some parasites produce no symptoms, even when present in large numbers. Others cause very serious symptoms. It is, however, impossible to make a sharp distinction between pathogenic and non-pathogenic varieties. Parasites which cause no apparent ill effects in one individual may, under certain conditions, produce marked disturbances in another. The disturbances are so varied, and frequently so indefinite, that diagnosis can rarely be made from the clinical symptoms. It must rest upon detection, by the naked eye or the microscope, of (*a*) the parasites themselves, (*b*) their ova or larvæ, or (*c*) some of their products.

Unlike bacteria, the great majority of animal parasites multiply by means of alternating and differently formed generations, which require widely different conditions for their development. The few exceptions are chiefly among the protozoa. Multiplication of parasites within the same host is thus prevented. In the case of the hookworm, for example, there is no increase in the number of worms in the host's intestine, except through reinfection from the outside. The ova are carried out of the intestine and the young must pass a certain period of development in warm, moist earth before they can again enter the human body and grow to maturity. This also explains the geographic distribution of parasites. The hookworm cannot flourish in cold countries; malaria can prevail only in localities in which the mosquito, Anopheles, exists, and then only after the mosquitoes have become infected from a human being.

In general, this alternation of periods of development takes place in one of three ways:

1. The young remain within the original host, but travel to other organs, where they do not reach maturity, but lie quiescent until taken in by a new host. A good example is *Trichinella spiralis*.

2. The young or the ova which subsequently hatch pass out of the host, and either (a) go through a simple process of growth and development before entering another host, as is the case with the hookworm, or (b) pass through one or more free-living generations, the progeny of which infect new hosts, as is the case with *Strongyloides intestinalis*.

3. The young or ova or certain specialized forms either directly (*e.g.*, malarial parasites) or indirectly (*e.g.*, tapeworms) reach a second host of different species, where a widely different process of development occurs. The host in which the adult or sexual existence is passed is called the *definitive* or final host; that in which the intermediate or larval stage occurs, the

Intermediate host. Man, for example, is the definitive host for *Tænia saginata*, and the intermediate host for the malarial parasites and *Tænia echinococcus*.

At this place a few words concerning the classification and nomenclature of living organisms in general will be helpful. Individuals which are alike *in all essential respects* are classed together as a *species*. Closely related species are grouped together to form a *genus;* genera that have certain characteristics in common make up a *family;* families are grouped into *orders;* orders, into *classes;* and classes, finally, into the *branches* or *phyla*, which make up the animal and vegetable *king-* *doms*. In some cases these groups are subdivided into intermediate groups—subphyla, subfamilies, etc., and occasionally slight differences warrant subdivision of the species into *varieties*.

The scientific name of an animal or plant consists of two parts, both Latin or Latinized words, and is printed in italics. The first part is the name of the genus and begins with a capital letter; the second is the name of the species and begins with a lower case letter, even when it was originally a proper name. When there are varieties of a species, a third part, the designation of the variety, is appended. The author of the name is sometimes indicated in Roman type immediately after the name of the species. Examples: *Spirochæta vincenti*, often abbreviated to *Sp. vincenti* when the genus name has been used just previously; *Staphylococcus pyogenes albus; Necator americanus*, Stiles.

At the present time there is great confusion in the naming and classification of parasites. Some have been given a very large number of names by different observers, and in many cases different parasites have been described under the same name. The alternation of generations and the marked differences in some cases between male and female have contributed to the confusion, different forms of the same parasite being described as totally unrelated species.

The number of parasites which have been described as occurring in man and the animals is extremely large. Only those which are of medical interest are mentioned here. They belong to four phyla—Protozoa, Platyhelminthes, Nemathelminthes, and Arthropoda.

PHYLUM PROTOZOA

These are unicellular organisms, the simplest types of animal life. There is very little differentiation of structure. Each contains at least one, and some several, nuclei. Some contain contractile vacuoles; some have cilia or flagella as special organs of locomotion. They reproduce by division, by budding, or by sporulation. Sometimes there is an alternation of generations, in one of which sexual processes appear, as is the case with the malarial parasites. The protozoa are very numerous, the subphylum Sarcodina alone including no less than 5000 species. Most of the protozoa are microscopic in size; a few are barely visible to the naked eye. The beginning student can gain a general idea of their appearance by examining water (together with a little of the sediment) from the bottom of any pond. Such water usually contains amebæ and a considerable variety of ciliated and flagellated forms.

The following is an outline of those protozoa which

are of medical interest, together with the subphyla and classes to which they belong:

PHYLUM PROTOZOA

SUBPHYLUM I. SARCODINA.—Locomotion by means of pseudopodia.

CLASS **Rhizopoda.**—Pseudopodia form lobose or reticulose processes.

Genus	Species
Endamœba.	E. histolytica.
	E. coli.
	E. gingivalis.

SUBPHYLUM II. **MASTIGOPHORA (FLAGELLATA).**—Locomotion by means of flagella.

CLASS **Zoömastigophora.**—Forms in which animal characteristics predominate.

Genus	Species
Spirochæta.	Sp. recurrentis.
	Sp. vincenti.
	Sp. buccalis.
	Sp. dentium.
	Sp. refringens.
Treponema.	T. pallidum.
	T. pertenue.
Trypanosoma.	T. gambiense.
	T. rhodesiense.
	T. cruzi.
	T. lewisi.
	T. evansi.
	T. brucei.
	T. equiperdum.
Leishmania.	L. donovani.
	L. tropica.
	L. infantum.
Cercomonas.	C. hominis.
Bodo.	B. urinarius.
Trichomonas	T. intestinalis.
	T. vaginalis.
	T. pulmonalis.
Lamblia.	L. intestinalis.

SUBPHYLUM III. **SPOROZOA.**—All members parasitic. Propagation by means of spores. No special organs of locomotion.
CLASS **Telosporidia.**—Sporulation ends the life of the individual.

Genus	Species
Coccidium.	C. cuniculi.
Plasmodium.	P. vivax.
	P. malariæ.
	P. falciparum.
Babesia.	B. bigeminum.

SUBPHYLUM IV. **INFUSORIA.**—Locomotion by means of cilia.
CLASS **Ciliata.**—Cilia present throughout life.

Genus	Species
Balantidium.	B. coli.
	B. minutum.

SUBPHYLUM SARCODINA
CLASS RHIZOPODA

These are protozoa the body substance of which forms changeable protoplasmic processes, or pseudopodia, for the taking in of food and for locomotion. They possess one or several nuclei.

1. Genus Endamœba.—1. **Endamœba histolytica.**—This organism is found, often in large numbers, in the stools of tropical dysentery and in the pus and walls of hepatic abscesses associated with dysentery. Infection is more common in this country than was at one time supposed and has even been reported in the Northern States. The parasite is a grayish or colorless, granular cell, usually between 25 and 40 μ in diameter (Fig. 155). Its appearance varies according to its stage of development. In the vegetative stage, which is found in acute dysentery, there is a distinct, homogeneous, refractile ectoplasm and a granular endoplasm containing one or more distinct vacuoles, a round nucleus which is ordinarily very indistinct, and, frequently, ingested red blood-corpuscles and bacteria. When at rest its shape is spheric, but upon a warm slide it exhibits the characteristic *ameboid motion*, constantly changing its shape or moving actively about by means of distinct pseudopodia. This motion is its most distinctive feature, and should always be seen to establish

FIG. 155.—*Endamœba histolytica* in intestinal mucus, with blood-corpuscles and bacteria (Lösch).

the identity of the organism in this stage. It is lost when the specimen cools, and can usually not be reestablished by warming. If neutral red in 0.5 per cent. solution be run under the cover-glass, it will be taken up by the endamebæ and other protozoa and will render them conspicuous without killing them ("vital staining").

In dysentery "carriers" and in chronic cases when the stools are formed and hard, most or all of the parasites may become encysted. Their appearance in this stage of development is given in the table on pages 456, 457. The structure of the cysts is best seen when a drop of Lugol's solution is mixed with the fecal matter on the slide.

When the presence of endamebæ is suspected, the stool should be passed into a warm vessel and kept warm until and during the examination. A warm stage can be improvised from a plate of copper with a hole cut in the center. This is placed upon the stage of the microscope, and one of the projecting ends is heated with a small flame. Endamebæ are most likely to be found in grayish or blood-streaked particles of mucus. Craig recommends the liquid stool following a saline cathartic. Favorable material for xamination can often be obtained at one's convenience by inserting into the rectum a large catheter with roughly cut· lateral openings. A sufficient amount of mucus or fecal matter will usually be brought away by it. .

No staining method is as useful in diagnosis as the study of the living and moving parasite. For more detailed study, Darling recommends the following method: Stain with Wright's (or Hastings' or Leishman's) stain in the usual way, and follow this with Giemsa's stain, diluted 1 : 10, until the film has a purple cast. Then plunge the preparation into a small beaker of 6q per cent. alcohol to which 10 to 20 drops of ammonia have been added and keep it in motion until the desired differentiation is obtained, when the film will have a violet color.

2. **Endamœba coli.**—This organism, which is frequently found in the stools of healthy persons, is simi-

lar to *E. histolytica*, but is smaller, rarely over 25 μ
in diameter. It has less distinct pseudopodia, less
sharp differentiation between ectoplasm and endoplasm,
less active motion, and more distinct nucleus, and does
not contain ingested red corpuscles or never more than
one or two. The vacuoles contain glycogen-granules
which stain brown with Lugol's solution; such granules
are rare in *E. histolytica*.

The principal points of distinction between *E. histolytica* and *E. coli* are included in the following table
which is slightly modified from Craig[1]:

VEGETATIVE STAGE

This stage of *E. histolytica* is found in acute dysentry.

Endamœba histolytica	*Endamœba coli*
Averages larger. Unimportant.	Averages smaller.
Actively motile. Characteristic. Often moves from place to place.	Sluggishly motile. Seldom moves from place to place.
Ectoplasm hyaline, glass-like, sharply differentiated from endoplasm. Characteristic.	Ectoplasm not glass-like, poorly differentiated from endoplasm.
Nucleus usually indistinct, often invisible. Changes position with motion of parasite.	Nucleus distinct. Located near center.
Red blood-cells present in endoplasm when stool contains blood. Very characteristic.	No red blood-cells (or never more than one or two) in endoplasm when stool contains blood.

PRECYSTIC STAGE

E. histolytica may be found in this stage when symptoms of dysentery have practically disappeared. The

[1] Craig: Archives of Internal Medicine, 1914, xiii, 917.

parasite is reduced in size, is sluggishly motile, and becomes practically indistinguishable from *E. coli*. The distinction must be based upon the vegetative or cystic forms, a few of which can usually be found in the same stool.

CYSTIC STAGE

In formed stools both endamebæ are commonly encysted. This, therefore, is the form of *E. histolytica* to be looked for between recurrences and in dysentery "carriers."

Endamœba histolytica	*Endamœba coli*
Cysts spheric or oval. Cyst wall single and delicate in young cysts; thicker and sometimes double outlined in older ones.	Similar, but double outline of wall more frequently observed and more distinct.
Diameter 10–20 μ; average, 12 μ.	Diameter 10–25 μ; average, 15 μ.
Cytoplasm of young cysts granular, often with a large vacuole. Presence of chromidia (brightly refractive, spindle-shaped or irregular masses of chromatin) characteristic.	Similar, but chromidia very rare.
Fully developed cysts contain four distinct nuclei seen by focusing at different levels. Very characteristic.	Fully developed cyst contains eight to sixteen nuclei, eight being the normal number.

3. **Endamœba gingivalis.**—That endamebas are common in the mouth and about the teeth has long been recognized, but they have generally been regarded as harmless or even as beneficial because they feed extensively upon bacteria. There is apparently only one species, which has been variously called *E. buccalis*, *E. dentalis*, and *E. gingivalis*, the last name being now

accepted as correct. Within the past few years it has attracted much attention as the possible cause of pyorrhea alveolaris. The organisms are found in the lesions of practically every case of pyorrhea, often in

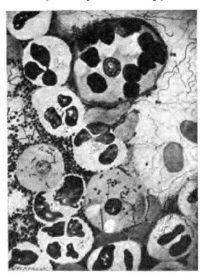

FIG. 156.—*Endamœba gingivalis* pus-corpuscles, red blood-cells, spirochetes, and bacteria in a smear from a lesion of pyorrhea alveolaris. Giemsa's stain, without alkali, twelve hours (× 850). The figure shows three endamebæ, each with one round nucleus (red). The cytoplasm (deep sky blue) contains vacuoles and bacteria. The largest parasite contains ten nuclei (blackish-purple) from ingested cells. A digestion vacuole is seen at each end of the long bacillus in the endameba near the bottom. The red corpuscles were salmon colored; nuclei of leukocytes, reddish-purple; spirochetes, bluish-purple.

large numbers. In some parts of the slide from which Fig. 156 was made, there were as many as 20 in a single field of the oil-immersion lens. Upon the other hand, a few are often found between the gums and teeth when no lesions are recognizable. The evidence at present available suggests that the organism is a factor in the etiology of pyorrhea, but the claim that it is the sole specific cause is not warranted.

Material is obtained for study by scraping between the teeth and the gum with a sterile wooden toothpick. When pus-pockets exist, the bottom and side of a pocket should be scraped with a dental scaler. This material may be examined in the fresh state by mixing it with a little saliva and placing on a warmed slide. The organism is less active than *E. histolytica*, more so than *E. coli*. Unless motion is seen it will be difficult to recognize. Individuals range in size from 10 to 35 μ.

In general, the endamebæ are more easily identified in stained smears. The smears are made by streaking the toothpick three or four times across the slide. Often one of the streaks will contain many of the parasites and the others only a few. Giemsa's solution, applied as described for blood (see p. 313) but allowed to act for three to twelve hours, is the most satisfactory stain. With this, the cytoplasm of endamebæ is blue and shows the vacuoles clearly, the small round nucleus is red, ingested bacteria purple and nuclei of ingested cells deep purple. In such preparations it is well-nigh impossible to mistake pus and epithelial cells for endamebæ. Wright's stain gives a similar picture but the differentiation is somewhat less sharp. The writer has found pyronin methyl green (see p. 642) be fairly satisfactory. It stains the cytoplasm of endamebæ red.

4. **Other Endamebæ.**—*E. tetragena*, which was described in 1907 by Viereck, is now regarded as identical with *E. histolytica*. A number of similar organisms have

been described as occurring in pus and in ascitic and other body fluids, but it is probable that in many cases, at least, the structures seen were ameboid body cells.

SUBPHYLUM MASTIGOPHORA (FLAGELLATA)
CLASS ZOÖMASTIGOPHORA

The protozoa of this subphylum are provided with one or several whip-like appendages with lashing motion, termed flagella, which serve for locomotion and, in some cases, for feeding. They generally arise from the anterior part of the organism. Some members of the group also possess an undulating membrane—a delicate membranous fold which extends the length of the body and somewhat suggests a fin. When in active motion this gives the impression of a row of cilia. The flagellata do not exhibit ameboid motion, and, in general, maintain an unchanging oval or spindle shape, and contain a single nucleus. The cytoplasm contains numerous granules and usually several vacuoles, one or more of which may be contractile. Encystment as a means of resisting unfavorable conditions is common.

1. Genus Spirochæta.—The spirochetes appear to occupy a position midway between the bacteria and protozoa, but are more frequently described with the latter. They are receiving much attention at the present time and the appreciation of their importance is growing rapidly.

1. Spirochæta recurrentis.—This spirochete was described by Obermeier as the cause of relapsing fever. It appears in the circulating blood during the febrile attack, and, unlike the malarial parasite, lives in the plasma without attacking the red corpuscles. The organism is an actively motile spiral, 15 to 20 μ long, with three to twelve wide, fairly regular turns. It can be seen in fresh unstained blood with a high dry lens, being located by the commotion which it creates among the red cells. For diagnosis, thin films, stained with Wright's or some similar blood-stain, are used (Fig. 157). In such preparations the spirals are not so regular.

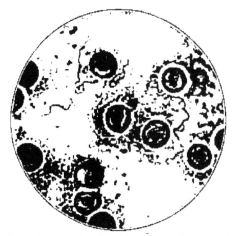

FIG. 157.—Spirochete of relapsing fever in blood (× 1000) (Karg and Schmorl).

It is generally believed that relapsing fever does not occur in the United States, but Meader has recently reported five cases which originated in Colorado. Spirochetes from one of these cases are shown in Plate VI.

Besides *Spirochæta recurrentis*, a number of distinct strains have been described in connection with different types of relapsing fever: *Sp. novyi, Sp. kochi, Sp. duttoni,* and *Sp. carteri.*

2. **Spirochæta vincenti.**--In stained smears from the ulcers of Vincent's angina (see p. 539) are found what appear to be two organisms. One, the "fusiform bacillus," is a slender rod, 4 to 8 μ long, pointed at both ends, and sometimes curved. The other is a slender spiral organism, 10 to 20 μ long, with three to ten comparatively shallow turns (see Fig. 216). These were formerly thought to be bacteria, a spirillum and a bacillus living in symbiosis. The present tendency is to regard them as stages or forms of the same organism, and to class them among the spirochetes. The same organisms are quite constantly present in large numbers in ulcerative stomatitis and in noma. They are not infrequently found in small numbers in normal mouths.

3. **Other Spirochetes.**—A number of harmless forms are of interest because of the possibility of confusing them with the more important pathogenic varieties. Of these, *Sp. buccalis* and *Sp. dentium* are inhabitants of the normal mouth. When the teeth and gums are not in good condition they are often found in immense numbers (see Fig. 156). The former is similar in morphology to *Sp. vincenti*. *Sp. dentium* (Fig. 158) is smaller (4 to 10 μ), more delicate, has deep curves, and may be easily mistaken for *Treponema pallidum*. It, also, stains reddish with Giemsa's stain. In suspected syphilitic sores of the mouth it is, therefore, important to make smears from the tissue juices rather than from the surface (see p. 550). *Sp. refringens* is frequently present upon the surface of ulcers, especially about the genitals, and has doubtless many times been mistaken for *Treponema pallidum*. It

can be avoided by properly securing the material for examination; but its morphology should be sufficient to prevent confusion. It is thicker than the organism of syphilis, stains more deeply, and has fewer and shallower curves (Figs. 158 and 222). Giemsa's stain gives it a bluish color.

Castellani has called attention to a bronchial spirochetosis and his observations have been confirmed by other workers in Europe, Asia, and the Philippines. When chronic the condition resembles tuberculosis but

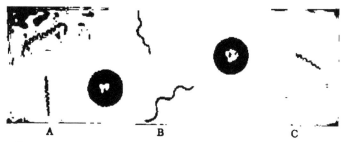

FIG. 158.—Spiral organisms: A, *Treponema pallidum:* B, *Spirochæta refringens*; C, *Spirochæta dentium.* Two red corpuscles are also shown (× 1200).

is distinguished by finding spirochetes in the sputum. Infectious jaundice is also now known to be caused by a spirochete, *Sp. icterohemorrhagicæ.* This parasite has recently been found in the blood and organs of wild rats in various parts of this country and Europe.

2. Genus Treponema.—1. **Treponema pallidum.** —This is the organism of syphilis. Its description and methods of diagnosis will be found on pp. 548–553.

2. **Treponema pertenue,** morphologically very similar to *Treponema pallidum,* was found by Castellani in yaws, a skin disease of the tropics.

3. Genus Trypanosoma.—Trypanosomes have been found in the blood-plasma of a great variety of vertebrates. Many of them appear to produce no symptoms, but a few are of great pathologic importance. As seen in the blood, they are elongated, spindle-shaped bodies, the average length of different species varying from 10 to 70 μ. Along one side there runs a delicate undulating membrane, the free edge of which appears to be somewhat longer than the attached edge, thus throwing it into folds. Somewhere in the body, usually near the middle, is a comparatively pale-staining nucleus; and near the posterior end is a smaller, more deeply staining chromatin mass, the micronucleus or blepharoplast. A number of coarse, deeply staining granules, chromatophores, may be scattered through the cytoplasm. A flagellum arises in the blepharoplast, passes along the free edge of the undulating membrane, and is continued anteriorly as a free flagellum. These details of structure are well shown in Plate VI.

The life history of the trypanosomes is not well known. In most cases there is an alternation of hosts, various insects playing the part of definitive host.

Trypanosomes have been much studied of late, and many species have been described. At least three have been found in man.

Trypanosoma gambiense is the parasite of African "sleeping sickness." Its detection in the blood is described on page 348. It is more abundant in the juice obtained by aspirating a lymph gland with a large hypodermic needle, and in the late stages is also found in the cerebrospinal fluid. A new species causing sleeping sickness in man has recently been described and has been named *T. rhodesiense*. The chief point of distinction from *T. gambiense* is the situation of the nucleus close to or even posterior to the blepharoplast. It is transmitted by the fly *Glossina morsitans*.

Trypanosoma cruzi is a small form which has been found in the blood of man in Brazil.

Trypanosoma lewisi, a very common and apparently harmless parasite of gray rats, especially sewer rats, is

Fig. 159.—*Trypanosoma lewisi* in blood of rat. The red corpuscles were decolorized with acetic acid (× 1000).

interesting because it closely resembles the pathogenic forms, and is easily obtained for study. Its posterior end is more pointed than that of *T. gambiense* (Fig. 159).

Trypanosoma evansi, *T. brucei*, and *T. equiperdum* produce respectively surra, nagana, and dourine, which are common and important diseases of horses, mules, and cattle in the Philippines, East India, and Africa.

30

4. Genus Leishmania. The several species which compose this genus are apparently closely related to the trypanosomes, but their exact classification is undetermined. They have been grown outside the body and their transformation into flagellated trypanosome-like structures has been demonstrated. Calkins places them in the genus *Herpetomonas*.

1. **Leishmania donovani** is the cause of kala-azar, an important and common disease of India. With Wright's stain the "Leishman-Donovan bodies" are round or oval, light blue structures, 2 to 3 μ in diameter, with two distinct reddish purple chromatin masses, one large and pale (trophonucleus), the other small

FIG. 160.—*Cercomonas hominis* (about × 500): A, Larger variety; B, smaller variety (Davaine).

and deeply staining (blepharoplast). The parasites are especially abundant in the spleen, splenic puncture being resorted to for diagnosis. They are readily found in smears stained by any of the Romanowsky methods. Then lie chiefly within endothelial cells and leukocytes. They are also present within leukocytes in the peripheral blood, but are difficult to find in blood-smears.

2. **Leishmania tropica** resembles the preceding. It is found, lying intracellularly, in the granulation tissue of Delhi boil or Oriental sore

3. **Leishmania infantum** has been found in an obscure form of infantile splenomegaly in Algiers.

5. Genus Cercomonas.—1. **Cercomonas hominis,** sometimes found in the feces, particularly in tropical regions, is probably harmless. The body is 10 to 12 μ long, is pointed posteriorly, and has a flagellum at the anterior end (Fig. 160). The nucleus is difficult to make out. The feces should be examined in the fresh state, and preferably while warm, in order to observe the active motion of the organism.

6. Genus Bodo.—1. **Bodo urinarius** is sometimes seen in the urine, darting about in various directions. It is probably an accidental contamination. It has a lancet-shaped body, about 10 μ long, and is somewhat twisted upon itself, with two flagella at the end.

7. Genus Trichomonas.—1. **Trichomonas intestinalis** is sometimes confused with the two flagellates just described but is more important and more definitely known than either. It is an oval or pear-shaped cell of somewhat changing shape. The average size is about 10 by 15 μ although there is considerable variation among individuals. An oval nucleus can sometimes be made out in the anterior half of the cell. At the anterior or blunt end there is a cluster of three—some say four—flagella of equal length, and along one side is an undulating membrane the thickened free edge of which is continued backward as a short flagellum. Owing to the active motion of the flagella and undulating membrane, these are not easily seen, and at first sight the parasite has much the appearance of a pus-corpuscle moving busily about among the fecal particles. According to Stitt the flagella can be seen more clearly if a drop of Gram's iodin solution be added to the preparation on the slide. The organism's usual habitat is the colon where, contrary to what is known of the similar or identical *T. vaginalis*, it is said to prefer an alkaline medium. Under conditions unfavorable to active life it becomes encysted.

Trichomonas intestinalis is common in the tropics, and from recent reports it appears to be widespread throughout the United States. While it was formerly believed to be non-pathogenic, it is now well-established that it may cause a diarrhea of the dysenteric type or that it at least may greatly aggravate an already existing in-

flammatory condition. The parasites are often so abundant that four or more may be seen in a single field of the high dry objective.

2. **Other Trichomonads.**— Various forms have been described, regarded by some as identical with

Fig. 161.—*Trichomonas vaginalis* (about + 1000) (after Kölliker and Scanzoni).

T. intestinalis, by others as distinct species. Among these are *T. pulmonalis*, which has been encountered in the sputum of persons suffering from pulmonary gangrene and putrid bronchitis, and *T. vaginalis* which is often found in the leucorrheal discharge of catarrhal vaginitis. The latter flourishes only in an acid medium and disappears when the discharge becomes alkaline as, for instance, during menstruation. In a case of coincident severe inflammation of the vagina and of the gums recently studied by Lynch, the parasites were very numerous in both situations. They averaged 22 by 26 μ in size, presented four flagella at the anterior end and had an undulating membrane but no posterior flagellum. Treatment with an alkaline wash was completely successful.

A few cases have been reported in which *T. vaginalis* was apparently the cause of a urethritis in the male.

8. Genus Lamblia.—1. **Lamblia intestinalis** is a very common parasite in the tropics, but is generally considered of little pathogenic importance. It appears, however, to be capable of producing a chronic diarrhea. In 6000 stool examinations at the Mayo

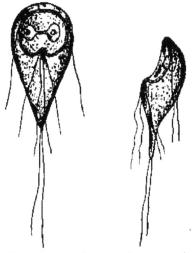

FIG. 162.—*Lamblia intestinalis* from the intestines of a mouse (about × 2000) (Grassi and Schweiakoff).

Clinic this parasite was found 66 times, the infected individuals coming chiefly from the Northern States and Canada. The organism is pear-shaped, measures 10 to 15 μ or more in length, and has a depression on one side of the blunt end, by which it attaches itself to the tops of the epithelial cells of the intestinal wall. Three pairs of flagella are arranged about the depression and one pair at the pointed end (Fig. 162). Its usual habitat is the upper part of the small intestine. Unless the stool is obtained by catharsis (see p. 443), encysted forms only may be found, and these may be difficult or impossible to recognize.

SUBPHYLUM SPOROZOA
Class Telosporidia

All the members of this class are parasitic, but only a few have been observed in man, and only one genus, *Plasmodium*, is of much importance in human pathology. Propagation is by means of spores, and sporulation ends the life of the individual. In some species there is an alternation of generations, in one of which sexual processes appear. In such cases the male individual may be provided with flagella. Otherwise, there are no special organs of locomotion.

1. Genus Coccidium.—1. Coccidium cuniculi. —This is a very common parasite of the rabbit and has been much studied; but extremely few authentic cases of infection in man have been reported. The parasite, which when fully developed is ovoid in shape and measures about 30 to 50 μ in length and has a shell-like integument, develops within the epithelial cells of the bile-passages. Upon reaching adult size it divides into a number of spores or merozoites which enter other epithelial cells and repeat the cycle. A sexual cycle outside the body, which suggests that of the malarial parasite but does not require an insect host, also occurs. Infection takes place from ingestion of the resulting sporozoites.

2. Genus Plasmodium.—This genus includes the malarial parasites which have already been described (see pp. 349–362).

3. Genus Babesia.—The proper position of this genus is uncertain. It is placed among the flagellates by some. The chief member is *Babesia bigeminum*, the cause of Texas fever in cattle. It is a minute, pear-shaped organism, lying in pairs within the red blood-corpuscles. An organism found in the red blood-corpuscles and certain tissue cells in Rocky Mountain spotted fever was at one time placed in this genus under the name *B.* (or *Piroplasma*) *hominis*, but its classification is uncertain.

SUBPHYLUM INFUSORIA
Class Ciliata

The conspicuous feature of this class is the presence of cilia. These are hair-like appendages which have a regular to-and-fro motion, instead of the irregular lashing motion of flagella. They are also shorter and more numerous than flagella. Most infusoria are of fixed shape and contain two nuclei. Contractile and food-vacuoles are also present. Encystment is common. Only one species is of medical interest. Certain ciliated structures, which have been described as infusoria, notably in sputum and nasal mucus, were probably ciliated body cells.

1. Genus Balantidium.—1. **Balantidium coli.**— This parasite, formerly called *Paramœcium coli*, is an occasional inhabitant of the colon of man, where it sometimes penetrates into the mucous membrane and produces a diarrheal condition resembling amebic dysentery. It is an actively moving oval organism, about 60 to 100 μ long and 50 to 70 μ wide, is covered with cilia, and contains a bean-shaped macronucleus, a globular micronucleus, two contractile vacuoles, and variously sized granules (Fig. 163).

Its ordinary habitat is the large intestine of the domestic pig, where it apparently causes no disturbance. It probably reaches man in the encysted condition.

2. **Balantidium minutum** resembles *B. coli* but is smaller, measuring 20 to 30 by 15 to 20 μ. It has been found a few times in diarrheal stools.

FIG. 163.—*Balantidium coli* (about × 350) (after Eichhorst).

PHYLUM PLATYHELMINTHES

The old phylum Vermidea has been subdivided into three phyla, those which are of interest here being the Platyhelminthes and Nemathelminthes, the flat worms and the round worms respectively. Of these, many species are parasitic in man and the higher animals. In some cases man is the regular host; in others the usual habitat is some one of the animals, and the occurrence of the worm in man is more or less accidental. Such are called *incidental parasites*. Only those worms that are found in man with sufficient frequency to be of medical interest are mentioned here.

The most important means of clinical diagnosis of

infection by either the flat worms or the round worms is the finding of ova. In many cases the ova are so characteristic that the finding of a single one will establish the diagnosis. In other cases they must be carefully studied and a considerable number measured. While ova from the same species will naturally vary somewhat, the average size of a dozen or more is pretty constant. The measurements given here are mainly those accepted by Stiles or Ward.

PHYLUM PLATYHELMINTHES
(Flat Worms)

CLASS Trematoda.—Flukes. Unsegmented, leaf-shaped.

Genus	*Species*
Fasciola.	F. hepatica.
Dicrocœlium.	D. lanceatum.
Opisthorchis.	Op. felineus.
	Op. sinensis.
Fasciolopsis.	F. buski.
Paragonimus.	P. westermani.
Schistosomum.	S. hæmatobium.
	S. mansoni.
	S. japonicum.

CLASS Cestoda.—Tapeworms. Segmented, ribbon-shaped.

Genus	*Species*
Tænia.	T. saginata.
	T. solium.
	T. echinococcus.
Hymenolepis.	H. nana.
	H. diminuta.
Dipylidium.	D. caninum.
Dibothriocephalus.	D. latus.

Class Trematoda

The trematodes, commonly known as "flukes," are flat, unsegmented, generally tongue- or leaf-shaped worms. They are comparatively small, most species averaging between 5 and 15 mm. in length. They possess an incomplete digestive tract, without anus, and are provided with one or more sucking disks by means of which they can attach themselves to the host. Some are also provided with hooklets. Nearly all species are

FIG. 164.—*Fasciola hepatica*, about two-thirds natural size (Mosler and Peiper).

hermaphroditic, and the eggs of nearly all are operculated (provided with a lid), the only important exception being the several species of *Schistosomum*. Development takes place by alternation of generations, the intermediate generation occurring in some water animal: mollusks, amphibians, fishes, etc. Trematode infection is uncommon in this country.

1. Genus Fasciola.—1. **Fasciola hepatica.**—The "liver fluke" inhabits the bile-ducts of numerous herbivorous animals, especially sheep, where it is an important cause of disease. It brings about obstruction of the bile-passages, with enlargement and degeneration of the liver—"liver rot." A species of snail serves as intermediate host. The worm is leaf-shaped, the average size being about 2.8 by 1.2 cm. The anterior end projects like a beak (head-cone 3 to 4 mm. long) (Fig. 164). Ova appear in the feces. They are yellowish brown, oval, operculated, and measure about 130 to 140 by 75 to 90 μ.

2. Genus Dicrocœlium.—1. **Dicrocœlium lanceatum** is often associated with the liver fluke in the bile-passages of animals, but is neither so common nor so widely distributed geographically. It has rarely been observed in man. It is smaller (length about 1 cm.) and more elongated. The eggs measure 38 to 45 μ long and 22 to 30 μ wide.

3. Genus Opisthorchis.—1. **Opisthorchis felineus** inhabits the gall-bladder and bile-ducts of the domestic cat and a few other animals. Infection in man has been repeatedly observed in Europe, and especially in Siberia. The body is flat, yellowish-red in color, and almost transparent. It measures 8 to 11 by 1.5 to 2 mm. The eggs, which are found in the feces, are oval, with a well-defined operculum at the narrower end, and contain a ciliated embryo when deposited. They measure about 30 by 11 μ.

2. **Opisthorchis sinensis,** like the preceding fluke, inhabits the gall-bladder and bile-ducts of domestic cats and dogs. It is, however, much more frequent in man, being a common and important parasite in certain parts of Japan and China. The number present may be very great; over 4000 were counted in one case. The parasite resembles *Op. felineus* in shape and color.

It is 10 to 14 mm. long and 1.5 to 4 mm broad. The eggs have a sharply defined lid and measure 25 to 30 by 15 to 17 μ. When they appear in the feces they contain a ciliated embryo. The intermediate host is unknown.

4. Genus Fasciolopsis.—1. Fasciolopsis buski.— This fluke is parasitic in the duodenum of man, and is widespread in the East, notably in India, China, and Japan. A few imported cases have been reported in this country. When in considerable numbers it causes a bloody diarrhea accompanied by high fever. The usual length is about 30 mm.; width, 10 to 12 mm.; thickness, 1.5 to 4 mm. The eggs are thin shelled, with granular contents, possess a minute operculum, and measure about 125 by 75 to 80 μ.

5. Genus Paragonimus.—1. Paragonimus westermannii, called the "lung fluke," is also a common parasite of man in Japan, China, and Korea. It inhabits the lung, causing the formation of small cavities. Moderate hemoptysis is the principal symptom. Ova are readily found in the sputum (Fig. 165); the worms themselves are seldom seen, except *postmortem*. The worms are faintly reddish-brown in color, egg shaped, with the ventral surface flattened, and measure 8 to 10 by 4 to 6 mm. The ova are thin shelled, operculated, brownish yellow, and measure about 87 to 100 by 52 to 66 μ.

The larval stage occurs in several species of freshwater crab which are common articles of food in Japan.

According to Ward, three distinct species have been confused under the name *P. westermani:* the original

form, *P. westermani*, found in the tiger; the American
lung fluke, *P. kellicotti*, thus far found only in cat, dog,
and hog; and the Asiatic lung fluke of man, *P. ringeri*,
described above.

6. Genus Schistosomum.—1. **Schistosomum
hæmatobium.**—This trematode, frequently called *Bil-*
harzia hæmatobia, is an extremely common cause of

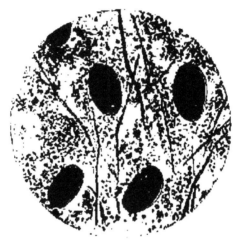

Fig. 165.—Sputum of man containing eggs of the lung fluke, greatly
enlarged (after Manson).

disease (bilharziasis or Egyptian hematuria) in northern Africa, particularly in Egypt.

Unlike the other flukes, the sexes are separate. The
male is 12 to 14 mm. long and 1 mm. broad. The body
is flattened and the lateral edges curl ventrally, forming a longitudinal groove, in which the female lies
(Fig. 166). The latter is cylindric in shape, about 20
mm. long and 0.25 mm. in diameter. The eggs are
an elongated oval, about 120 to 190 μ long and 50 to

73 μ broad, yellowish in color, and slightly transparent. They possess no lid, such as characterizes the eggs of most of the trematodes, but are provided with a thorn-like spine which is placed at one end (Fig. 167). Within is a ciliated embryo.

FIG. 166.—*Schistosomum hæmatobium,* male and female (about × 4), with egg (about × 70) (von Jaksch).

In man the worm lives in the veins, particularly the portal vein and the veins of the bladder and rectum, leading to obstruction and inflammation. The eggs penetrate into the tissues and are present in abundance

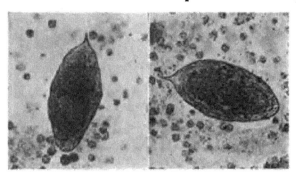

FIG. 167.—Ova of *Schistosomum hæmatobium* with pus corpuscles in urine (× 250).

in the mucosa of the bladder and rectum. They also appear in the urine and, less commonly, in the feces.

A species of snail serves as intermediate host, and infection in man apparently takes place both by mouth and through the skin.

2. **Schistosomum mansoni.**—It has long been observed that schistosomum eggs in the urine have usually a terminal spine, while in the feces the lateral spine is more common. It is now known that the lateral-

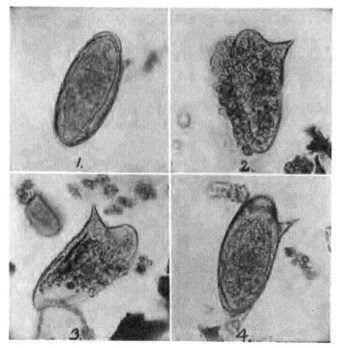

FIG. 168.—Ova of *Schistosomum mansoni*: 1, With spine out of focus; 2, in a clump of red blood-cells; 3, apparently unfertilized; 4, usual appearance (× 250).

spined egg is that of a distinct species, to which the name *Schistosomum mansoni* has been given. It is found in Africa along with *Schistosomum hæmatobium*, but is especially prevalent in the West Indies and Central America. The adult worms closely resemble

the male and female of *S. hæmatobium.* They inhabit
the rectal and portal veins, and ova appear in the
feces, where they are very easily recognized from their
size and the characteristic spine (see Figs. 168 and 182).
They are light yellow in color, measure 112 to 162 by
60 to 70 μ, and are provided with a cleanly cut, sharply
pointed spine, which is situated at the juncture of the
last and third quarter of the egg and is directed backward. Within the egg is a ciliated embryo (mira-
cidium) which can be seen without difficulty.

3. **Schistosomum japonicum** resembles *S. hæmato-*
bium morphologically, but both the male and female are
smaller. The ova, which appear in the feces, are ovoid,
thin shelled, and without lid or spine. They average
83 by 62 μ in size, and contain a ciliated embryo. The
worm inhabits the portal and probably also other veins.

Class Cestoda

The cestodes, or tapeworms, are very common parasites of both man and the animals. In the adult stage
they consist of a linear series of flat, rectangular segments (proglottides), at one end of which is a smaller
segment, the scolex or head, especially adapted by
means of sucking disks and hooklets for attachment
to the host. The series represents a colony, of which
the scolex is ancestor. The proglottides are sexually
complete individuals (in most cases hermaphroditic)
which are derived from the scolex by budding. With
the exception of the immature segments near the
scolex, each contains a uterus filled with ova.

The large tapeworms, *Tænia saginata, T. solium,* and

Dibothriocephalus latus, are distinguished from one another mainly by the structure of the scolex and of the uterus. The scolex should be studied with a low-power objective or a hand lens. The uterus is best seen by pressing the segment out between two plates of glass.

All the tapeworms pass a larval stage in the tissues of an intermediate host, which is rarely of the same species as that which harbors the adult worm. Within the ova which have developed in the proglottides of the adult worm, and which pass out with the feces of the host, there develop embryos, or *oncospheres*, each provided with three pairs of horny hooklets. When the egg is taken into the intestines of a suitable animal, the oncosphere is liberated and penetrates to the muscles or viscera and there, in the case of most of the tapeworms, forms a cyst in which develop usually one, but sometimes many, scolices, which are identical with the head of the adult worm. When the flesh containing this cystic stage is eaten without sufficient cooking to destroy the scolices, the latter attach themselves to the intestinal wall and produce adult tapeworms by budding. The oncosphere of some of the tapeworms leaves the egg in the open and exists for a time as a free-living larva before entering the intermediate host.

Ordinarily, only the adult stage occurs in man. In the case of *Tænia echinococcus* only the larval stage is found. *T. solium* may infect man in either stage, although the cystic stage is rare.

Since the head, or scolex, is the ancestor from which the worm is formed in the intestine, it is important, after giving a vermifuge, to make certain that the head

31

has been passed with the worm. Should it remain, a new worm will develop.

The principal tapeworms found in man belong to the genera Tænia, Hymenolepis, and Dibothriocephalus.

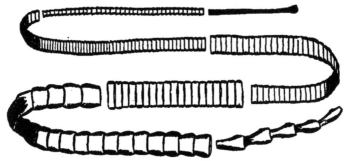

FIG. 169.—*Tænia saginata* (Eichhorst)

1. Genus Tænia.—1. **Tænia saginata** (Fig. 169).— This, the beef tapeworm, is the common tapeworm of

FIG. 170.—Head of *Tænia saginata* (Mosler and Peiper).

the United States, and is widely distributed over the world. Its length is generally about 4 to 8 meters. The scolex is about the size of a large pin-head (1.5 to

2 mm. in diameter), and is surrounded by four sucking disks, but has no hooklets (Fig. 170). The neck is about 1 mm. wide. The terminal segments, which become detached and appear in the feces, measure about 18 to 20 mm. long by 4 to 7 mm. wide. The uterus extends along the midline of the segment and gives off twenty to thirty branches upon each side (see Fig. 180, 1).

The larval stage is passed in the muscles of various animals, especially cattle. It rarely or never occurs in

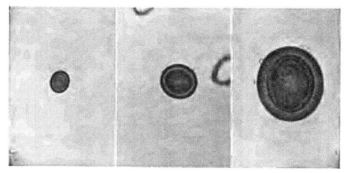

FIG. 171.—Eggs of *Tænia saginata*, magnifications 100, 250, and 500 diameters.

man, hence there is little or no danger of infection from examining feces.

The scolex is ingested with the meat, its capsule is dissolved by the digestive juices, and it attaches itself to the intestinal wall by means of its suckers. It then develops into the mature worm, which may grow very rapidly, even as many as thirteen or fourteen segments being formed in a day.

The ova are present in the feces of infected persons, sometimes in great numbers. When, however, segments are passed, the ova for the most part remain within them and comparatively few are found free in the feces. They are spheric or ovoid, yellow to brown in color, and have a thick, radially striated shell (Fig. 171). Within them the six hooklets of the embryo (oncosphere) can usually be made out as three pairs of parallel lines. The size of the ova varies from 20 to 30 μ wide and 30 to 40 μ long. Vegetable cells, which are generally present in the feces, are often mistaken for them, although there is no great resemblance.

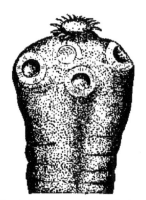

FIG. 172.—Head of *Tænia solium* (Mosler and Peiper).

2. **Tænia solium,** the pork tapeworm is very rare in this country. It is usually much shorter than *Tænia saginata*. The scolex is about 0.6 to 1 mm. wide, is surrounded by four sucking disks, and has a projection, or rostellum, with a double row of horny hooklets, usually twenty-six to twenty-eight in number (Fig. 172). The terminal segments measure about 5 to 6 by 10 to 12 mm. The uterus has only seven to fourteen branches on each side (see Fig. 180, 3).

The cysticercus stage occurs ordinarily in the muscles of the pig, but is occasionally seen in man, most frequently affecting the brain and eye (*Cysticercus cellulosæ*). There is, therefore, danger of infection from handling feces.

The ova so closely resemble those of *Tænia saginata* as to be practically indistinguishable. They average about 31 to 36 μ in diameter and are usually spheric.

3. **Tænia echinococcus.**—The mature form of this tapeworm inhabits the intestines of the dog and wolf. The larvæ develop in cattle and sheep ordinarily, but are sometimes found in man, where they give rise to echinococcus or "hydatid" disease. The condition is unusual in America, but is not infrequent in Central Europe and is common in Iceland and Australia.

FIG. 173.—*Tænia echinococcus;* enlarged (Mosler and Peiper).

The adult parasite is 2.5 to 5 mm. long and consists of only four segments (Fig. 173). It contains many ova. When the ova reach the digestive tract of man the embryos are set free and find their way to the liver, lung, or other organ, where they develop into cysts, thus losing their identity. The cysts may attain the size of a child's head. Other cysts, called "daughter-cysts," are formed within these. The cyst-wall is made up of two layers, from the inner of which (the so-called "brood membrane") there develop larvæ which are identical with the head, or scolex, of the mature parasite. These are ovoid structures 0.2 to 0.3 mm. long. Each has four lateral suckers and a rostellum surmounted by a double circular row of horny hooklets. The rostellum with its hooklets is frequently invaginated into the body.

Diagnosis of echinococcus disease depends upon detection of scolices, free hooklets which have fallen off from degenerated scolices, or particles of cyst-wall which

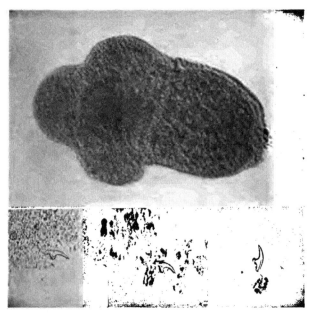

FIG. 174.—Degenerated scolex without hooklets and free hooklets of *Tænia echinococcus* in fluid from hepatic cyst (× 300).

are characteristically laminated and usually have curled edges. The lamination is best seen at the torn edge of the membrane. All of these structures can be found in fluid withdrawn from the cysts or, less frequently, in the sputum or the urine, when the disease involves the lung or kidney (see Figs. 75, 174, 175). In such material the scolices are usually much degenerated and many of them have entirely lost the hooklets. The cysts are sometimes "barren," growing to a considerable size without producing scolices.

The cyst fluid is clear, between 1.009 and 1.015 in specific gravity, and contains a notable amount of sodium chlorid, but no albumin. Recently diagnosis of echinococcus disease has been made by the complement fixation method.

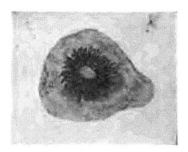 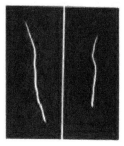

FIG. 175.—Portion of a degenerated scolex of *Tænia echinococcus*, showing circle of hooklets. From a hepatic cyst (× 250).

FIG. 176.—Dwarf tapeworm (*Hymenolepis nana*) adults. From photographs. Natural size.

2. Genus Hymenolepis.—1. **Hymenolepis nana,** the dwarf tapeworm (Figs. 176 and 177), is 1 to 4.5 cm. in length and 0.4 to 0.7 mm. in breadth at the widest part. The head has a rostellum with a crown of 24 to 30 hooklets. There are about 150 segments. Diagnosis must, in general, depend upon the discovery of ova in the feces since the worms themselves are usually partly disintegrated when they leave the body and are recognized with difficulty. The ova are nearly spheric and contain an embryo surrounded by two distinct membranous walls, between which is a

The eggs are grouped in packets of eight to fifteen and are usually passed from the bowel within the proglottids.

The intermediate host is the flea or louse. Infection of human beings is rare, and is mostly confined to children, who are probably infected from getting lice or fleas of dogs or cats into their mouths.

4. Genus Dibothriocephalus.—1. **Dibothriocephalus latus,** the fish tapeworm, sometimes reaches 20 meters in length, although it is generally not more than one-half or one-third as long. When several worms are present, they are much shorter, often only 1.5 or 2 meters. The head is a flattened ovoid, about 1 mm. broad

FIG. 179.—Head of *Dibothriocephalus latus* seen from the narrow side, showing one of the two grooves (× 15).

and 1.5 mm. long. It is unprovided with either suckers or hooklets, but has two longitudinal grooves which serve the same purpose (Fig. 179). The length of the segments is generally less than their breadth, mature segments measuring about 3 by 10 or 12 mm. The uterus, which is situated in the center of the segment, is roset shaped (Fig. 180, 2) and brown or black in color.

The number of segments sometimes exceeds 3000. As a rule they do not appear in the feces singly, but in chains of considerable length.

The larvæ, which do not form cysts but live as worm-like structures 2 to 3 cm. long (plerocercoids), are found in various fish, notably the pike, burbot, grayling, and certain trout. Infection of man prevails only in regions where these fish are found. It is very common in Japan and in various countries of Europe, especially Ireland and the Baltic provinces of Russia. A number of cases of infection have been reported in this country, a few of which were undoubtedly acquired here. Any

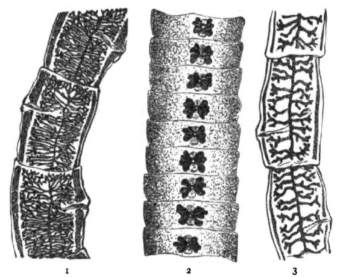

FIG. 180.—Segments of—(1) *Tænia saginata;* (2) *Dibothriocephalus latus;* (3) *Tænia solium,* showing arrangement of uterus.

locality in which favorable fish are native becomes a possible center of infection if the worm is introduced by infected immigrants.

The ova are characteristic. They measure about 45 by 70 μ, are brown in color, and are filled with small spherules. The shell is thin and has a small hinged lid at one end. As the eggs appear in the feces the lid is not easily seen, but it may be demonstrated by sufficient pressure upon the cover-glass to force it open (Figs. 181, 182). The only other operculated eggs met with in man are those of the fluke-worms.

Dibothriocephalus latus is interesting clinically because in many cases it causes a very severe grade of anemia, which may be indistinguishable from pernicious anemia.

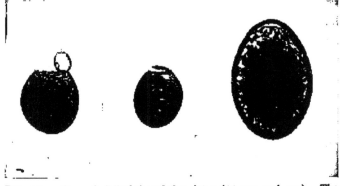

FIG. 181.—Ova of *Dibothriocephalus latus* (× 250 and 500). The lids were forced open by pressure upon the cover-glass.

PHYLUM NEMATHELMINTHES
(Round Worms)

CLASS **Nematoda.**—Unsegmented, cylindric or fusiform.

Genus	Species
Anguillula.	A. aceti.
Ascaris.	A. lumbricoides.
	A. canis.
Oxyuris.	O. vermicularis.
Filaria.	F. bancrofti.
	F. philippinensis.
	F. perstans.
	F. diurna.
	F. medinensis.

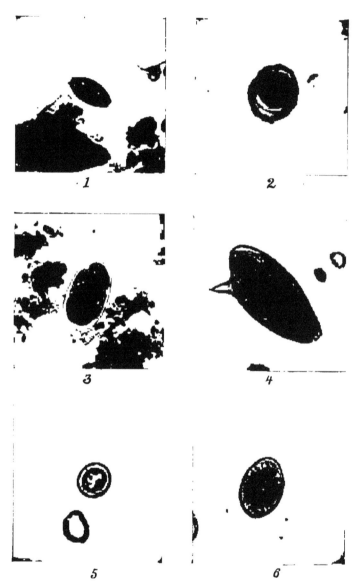

FIG. 182.—Showing comparative size of ova found in the feces: 1, *Trichocephalus trichiurus;* 2, *Ascaris lumbricoides;* 3, *Necator americanus,* four-cell stage; 4, *Schistosomum mansoni;* 5, *Tænia saginata;* 6, *Dibothriocephalus latus,* the line of the lid being out of focus (× 250).

Ancylostoma.	A. duodenale.
Necator.	N. americanus.
Strongyloides.	S. intestinalis.
Trichinella.	T. spiralis.
Trichocephalus.	T. trichiurus.

Class Nematoda

The nematodes, or round-worms, are cylindric or fusiform worms, varying in length, according to species, from 1 mm. to 40 or 80 cm. As a rule, the sexes are separate. The male is smaller and more slender than the female. In a few cases the female is viviparous; in most cases she deposits ova which are characteristic, so that the finding of a single egg may establish the diagnosis. Except in a few instances the young are different from the adult, and must pass a certain larval stage of development before again reaching a host. An intermediate host is, however, necessary with only a few species.

1. Genus Anguillula.—1. Anguillula aceti.— This worm, commonly called the "vinegar eel," is usually present in vinegar. A drop of the vinegar, particularly of the sediment, will frequently show great numbers, all in active motion: males, about 1 or 1.5 mm. long; females, somewhat larger and frequently containing several coiled embryos; and young, of all sizes up to the adult (see Fig. 76).

The vinegar eel is never parasitic, but is occasionally met with as a contamination in the urine (see p. 237), and has there been mistaken for the larva of filaria or strongyloides.

2. Genus Ascaris.—1. **Ascaris Lumbricoides.**—
The female is 20 to 40 cm. long and about 5 mm. thick
(Fig. 183); the male, 15 to 17 cm. long and 3 mm. thick.
They taper to a blunt point anteriorly and posteriorly.
At the anterior end are three small lips which can easily
be seen with a hand lens.
Their color is reddish or
brown. They are the com-
mon "round-worms" so fre-
quently found in children.
Their habitat is the small
intestine. Usually several
individuals are present and
sometimes many.

The diagnosis is made by
detection of the worms or
ova in the feces. The latter
are generally numerous and
are easily recognized.
They are elliptic, measuring
about 50 by 65 to 80 μ, and
have an unsegmented pro-
toplasm. There is usually
a crescentic clear space at
each pole between the contents and the shell (Fig.

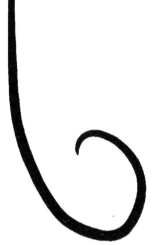

FIG. 183.—*Ascaris lumbricoides*
(female) (Mosler and Peiper).

184). The shell is thick and has a roughly mammillated
or sculptured surface (Fig. 185). When only females
are present in the intestine, and occasionally at other
times, one finds unfertilized eggs. These are generally
much more elongated, have a thinner and smoother

shell, have coarsely granular contents, and lack the crescentic clear spaces.

The eggs do not hatch in the intestine of the original host. They pass out in the feces and, after a variable period, usually about five weeks, come to contain an embryo which remains within the shell until ingested by a new host. The embryo is very resistant and may remain alive within the shell for years, even, according to Morris, when preserved in 2 per cent. formalin. Upon reaching the intestine of the new host it hatches

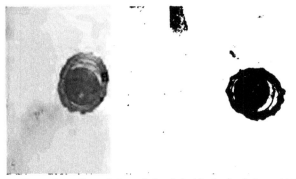

FIG. 184.—Ova of *Ascaris lumbricoides* in fresh feces (× 250).

out and develops into the adult worm in about a month.

2. **Ascaris canis** is the very common "stomach worm" of cats and dogs. It is also known as *Ascaris mystax* and *Toxocara canis*. It is rare as a human parasite. The male is 4 to 9 cm. long; the female, 12 to 20. Individuals from dogs are generally larger than those from cats. The egg is spheric, 68 to 70 μ in diameter, and has a thin shell with comparatively smooth surface.

3. Genus Oxyuris.—1. Oxyuris vermicularis.—
This is the "thread-worm" or "pin-worm" which
matures in the small intestine and cecum and in the
adult stage inhabits the colon and rectum, especially of

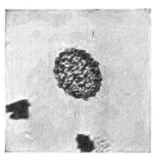

FIG. 185.—Egg of *Ascaris lumbricoides*, surface view (× 250).

young children. Its presence should be suspected in
all unexplained cases of pruritus ani. The female is
about 9 to 12 mm. long; the male, about 3 to 5 mm.

FIG. 186.—*Oxyuris vermicularis*, male and female, natural size (after
Heller).

The worms are not infrequently found in the feces
particularly after an enema; the ova, rarely. The
latter are best found by scraping the skin with a dull
knife at the margin of the anus, where they are deposited by the female, who wanders out from the rec-

32

tum for this purpose, thus producing the troublesome itching. They are asymmetrically oval with one flattened side, are about 50 μ long by 16 to 25 μ wide, have a moderately thin, double-contoured shell, and when deposited contain a partially developed embryo (Fig. 187). The diagnosis is best made by giving a purgative and searching the stool for the adult worms (see p. 427).

Infection takes place through swallowing the ova. Auto-infection is likely to occur in children; the ova cling to the fingers after scratching and are thus carried

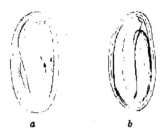

a   *b*

FIG. 187.—Eggs of *Oxyuris vermicularis: a*, freshly deposited, with tadpole-like embryo; *b*, twelve hours after deposition, with nematode-like embryo (× 500) (after Fantham, Stephens and Theobald).

to the mouth. Diagnosis can sometimes be made by finding the ova in the dirt beneath the finger-nails.

4. Genus Filaria.—1. **Filaria bancrofti.**—The adults are thread-like worms, the male about 4 cm., the female about 8 cm., long. They live in pairs in the lymphatic channels and glands, especially those of the pelvis and groin, and often occur in such numbers as to obstruct the flow of lymph. This is the most common cause of elephantiasis and chyluria. Infection is very common in tropical and sub-tropical countries, where in some regions as high as 50 per cent. of the natives harbor microfilariæ. Even as far north as

Charleston, S. C., Johnson has found over 19 per cent. of infection among the poorer classes. Of these only one-fourth showed any symptoms referable to the filariæ.

The female is viviparous, and produces vast numbers of larvæ, which appear in the circulating blood. These are conveniently called microfilariæ; the name *Filaria sanguinis hominis*, which was formerly applied to them, is incorrect, since they do not constitute a species. These larvæ are slender, being about as wide as a red corpuscle and 0.2 to 0.4 mm. long (see Fig. 139), and are very active, although, owing to the fact that they are inclosed in a loose transparent sheath, they do not move about from place to place. They are found in the peripheral blood chiefly at night, being usually easily demonstrable by eight o'clock and reaching their maximum number—which may be enormous—about 2 a.m. By the concentration method given on page 363 a few can usually be demonstrated during the day. In the case of a medical student from Porto Rico with no symptoms, Smith and Rivas found 30 microfilariæ per cubic centimeter of blood at 4 p. m. and 6500 per cubic centimeter at 2 a.m. If the patient change his time of sleeping, they will appear during the day. Infection is carried by a variety of mosquito, which acts as intermediate host.

Diagnosis rests upon detection of larvæ in the blood, as described on page 362, but the number of larvæ found bears little relation to the severity of the symptoms, since the symptoms are largely mechanical and depend upon the localization of the adults within the body.

The larvæ are sometimes found in urine and chylous fluids from the serous cavities. Their motion is then usually less active than when in blood. That shown in Fig. 188 was alive sixty hours after removal of the fluid. Larvæ were present in the blood of the same patient.

A number of other filariæ whose larvæ appear in the blood are known, some of them only in the larval stage.

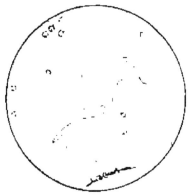

Fig. 188.—Larva of *Filaria bancrofti* in chylous hydrocele fluid; length, 300 μ; width, 8 μ. The sheath is not shown. A number of red blood-corpuscles also appear (studied through courtesy of Dr. S. D. Van Meter).

Among these are *Filaria philippinensis* and *F. perstans*, which exhibit no periodicity, and *F. loa*, whose larvæ appear in the blood during the day. The adult of the last named is especially frequent in the orbit and beneath the conjunctiva.

2. **Filaria medinensis** the "guinea-worm," is a very interesting and important worm of Africa and southern Asia. It has been thought to be the "fiery

serpent" which molested the Children of Israel in the Wilderness.

The larva probably enters the body through the skin or gastro-intestinal tract. It wanders about in the sub-cutaneous tissues until maturity, producing slight, if any, symptoms. The male, which is very rarely seen, is only about 4 cm. long. It dies soon after the female is impregnated. The adult female is a very slender, yellowish worm, about 50 to 80 cm. long, its appear-ance somewhat suggesting a catgut suture. When gestation is complete the greater part of the female's body consists of a uterus filled with embryos. The female then travels to the feet or ankles of the host and there causes the formation of a red nodule and, finally, an ulcer, from the center of which her head protrudes. Through this great numbers of larvæ are discharged whenever it comes in contact with water. Little dam-age is done unless the worm is pulled out, when the larvæ are set free in the tissues and cause serious disturbances.

When discharged the larvæ seek out a small crus-tacean, cyclops, which serves as intermediate host.

5. Ancylostoma duodenale and Necator ameri-canus.—These, the Old and the New World hookworm, respectively, are among the more harmful of the animal parasites. They inhabit the small intestine, often in great numbers, and commonly produce an anemia which is often severe and sometimes fatal. The presence of a few, however, may cause no disturbance.

Ancylostoma duodenale is common in southern Europe and in Egypt. The body is cylindric, reddish in color, and the head is bent sharply. The oral cavity

has six hook-like teeth. The female is 12 to 18 mm.
long and the tail is pointed. The male is 8 to 10 mm.
long and the posterior end is expanded into an umbrella-
like pouch, the caudal bursa. The eggs are oval and
have a thin, smooth, transparent shell. As they appear
in the feces the protoplasm is divided into 2, 4, 8, or
more rounded segments (Fig. 189). They measure 32
to 38 by 52 to 61 μ.

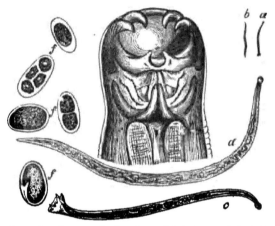

FIG. 189.—*Ancylostoma duodenale:* a, Male (natural size); b, female
(natural size); c, male (enlarged); d, female (enlarged); e, head; f, f, f,
eggs (after v. Jaksch).

Necator americanus is very common in subtropical
America, including the southern part of the United
States and the West Indies. In Porto Rico 90 per cent.
of the rural population is infected. Isolated cases,
probably imported, have been seen in most of the
Northern States. The American hook-worm is smaller
than the Old World variety, the male being 7 to 9 mm.
long, the female 9 to 11 mm. The four ventral hook-

like teeth are replaced by chitinous plates. There are also differences in the caudal bursa of the male, and in the situation of the vulva in the female. The ova (Fig. 190) resemble those of *Ancylostoma duodenale*, but are larger, 36 to 45 by 64 to 75 μ.

The life-history of the two worms is probably the same. The ova pass out with the feces, and, under favorable conditions of warmth and moisture, develop an embryo which hatches within a few days. The resulting larvæ pass through a stage of development in warm moist earth, growing to a length of 0.5 to 0.6 mm., and moulting twice. They are then ready to infect a new host. In some cases they probably reach the host's intestine by way of the mouth, with food or water; but the usual route is probably that established by Loos. When moist earth containing the larvæ comes in contact with the skin, they penetrate into the subcutaneous tissues. This is favored by retention of mud between the toes of those who go barefooted. When the larvæ are abundant a dermatitis is induced ("ground itch"). From the subcutaneous tissue they pass by way of lymph- and blood-streams to the lungs. Here they make their way into the smaller bronchi, are carried by the bronchial mucus to the pharynx, and are swallowed. They thus ultimately reach the small intestine, where they develop into mature worms.

The diagnosis of hookworm infection, which is assuming increasing importance in this country, must rest upon detection of ova in the feces. The worms themselves seldom appear except after thymol and a cathartic. A small portion of the feces, diluted with water if necessary, is placed upon a slide and the larger par-

ticles removed. The material is covered and searched
with a 16-mm. objective. A higher power may rarely
be necessary to positively identify an egg, but should
not be used as a finder. The eggs (see Figs. 182, 190)

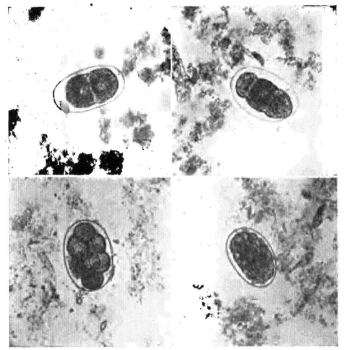

FIG. 190.—Ova of *Necator americanus* in feces. The egg, showing
three cells, is a lateral view of a four-cell stage (× 250).

are nearly always typical, showing a thin but very distinct shell, a clear zone, and a finely granular segmented
protoplasm. A light spot, representing the nucleus,
can usually be made out in each segment. After having
once been seen the eggs are not easily mistaken.

In heavy infections they may be found in nearly every microscopic field; in most cases, even when so mild as to cause no symptoms, they can be found on the first slide examined. It is seldom necessary to search more than half a dozen slides. From the estimate of Dock and Bass it seems probable that ova will average at least one to the slide if ten or more laying females are present in the intestine. Very old females may fail to produce

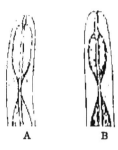

FIG. 191.—Diagram showing the differences in the mouth cavities of the larvæ of A, *Strongyloides*, and B, *Necator*.

eggs. When they are scarce, some method of sedimenting the feces should be tried (see p. 443).

Pepper's method of concentration is simple, but is not applicable to other ova than those of the hookworm. It is best first to sediment the feces. A layer of the diluted feces is placed on a slide and allowed to remain for some minutes. The slide is then gently immersed in water. The ova, which have settled to the bottom, cling to the glass and are not washed away as is other material. This may be repeated several times and numerous eggs collected.

Hookworm larvæ are not found in fresh feces, but may

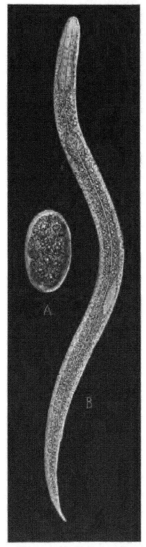

hatch within twenty-four to forty-eight hours after stool is passed. They are then easily mistaken for the larvæ of *Strongyloides intestinalis*, but can be distinguished by the depth of the mouth cavity, which is about equal to the diameter of the larvæ at the posterior end of the cavity. In *St. intestinalis* the mouth is about one-half as deep (see Fig. 191).

6. Genus Strongyloides. —1. Strongyloides intestinalis.—Infection with this worm is by no means so rare in this country as the few clinical reports would indicate. It is apparently widespread in the Southern States and is very common in subtropical countries, notably in Italy and in southern China. It seems probable that the parasite is the cause of "Cochin China diarrhea," although some authorities regard it as harmless.

FIG. 192.—A, Egg of *Strongyloides intestinalis* (parasitic mother worm) found in stools of case of chronic diarrhea; B, Rhabditiform larva of *Strongyloides intestinalis* from the stools. (William Sydney Thayer, in Journal of Experimental Medicine.)

The adult female, which reproduces by partheno-
genesis and is about 2 mm. long, inhabits the upper
portion of the small intestine, but neither it nor the
ova appear in the stool unless an active diarrhea exists.
Ordinarily they hatch in the intestines, and when infection is severe larvæ can be found in the feces in large
numbers.   These are the "rhabditiform larvæ," which
measure 450 to 600 μ by 16 to 20 μ (Figs. 192, 193).
They are actively motile, with a striking "wriggling"

FIG. 193.—Rhabditiform larva of *Strongyloides intestinalis* in feces
(× 150).

motion, and, when the stool is solid, are best found by
making a small depression in the fecal mass, filling it with
water, and keeping in a warm place (preferably an incubator) for twelve to twenty-four hours.   The larvæ
will collect in the water, and can be easily found by
transferring a drop to a slide and examining with a
16-mm. objective.   The inexperienced worker should
make sure that the worms move, or he may be misled
by the vegetable hairs which are generally present in

the feces. Certain of these hairs (notably those from the skin of a peach) closely resemble small worms (see p. 438).

Outside the body the rhabditiform larvæ develop into a free-living, sexually differentiated generation. The young of this generation are the more slender "filariform larvæ," which constitute the infective form. Direct transformation of rhabditiform into filariform larvæ also occurs. Infection takes place by ingestion or by way of the skin.

7. Genus Trichinella.—1. Trichinella spiralis.— This is a very small worm—adult males, 1.5 to 1.6 by 0.04 mm.; female, 3 to 4 by 0.06 mm. Infection in man occurs from eating of pork which contains encysted larvæ and is insufficiently cooked. Ordinary "curing" of pork does not kill them. According to Winn heating to 55°C. for fifteen minutes for each pound of meat is sufficient to kill all larvæ. Six days' refrigeration at 0°F. (−17.7°C.) is also effective. Protection against infection must be secured through such measures as these; meat inspection is of little value unless every part of the carcass is examined and this, of course, is impracticable. When the larvæ reach the stomach the capsule surrounding them is digested away and they grow to maturity in the small intestine. Soon after copulation the males die, and the females penetrate into the mucous membrane where they live for about six weeks, giving birth to great numbers of young, averaging as high as 1500 from a single female. The larvæ migrate to the striated muscles, chiefly near the tendinous insertions, where they grow to a length of about 0.8 mm., and finally become encysted. In this condition they may remain alive and capable of development for as long as twenty-five years.

Trichinella is widespread throughout the world and is more abundant in the United States than the reported cases of human infection would lead one to expect. It is capable of living in many animals but is most common in the pig and the rat. Rats when once infected continue the infection through cannibalism. A convenient means of finding whether the parasite is com-

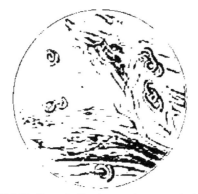

FIG. 194.—*Trichinella spiralis* (larvæ) from head of right gastrocnemius muscle; seventh week of disease (16-mm. objective; eye-piece 4) (Boston).

mon in a given community is to examine a series of slaughter-house rats.

Excepting during the acute stage trichiniasis is generally accompanied by a marked eosinophilia. The diagnosis is made by teasing out upon a slide a bit of muscle, obtained in man preferably from the outer head of the gastrocnemius, the insertion of the deltoid, or the lower portion of the biceps. In the case of rats the diaphragm, which is the most likely site, is pressed out between two glass slides. The colled larvæ can easily be seen with a 16-mm. objective (Fig. 194). The larvæ can usually be found in the spinal fluid and the blood (see p. 363) before they have reached their final resting-place in the muscles. During the diarrheal stage adults may be present in the feces, and are found by diluting with water, decanting several times and examining the sediment in a very thin layer in clean water with a hand lens.

8. Genus Trichocephalus.—1. **Trichocephalus trichiurus.**—This, the "whip-worm," is 3.5 to 5 cm. long. Its anterior portion is slender and thread-like,

FIG. 195.—Whip-worms (*Trichocephalus trichiurus*). A, females; B, males. The posterior portion of the male is usually coiled as is shown at the right. Photographs of mounted specimens. Natural size.

while the posterior portion is thicker (Fig. 195). It is widely distributed geographically, and is one of the most common of intestinal parasites in this country. It lives in the large intestine, especially the cecum, with its slender extremity embedded in the mucous membrane. Whip-worms do not, as a rule, produce any symptoms, although gastro-intestinal disturbances, nervous symptoms, and anemia have been ascribed to them. They, as well as many other intestinal parasites, are probably an important factor in the etiology of appendicitis, typhoid fever, and other intestinal infections. The damage which they do to the mucous membrane favors bacterial invasion.

The number present is usually small. The worms themselves are rarely found in the feces. The ova, which are not often abundant, are easily recognized with the 16-mm. objective. Although they are comparatively small, their appearance is striking. They are brown, ovoid in shape, 50 to 54 μ long by about 23 μ wide, and have a button-like projection at each end (Fig. 196). ·

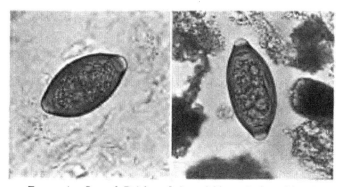

FIG. 196.—Ova of *Trichocephalus trichiurus* in feces (× 500).

PHYLUM ARTHROPODA

The arthropoda which are parasitic for man belong to the classes Arachnoidea and Insecta. They are nearly all external parasites, and the reader is referred to the standard works upon diseases of the skin for descriptions. The several species of the louse (*Pediculus capitis, P. vestimenti, Phthirius pubis*) (Figs. 197, 198, 199, 200), the itch mite (*Sarcoptes scabiei*), and the small organism (*Demodex folliculorum*) which lives in the sebaceous glands, especially about the face, are the most common members of this group.

A number of flies may deposit their ova in wounds or

FIG. 197.—Body louse (*Pediculus vestimenti*), female (× 15). The male is a little smaller and the posterior end of the abdomen has no notch. The body louse is distinguished from the head louse by its larger side and by the relative widths of thorax and abdomen.

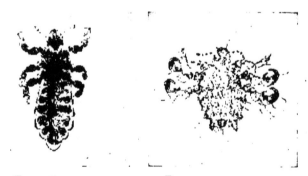

FIG. 198.   FIG. 199.

FIG. 198.—Head louse (*Pediculus capitis*), male (× 15). The female is a little larger and the posterior end of the abdomen is notched.
FIG. 199.—Pubic louse (*Phthirius pubis*) (× 15).

in such of the body cavities as they can reach, and the resulting maggots may cause intense irritation. To this condition the general term myiasis is applied. Ova may be swallowed with the food and the maggots appear in the feces. A few cases in which larvæ have been found in the urinary passages have been reported. Probably most important is infection with the "screw worm," the larva of *Chrysomyia macellaria*, which is not rare in some parts of the United States, particularly west of the Mississippi. The ova are most commonly deposited in

the nasal passages, and the larvæ, which may be present in great numbers, burrow through the soft parts, cartilage, and even bone,

FIG. 200.—Empty egg of *Pediculus capitis* on a hair (× 15).

FIG. 201.—Larva of *Linguatula serrata* (de Faria and Travassos).

always with serious and often with fatal results.

A few cases of human infection with *Linguatula serrata* have been reported from the Panama Canal Zone and from Brazil, and it may prove to be more common than has been recognized. The parasite belongs to the class Arachnoidea, which includes spiders,

33

mites, ticks, etc.' It is not at all rare in Europe. Related species are common in certain birds in North America. Man may be infected with either adult or larval stages, the former living in the nasal and accessory passages, the latter, encysted, in the internal organs, particularly the liver. The larvæ may be found in the feces, and, because of their serrations, may be mistaken for minute tapeworms (Fig. 201). They are white in color and measure about 4 to 6.5 mm. long and 0.9 to 1.5 mm. broad at the widest (anterior) part.

CHAPTER VII

MISCELLANEOUS EXAMINATIONS

PUS

Pus contains much granular débris and numerous more or less disintegrated cells, the great majority being polymorphonuclear leukocytes—so-called "pus-corpuscles." Eosinophilic leukocytes are common in gonorrheal pus and in asthmatic sputum. Examination of pus is directed chiefly to detection of bacteria.

When very few bacteria are present, culture methods, which are outlined in Chapter VIII, must be resorted to. When considerable numbers are present, they can be detected and often identified in cover-glass smears. Several smears should be made, dried, and fixed, as described on page 572.

One of these should be stained with a bacterial stain. Löffler's methylene-blue and Pappenheim's pyronin-methyl-green are especially satisfactory for this purpose. These stains are applied for one-half minute to two minutes or longer, without heating; the preparation is rinsed in water, dried, mounted, and examined with an oil-immersion lens. Another smear should be stained by Gram's method (see p. 572). These will give information concerning all bacteria which may be present, and frequently no other procedure will be necessary for their identification.

515

The most common pus-producing organisms are
staphylococci and streptococci. They are both cocci,

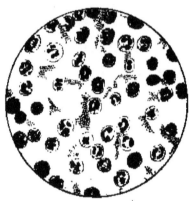

Fig. 202.—*Staphylococcus pyogenes albus* from an abscess of the
parotid gland (Jakob).

or spheres, their average diameter being about 0.7 μ.
Staphylococci are commonly grouped in clusters, often

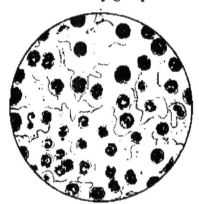

Fig. 203.—*Streptococcus pyogenes* from a case of empyema (Jakob).

compared to bunches of grapes (Fig. 202). There are
several varieties which can be distinguished only in

cultures. Streptococci are arranged side by side, forming chains of variable length (Fig. 203). Sometimes there are only three or four individuals in a chain; sometimes a chain is so long as to extend across several microscopic fields. Streptococci are more virulent than staphylococci, and are less commonly met. Both are Gram-positive. Their cultural characteristics are given on page 581.

Should bacteria resembling **pneumococci** be found, Rosenow's or Smith's method for capsules (see p. 87)

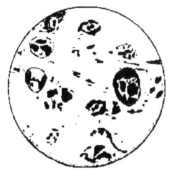

Fig. 204.—*Diplococcus pneumoniæ* from ulcer of cornea (oil-immersion objective) (study through courtesy of Dr. C. A. Oliver) (Boston).

should be tried. When these are not available, capsules can usually be shown by the method of Hiss. The dried and fixed smear is covered with a stain composed of 5 c.c. saturated alcoholic solution gentian-violet and 95 c.c. distilled water, and heated until steam rises. The preparation is then washed with 20 per cent. solution of copper sulphate, dried, and mounted in Canada balsam.

Pneumococci may give rise to inflammation in many

locations (see p. 8₅). When they form short chains, demonstration of the capsule or cultural methods (see p. 581) may be necessary to distinguish them from streptococci.

If tuberculosis be suspected, the smears should be stained by one of the methods for the **tubercle bacillus** (see pp. 77 to 79), or guinea-pigs may be inoculated. The bacilli are generally difficult to find in pus, and bacteria-free pus would suggest tuberculosis.

Gonococci, when typical, can usually be identified with sufficient certainty for clinical purposes in the

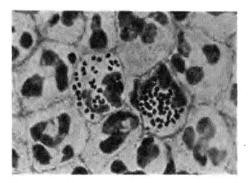

FIG. 205.—Gonococci in urethral pus (× 1500).

smear stained with Löffler's methylene-blue or, much better, Pappenheim's pyronin-methyl-green. They are ovoid or coffee-bean-shaped cocci which lie in pairs with their flat surfaces together (Fig. 205). They lie for the most part within pus-cells, an occasional cell being filled with them, while the surrounding cells contain few or none. Their intracellular position and their appearance in clusters are very important points in their identification. While a few are often found outside of cells, one should hesitate to accept them as gonococci unless further search reveals intracellular organisms. It is usually difficult to find gonococci when many other bacteria are present, even though the pus is primarily of gonorrheal origin. Whenever the identity of the organism is at all questionable, Gram's method should be tried. In rare instances it may be necessary to resort to cultures. The gonococcus is distinguished by its failure to grow upon ordinary media (see p. 582).

Gonococci are generally easily found in pus from untreated acute and subacute gonorrheal inflammations— conjunctivitis, urethritis, etc.—but are found with difficulty in pus from chronic inflammations and abscesses, and in urinary sediments.

In the urine gonococci are most likely to be present in the well-known "gonorrheal threads" or "floaters," which consist of strands of mucus with entangled pus corpuscles and are suggestive of chronic gonorrhea, but are by no means diagnostic of it. These are fished out with a platinum wire, spread upon slides, fixed, and stained. When floaters are absent it may be necessary to examine the sediment obtained by thorough centrifugation. In order to remove urea, which prevents proper drying of the smear, the sediment may be washed once with water or normal salt. Smears should be thin and quickly dried in order that the pus-corpuscles may be as well preserved as possible. Very often the pus-cells are so shrunken that the contained gonococci are difficult to recognize. There is likewise difficulty in finding gonococci in vaginal discharges unless comparatively pure pus from the suspected lesion can be obtained; otherwise the organisms sought are to a great extent lost among the myriads of bacteria and the epithelial and pus-cells of the leukorrheal discharge. Also, it should be borne in mind that the female genitals frequently harbor a non-pathogenic Gram-negative diplococcus which closely resembles the gonococcus.

PERITONEAL, PLEURAL, AND PERICARDIAL FLUIDS

The serous cavities contain very little fluid normally, but considerable quantities are frequently present as a result of pathologic conditions. The pathologic fluids are classed as transudates and exudates.

Transudates are non-inflammatory in origin. They contain only a few cells, and less than 2.5 per cent. of albumin, and do not coagulate spontaneously. The specific gravity is below 1.018. Micro-organisms are seldom present.

Exudates are of inflammatory origin. They are richer in cells and albumin, and tend to coagulate upon standing. The specific gravity is above 1.018. The amount of albumin is estimated by Esbach's method, after diluting the fluid, if much albumin is present. A mucin-like substance, called serosomucin, is likewise found in exudates. It is detected by acidifying with a few drops of 5 per cent. acetic acid, when a white cloudy precipitate results. This reaction is very helpful in distinguishing between transudates and exudates, although some transudates give a slight turbidity with acetic acid. Bacteria are generally present and often numerous. When none are found in stained smears or cultures, tuberculosis is to be suspected, and animal inoculation should be resorted to.

Exudates are usually classed as serous, serofibrinous, seropurulent, purulent, putrid, and hemorrhagic, which terms require no explanation. In addition, chylous and chyloid exudates are occasionally met, particularly in the peritoneal cavity. In the chylous form the milkiness is due mainly to the presence of minute fat-droplets, and is the result of rupture of a lymph-vessel, usually from obstruction of the thoracic duct. Chyloid exudates are milky chiefly from proteins in suspension, or fine débris from broken-down cells. These exudates are most frequently seen in carcinoma and tuberculosis of the peritoneum.

Cytodiagnosis.—This is diagnosis from a differential count of the cells in a transudate or exudate, particularly one of pleural or peritoneal origin.

The fresh fluid, obtained by aspiration, is centrifugalized for at least five minutes; the supernatant liquid is poured off; and smears are made from the sediment and dried in the air. The fluid must be very fresh, and the smears must be thin and quickly dried, otherwise the cells will be small and shrunken and hence difficult to identify. The smears are then stained with Wright's blood-stain which has been previously diluted with one-third its volume of pure methyl alcohol, mounted, and examined with an oil-immersion objective.

Predominance of *polymorphonuclear leukocytes* (pus-corpuscles) points to an acute infectious process (Fig. 206). These cells are the neutrophiles of the blood. Eosinophiles and mast-cells are rare. In thin smears they are easily recognized, the cytoplasmic granules often staining characteristically with polychrome-methylene-blue-eosin stains. In thick smears, upon

FIG. 206.—Cytodiagnosis. Polymorphonuclear leukocytes and swollen endothelial cells from acute infectious non-tuberculous pleuritis (Percy Musgrave; photo by L. S. Brown).

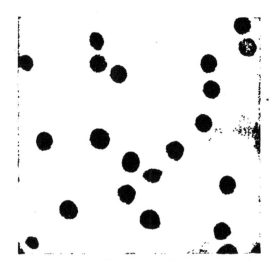

FIG. 207.—Cytodiagnosis. Lymphoid cells from pleural fluid; case of tuberculous pleuritis (Percy Musgrave; photo by L. S. Brown).

the other hand, they are often small and shrunken, and may be identified with difficulty, being easily mistaken for lymphocytes.

A large number, or even a preponderance, of eosinophiles is occasionally seen in the effusions following artificial pneumothorax and in those of early tuberculous pleuritis. In the latter case the eosinophiles gradually give place to lymphocytes as the disease progresses.

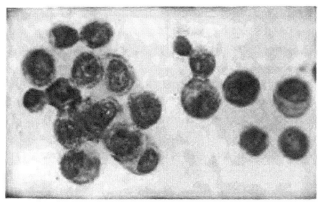

Fig. 208.—Cytodiagnosis. Mesothelial cells from transudate or mechanical effusion (Percy Musgrave; photo by L. S. Brown).

Predominance of *lymphocytes* (Fig. 207) generally signifies tuberculosis. They are the same as found in the blood. The cytoplasm is usually scanty, is often ragged, and sometimes is apparently absent entirely. Tuberculous pleurisy due to direct extension from the lung may give excess of polymorphonuclears owing to mixed infection.

Predominance of *mesothelial cells*, few cells of any kind being present, indicates a transudate (Fig. 208). These cells are large, with relatively abundant cytoplasm, and contain one, sometimes two, round or oval, palely staining nuclei. Mesothelial cells generally predominate in carcinoma, but are accompanied by considerable numbers of lymphocytes and red blood-corpuscles. Cancer cells cannot be recognized as such, although the presence of mitotic figures would suggest malignant disease.

CEREBROSPINAL FLUID

Examination of the fluid obtained by lumbar puncture has of recent years become a very important aid in diagnosis, particularly in syphilitic conditions of the nervous system.

1. Macroscopic Examination.—The amount obtainable varies from a few drops to 100 c.c. Normally, the fluid is clear and limpid, resembling water. The reaction is alkaline. The specific gravity is 1.003 to 1.008. Not infrequently it is tinged with fresh blood from a punctured vessel. This should not be confused with the dull-red or brown color which is seen in hemorrhagic conditions like intraventricular and subdural hemorrhage and hemorrhagic meningitis. When the bleeding is extensive and recent it may give the appearance of practically pure blood.

In purulent meningitis the fluid may exhibit varying degrees of cloudiness, from slight turbidity to almost pure pus. In the less acute stage of the epidemic form it is sometimes quite clear.

After standing for twelve to twenty-four hours the fluid will often coagulate   This occurs especially in the various forms of meningitis, rarely in non-inflammatory conditions. In tuberculosis the coagulum is usually very delicate and cobweb-like and is not easily seen.

(2) Chemical Examination.—Only a few constituents are of clinical importance.

(1) **Globulin.**—Traces are present normally. A notable increase occurs in acute inflammations and in syphilis and parasyphilitic affections. The three tests for globulin which follow are positive in 93 to 95 per cent. of all cases of paresis, and are, therefore, an important diagnostic consideration. When acute inflammation is excluded, they run practically parallel with the Wassermann reaction when the latter is applied to the spinal fluid. They must not be applied to fluid containing blood, owing to the presence of serum-globulin.

Noguchi's Butyric Acid Test.—In a small test-tube take 1 to 2 c.c. of the fluid and 5 c.c. of a 10 per cent. solution of butyric acid in normal salt solution. The original test calls for one-tenth these quantities, but they are too small for convenient manipulation. Heat to boiling and immediately add 1 c.c. of normal sodium hydroxid solution and boil again for a few seconds. A positive reaction, corresponding to a pathologic amount of globulin, varies from a distinct cloudiness to a heavy flocculent precipitate which generally appears within twenty minutes, but may be delayed for two hours. A slight opalescence may be seen in normal fluids.

Ammonium Sulphate Test.—Globulin is precipitated by strong solutions of ammonium sulphate. Ross and Jones apply the test after the manner of the ring tests for albumin in the urine. In a test-tube or horismascope take a few cubic centimeters of a completely saturated solution of ammonium sulphate and overlay with the suspected fluid. In the presence of an excess of globulin, a clear-cut, thin, grayish-white ring appears at the zone of contact of the two fluids within five minutes to three hours. This test appears to be fully as reliable as the butyric acid test.

Pandy's test is said to be more definite and more sensitive. The reagent consists of a saturated aqueous solution of phenol crystals. To 1 c.c. of the reagent add 1 drop of the cerebrospinal fluid. A bluish-white cloud indicates an abnormal amount of globulin.

(2) **Colloidal Gold Test.**—Lange's colloidal gold test, introduced in 1912 and now very widely used, con-

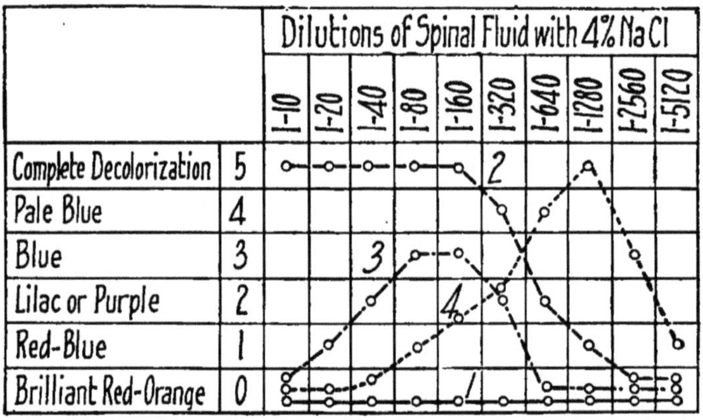

		Dilutions of Spinal Fluid with 4% NaCl									
		1-10	1-20	1-40	1-80	1-160	1-320	1-640	1-1280	1-2560	1-5120
Complete Decolorization	5						2				
Pale Blue	4										
Blue	3			3							
Lilac or Purple	2					4					
Red-Blue	1										
Brilliant Red-Orange	0						1				

FIG. 209.—Types of reactions in colloidal gold test: 1, normal cerebrospinal fluid, no reaction; 2, paretic type; 3, luetic type; 4, meningitic type.

sists in mixing cerebrospinal fluid in certain proportions with a solution of colloidal gold. Normal cerebrospinal fluid causes no change in color. Fluids from cases of syphilis and certain pathologic conditions of the nervous system induce changes in the color of the gold solution from red to purple, deep blue, pale blue, or colorless. Moreover, the dilution at which the maximum color change occurs is more or less characteristic of the different pathological conditions. The typical "curves" are shown in Fig. 209. The test gives its most consistent and valuable results in cases of general paresis. The exact explanation of the reaction is still uncertain.

The test itself is relatively simple, and any difficulty may be attributed to a faulty reagent, the preparation of which is time-consuming and uncertain. The reagent can be purchased ready prepared.

Preparation of Reagent.—Lange's Method (*modified by Miller, Brush, Hammers and Felton*).—It is imperative that all water used be triply distilled with avoidance of rubber connections in the still, that the beaker used for heating the solution be of Jena or Nonsol glass, and that all glassware be absolutely clean. It is recommended that the glass be boiled in a solution of Ivory soap, brushed thoroughly under the tap, rinsed well, soaked for one-half hour or longer in hot bichromate cleaning fluid (see p. 563), and immediately before use rinsed thoroughly with distilled water and finally with triply distilled water.

Heat slowly 1000 c.c. triply distilled water in a beaker. When the temperature reaches 60°C. add 10 c.c. of a 1 per cent. solution of chlorid of gold (Merck's yellow crystals in sealed ampoules) and 7 c.c. of a fresh 2 per cent. solution of potassium carbonate (Merck's "Blue Label") using a clean thermometer as a stirring rod. At 80°C. slowly add 10 drops of a 1 per cent. solution of oxalic acid (Merck's "Blue Label"), stirring briskly meanwhile. At 90°C. turn out the fire and add 5 c.c. of Merck's formaldehyd, "40 per cent., highest purity," drop by drop under constant stirring. Should a pink color develop before all the formaldehyd is added, stop at once, for this will slowly deepen to the brilliant orange-red of the finished solution. This solution should be absolutely transparent and free from any blue color; otherwise it is worthless.

Before use the solution must be neutralized with $\frac{n}{50}$ hydrochloric acid or $\frac{n}{50}$ sodium hydroxid, as the case may be. The amount to be added is found by titrating a small portion removed for the purpose, using a 1 per cent. solution of alizarin red in 50 per cent. alcohol as indicator. With an acid reaction this indicator gives a lemon-yellow color; with neutral reaction, brownish-red; with alkaline, red-purple.

Technic of Test.—Arrange a series of eleven clean test-tubes. Place 1.8 c.c. fresh sterile 0.4 per cent. solution of sodium chlorid in the first test-tube and 1 c.c. in each of the others. To the first tube add 0.2 c.c. of the spinal fluid, which must be free from any trace of blood. Mix well by sucking the fluid up into the pipet and expelling it, and then transfer 1 c.c. to the second tube. Mix and transfer 1 c.c. to the third tube, repeating this down the row to the tenth tube and discarding the last 1-c.c. portion. This leaves the eleventh tube with salt solution, only, to serve as a control. To each of these eleven tubes add 5 c.c. of the colloidal gold solution. Let stand at room temperature for an hour or longer, at the end of which time, in the case of a positive reaction, the solution in some of the tubes will have changed from red to purple, deep blue, pale blue, or colorless. In the case of normal fluids no change will occur. The results are usually charted as shown in Fig. 209. For the purpose of brevity the colors may be indicated by the corresponding numbers, which are placed in the same order as the tubes. Thus the "paretic reaction" in Fig. 209 may be expressed as 5555542100.

(3) **Mastic Test.**—Because of the many difficulties in the way of preparing satisfactory and uniform colloidal gold solutions, the mastic test has been proposed as a

satisfactory substitute for the gold test. The reagent is inexpensive and easily made, and the test is easily carried out. Results appear to parallel those obtained with colloidal gold, being almost uniformly positive in paresis, cerebrospinal syphilis, and tabes; but there does not appear to be the same opportunity to differentiate various types of reactions which the gold test offers. The method which follows is that used by Cutting.

FIG. 210.—The mastic reaction in cerebrospinal fluid. A, from a case of dementia præcox, negative; B, from a case of paresis, positive. (Courtesy of Jas. A. Cutting).

Preparation of Solutions.—(a) *Mastic Solution.*—Make a stock solution by completely dissolving 10 Gm. of gum mastic, U. S. P., in 100 c.c. of absolute alcohol and filter. To 2 c.c. of this stock solution add 18 c.c. of absolute alcohol, mix well, and pour rapidly into 80 c.c. of distilled water.

(b) *Alkaline-saline Solution.*—Make a 1.25 per cent. solution of sodium chlorid (C. P.) in distilled water, and to each 99 c.c. of this solution add 1 c.c. of a 0.5 per cent. solution of potassium carbonate in distilled water.

34

Technic of Test —Arrange a series of six small test-tubes. In the first place 1.5 c.c. of the alkaline-saline solution and in each of the others place 1 c.c. To the first tube add 0.5 c.c. of the spinal fluid, which must be completely free from blood. Mix by sucking the fluid up into the pipet and expelling it, and transfer 1 c.c. to the second tube. Again mix and transfer 1 c.c. to the third tube and continue down the line to the fifth tube, discarding the 1-c.c. portion which is removed from this and leaving the sixth tube with alkaline saline solution alone to serve as a control. Finally add 1 c.c. of the mastic solution to each tube. Mix well and set aside at room temperature for twelve to twenty-four hours, or in the incubator for six to twelve hours. Tubes in which the reaction is complete will show a heavy precipitate with clear supernatant fluid (Fig. 210).

(4) **Sugar.**—The normal cerebrospinal fluid gives a distinct reaction with the copper tests (see pp. 162, 163), apparently due to glucose, but it is usually necessary to use at least twice as much of the fluid as is recommended for the urine. A number of writers lay stress upon the absence of this reduction in meningitis. From a study of a series of cases, Jacob finds that: (1) No reduction of copper occurs in pyogenic meningitis (pneumococcus, streptococcus, etc.) or in acute meningococcic meningitis; (2) reduction occurs, but may be diminished in tuberculosis and in the more chronic cases of meningococcic meningitis; (3) reduction is normal in poliomyelitis.

(5) **Antimeningococcus-serum Test.**—Vincent and other French investigators have developed the following test, which they believe to be specific for epidemic cerebrospinal meningitis:

To a few cubic centimeters of the spinal fluid, which has been cleared by thorough centrifugation, are added a few drops of antimeningococcus serum. The tube, along with a control tube of the untreated fluid, is then placed in an incubator at 52°C. for a few hours. A positive reaction consists in the appearance of a white cloud. The test is said to be reliable even when meningococci cannot be found. The serum must be free from phenol and other interfering substances.

3. Microscopic Examination.—This consists in a study of the bacteria, and of the number and kind of cells.

(1) **Bacteria.**—*Tubercle bacilli* can be found in the great majority of cases of tuberculous meningitis. The delicate coagulum which forms when the fluid is allowed to stand in a cool place for twelve to twenty-four hours will entangle any bacilli which may be present. This clot may be removed, spread upon slides, and stained by one of the methods already given (see pp. 77 to 79). If desired, the coagulum may be treated with antiformin (see p. 80). In case no coagulum forms, the fluid should be thoroughly centrifugalized and the sediment stained, or, if much protein be present, it may be coagulated by heat, precipitated by the centrifuge, and treated with antiformin. It may be necessary to examine a considerable number of smears. In doubtful cases inoculation of guinea-pigs must be resorted to.

The *Diplococcus intracellularis meningitidis* is recognized as the cause of epidemic cerebrospinal fever, and can be détected in the cerebrospinal fluid of most cases, especially those which run an acute course. Cover-glass smears from the sediment should be stained by a simple bacterial stain and by Gram's method. The meningococcus is an intracellular diplococcus which often cannot be distinguished from the gonococcus in stained smears (Fig. 211). It also decolorizes by Gram's method. The presence of such a diplococcus in meningeal exudates is, however, sufficient for its identification in clinical work.

Various organisms have been found in other forms of meningitis—the pneumococcus most frequently, the

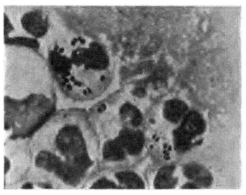

Fig. 211.—Meningococci in cerebrospinal fluid from a case of epidemic spinal meningitis. Gram's method and carbol-fuchsin (× 1500).

influenza bacillus (Fig. 212) rarely. When the pneumococcus is present, it is usually very abundant. In some cases no micro-organisms can be detected even by culture methods.

(2) **Cytology.**—The fluid must be as fresh as possible as the cells tend to degenerate. The routine examination should include both a total and a differential count.

The total number of cells may be counted with the hemacytometer, but the Fuchs-Rosenthal counting chamber which is 0.2 mm. deep is more convenient. Unna's polychrome methylene-blue or a solution of methyl-violet or other nuclear dye is drawn into the leukocyte pipet to the 0.5 mark, and the fresh spinal fluid, which has been well shaken, is drawn up to the mark 11. After mixing, a drop is placed on the counting slide and covered. If one is certain of recognizing the cells, the dye may be omitted and a small drop of

FIG. 212.—Influenza bacilli in spinal fluid. Case of meningitis (× 1000).

the well shaken fluid placed directly upon the counting slide. To reduce the error arising from the small number of cells present, it is necessary to count a large area on several slides. Normally, the cells rarely exceed 5 or 7 per cubic millimeter; 10 is perhaps the maximum. The differential count is made as described on page 521. Ordinarily, only two kinds of cells are seen: lymphocytes and polymorphonuclear neutrophiles.

Lymphocytes predominate normally. An increase

in the total count, together with predominance of lymphocytes (over 70 per cent.), strongly suggests tuberculosis or syphilitic disease of the nervous system, such as
paresis. It has been observed in the more chronic type
of epidemic cerebrospinal meningitis, but not to the
same extent.

In acute meningitis the total count is high and polymorphonuclears prevail.

ANIMAL INOCULATION

Inoculation of animals is one of the most reliable
means of verifying the presence of certain micro-organisms in fluids and other pathologic material, and is
helpful in determining the species of bacteria which have
been isolated in pure culture.

Clinically, it is applied most frequently to demonstration of the tubercle bacillus when other means have
failed or are uncertain. The guinea-pig is the most
suitable animal for this purpose. When the suspected
material is fluid and contains pus, it should be well
centrifugalized, and 1 or 2 c.c. of the sediment injected,
by means of a large hypodermic needle, into the peritoneal cavity or underneath the loose skin of the groin.
Fluids from which no sediment can be obtained must
be injected directly into the peritoneal cavity, since
at least 10 c.c. are required, which is too great an
amount to inject hypodermically. Solid material
should be placed in a pocket made by snipping the skin
of the groin with scissors, and freeing it from the underlying tissues for a short distance around the opening.
When the intraperitoneal method is selected, several
animals must be inoculated, since some are likely to die

from peritonitis caused by other organisms before the tubercle bacillus has had time to produce its characteristic lesions.

The animals should be killed at the end of six or eight weeks, if they do not die before that time; and a careful search should be made for the characteristic pearl-gray or yellow tubercles scattered over the peritoneum and through the abdominal organs, particularly the spleen and liver, and for caséous inguinal and retroperitoneal lymph-glands. The tubercles and portions of the caseous glands should be crushed between two slides, dried, and stained for tubercle bacilli. The bacilli are difficult to find in the caseous material.

It has recently been shown that exposure of the guinea-pigs to strong x-ray for about ten minutes so lowers their resistance that inoculation of tuberculous material is followed by recognizable tuberculosis within about two weeks. Susceptibility to other bacteria is also increased; therefore if the material contains pyogenic organisms it should first be treated with antiformin and washed with water.

THE MOUTH

Micro-organisms are always present in large numbers. Among these is *Leptothrix buccalis* (Fig. 213), which is especially abundant in the crypts of the tonsils and the tartar of the teeth. The whitish patches of *Pharyngomycosis leptothrica* are largely composed of these fungi. They are slender, segmented threads, which generally, but not always, stain violet with Lugol's solution, and are readily seen with a 4-mm. objective. At times they are observed in the sputum and stomach fluid. In the former they might be mistaken for elastic fibers; in the latter, for Boas-Oppler bacilli. In either case, the reaction with iodin will distinguish them.

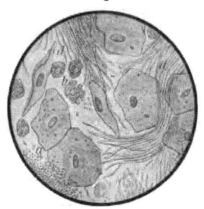

FIG. 213.—Gingival deposit (unstained): *a*, Squamous epithelial cells; *b*, leukocytes; *c*, bacteria; *d*, *Leptothrix buccalis* (Jakob).

FIG. 214.—Thrush fungus (*Endomyces albicans*) (Kolle and Wassermann).

The prevalence of endamebæ and spirochetes in the mouths of normal persons and of those suffering from pyorrhea alveolaris has already been mentioned (see pp. 457 and 462).

Thrush is a disease of the mouth seen most often in children, and characterized by the presence of white patches upon the mucous membrane. It is caused by the thrush fungus, *Endomyces albicans*. When a bit from one of the patches is pressed out between a slide and cover and examined with a 4-mm. objective, the

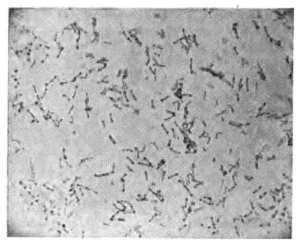

FIG. 215.—*Bacillus diphtheriæ* stained with methyl-green; culture from throat (× 1000).

fungus is seen to consist of a network of branching segmented hyphæ with numerous spores, both within the hyphæ and in the meshes between them (Fig. 214). The meshes also contain leukocytes, epithelial cells, and granular débris.

Acute pseudomembranous inflammations, which occur chiefly upon the tonsils and nasopharynx, are generally caused by the diphtheria bacillus, but may result from streptococcic infection. In many cases diphtheria bacilli can be demonstrated in smears made from the membrane and stained with Löffler's methylene-blue or 2 per cent. aqueous solution of methylgreen. They are straight or curved rods, which vary markedly in size and outline, and stain very irregularly. A characteristic form is a palely tinted rod with several deeply stained granules (metachromatic bodies), or with one such granule at each end (Fig. 215). They stain by Gram's method. It is generally necessary, and always safer, to make a culture upon blood-serum, incubate for twelve hours, and examine smears from the growth.

Neisser's stain has long been the standard differential stain for the diphtheria bacillus. It colors the bodies of the bacilli brown and the metachromatic bodies blue.

1. Make films and fix as usual.

2. Apply the following solution, freshly filtered, for about one-half minute:

> Methylene-blue.......................... 0.1 Gm.;
> Alcohol (96 per cent.)..................... 2.0 c.c.;
> Glacial acetic acid....................... 5.0 c.c.;
> Distilled water.......................... 95.0 c.c.

3. Rinse in water.

4. Apply a saturated aqueous solution of Bismarck brown one-half minute.

5. Rinse, dry, and mount.

Ponder's Stain.—This comparatively new stain is preferred by many to Neisser's:

> Toluidin blue (Gruebler).................. 0.02 Gm.;
> Glacial acetic acid....................... 1.00 c.c.;
> Absolute alcohol......................... 2.00 c.c.;
> Distilled Water to.......................100.00 c.c.

Cover the fixed film with the stain; turn the cover-glass over and examine as a hanging-drop preparation. Diphtheria bacilli are blue, with red granules.

Vincent's angina is a pseudomembranous and ulcerative inflammation of mouth and pharynx, which when acute may be mistaken for diphtheria, and when

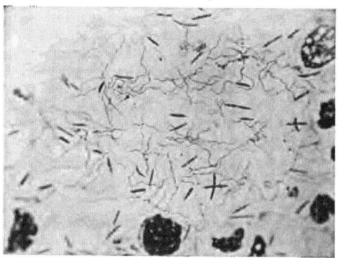

FIG. 216.—*Spirochæta vincenti* from case of ulcerative stomatitis stained with gentian-violet (× 1200).

chronic is very apt to be mistaken for syphilis. Stained smears from the ulcers or membrane show large numbers of spirochetes and "fusiform bacilli," giving a striking and characteristic picture (Fig. 216). Before making the smears the surface of the lesion should be gently cleaned by swabbing, otherwise so many miscellaneous bacteria may be present that the characteristic picture is obscured. The "bacillus" is spindle-shaped, more or less pointed at the ends, and about 4 to 8 μ long. The spirillum is a very slender, wavy thread, about 10 to 20 μ long, and stains feebly. Diluted formalin-gentian-violet makes a satisfactory stain. With methylene-blue the palely staining spirillum may easily be overlooked. Further description is given on page 462.

Tuberculous ulcerations of mouth and pharynx can generally be diagnosed from curetings made after careful cleansing of the surface. The curetings are well rubbed between slide and cover, and the smears thus made are dried, fixed, and stained for tubercle bacilli. Since there is much danger of contamination from tuberculous sputum, the presence of tubercle bacilli is significant only in proportion to the thoroughness with which the ulcer was cleansed. The diagnosis is certain when the bacilli are found within groups of cells which have not been dissociated in making the smears.

THE EYE

Staphylococci, pneumococci, and **streptococci** are probably the most common of the bacteria to be found in non-specific conjunctivitis and keratitis. Serpiginous ulcer of the cornea is generally associated with the pneumococcus (see Fig. 204).

The usual cause of acute infectious conjunctivitis ("pink-eye"), especially in cities, seems to be the **Koch-Weeks bacillus.** This is a minute, slender rod, which lies within and between the pus-corpuscles (Fig. 217), and is negative to Gram's stain. In smears it cannot be distinguished from the influenza bacillus, although its length is somewhat greater.

The **diplobacillus of Morax and Axenfeld** gives rise to an acute or chronic blepharoconjunctivitis without

follicles or membrane, for which zinc sulphate seems to
be a specific. It is widely distributed geographically

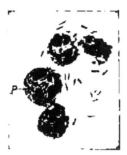

FIG. 217.—Conjunctival secretion from acute contagious conjunctivitis; polynuclear leukocytes with the bacillus of Weeks; P, phagocyte
containing bacillus of Weeks (oil-immersion objective, ocular III)
(Morax).

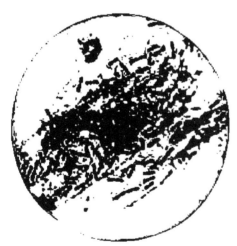

FIG. 218.—The diplobacillus of Morax and Axenfeld (from a preparation by Dr. Harold Gifford).

and is common in many regions. The organism is a
short, thick diplobacillus, is frequently intracellular, and

is Gram-negative (Fig 218) A delicate capsule can sometimes be made out.

Early diagnosis of gonorrheal ophthalmia is extremely important, and can be made with certainty only by detection of **gonococci** in the discharge. They are easily found in smears from untreated cases. After treatment is begun they soon disappear, even though the discharge continues.

Pseudomembranous conjunctivitis generally shows either **streptococci** or **diphtheria bacilli.** In diagnosing diphtheric conjunctivitis, one must be on his guard against the **Bacillus xerosis,** which is a frequent inhabitant of the conjunctival sac in healthy persons, and which is identical morphologically with the diphtheria bacillus. Hence the clinical picture is more significant than the microscopic findings.

Various micro-organisms—bacteria, molds, protozoa—have been described in connection with trachoma, but the more recent work points to certain minute intracellular bodies as the causative agents. These are best seen in smears stained with Giemsa's stain and appear as minute blue dots usually grouped in clusters in the cytoplasm of epithelial cells. A red-staining granule can be seen in many of the blue bodies. The nature of these "**trachoma bodies**" is not yet settled. They are thought by many to belong to the chlamydozoa.

Herbert has called attention to the abundance of eosinophilic leukocytes in the discharge of **vernal catarrh.** He regards their presence in considerable numbers as very helpful in the diagnosis of this disease.

THE EAR

By far the most frequent exciting causes of acute otitis media are the pneumococcus and the streptococcus. The finding of other bacteria in the discharge generally indicates a secondary infection, except in cases complicating infectious diseases, such as typhoid fever, diphtheria, and influenza. Discharges which have continued for some time are practically always contaminated with the staphylococcus. The presence of the streptococcus should be a cause of uneasiness, since it much more frequently leads to mastoid disease and meningitis than does the pneumococcus. The staphylococcus, bacillus of Friedländer, colon bacillus, and *Bacillus pyocyaneus* may be met in chronic middle-ear disease.

In tuberculous disease the tubercle bacillus is present in the discharge, but its detection offers some difficulties. It is rarely easy to find, and precautions must always be taken to exclude the smegma and other acid-fast bacilli (see p. 82), which are especially liable to be present in the ear. Rather striking is the tendency of old squamous cells to retain the red stain, and fragments of such cells may mislead the unwary.

PARASITIC DISEASES OF THE SKIN

Favus, tinea versicolor, and the various forms of ringworm are caused by members of the fungus group. To demonstrate them, a crust or a hair from the affected area is softened with a few drops of 20 per cent. caustic soda solution, pressed out between a slide and cover, and examined with a 4-mm. objective. They consist of a more or less dense network of hyphæ and numerous round or oval refractive spores. The cuts in standard works upon diseases of the skin will aid in differentiating the members of the group.

MILK

A large number of analyses of human and cows' milk are averaged by Holt as follows, Jersey milk being excluded because of its excessive fat:

	HUMAN MILK		COWS' MILK
	Normal variations, per cent.	Average, per cent.	Average, per cent.
Fat......................	3.00 to 5.00	4.00	3.50
Sugar....................	6.00 to 7.00	7.00	4.30
Proteins..................	1.00 to 2.25	1.50	4.00
Salts.....................	0.18 to 0.25	0.20	0.70
Water....................	89.82 to 85.50	87.30	87.50
	100.00	100.00 100.00	100.00

.The reaction of human milk is slightly alkaline; of cows', neutral or slightly acid. The specific gravity of each is about 1.028 to 1.032. Human milk is sterile when secreted, but derives a few bacteria from the lacteal ducts. Cows' milk, as usually sold, contains large numbers of bacteria, the best milk rarely containing fewer than 10,000 per cubic centimeter. Microscopically, human milk is a fairly homogeneous emulsion of fat, and is practically destitute of cellular elements. Any notable number of leukocytes indicates infection of the mammary gland.

Chemical examination of milk is of great value in solving the problems of infant feeding. The sample examined should be the middle milk, or the entire quantity from one breast. The fat and protein can be estimated

roughly, but accurately enough for many clinical purposes by means of Holt's apparatus, which consists of
a 10-c.c. cream gage and a small hydrometer (Fig. 219).
The cream gage is filled to the o mark with milk,

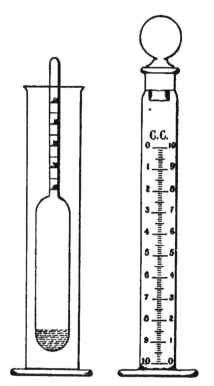

FIG. 219.—Holt's milk-testing apparatus.

allowed to stand for twenty-four hours at room temperature, and the percentage of cream then read off.
The percentage of fat is three-fifths that of the cream.
The protein is then approximated from a consideration
of the specific gravity and the percentage of fat. The

35

salts and sugar very seldom vary sufficiently to affect the specific gravity, hence a high specific gravity must be due to either an increase of protein or decrease of fat, or both, and *vice versâ*. With normal specific gravity the protein is high when the fat is high, and *vice versâ*. The method is not accurate with cows' milk.

For more accurate work the following methods, applicable to either human or cow's milk, are simple and satisfactory:

Fat.—*Leffmann-Beam Method.*—This is essentially the widely used Babcock method, modified for the small quantities of milk obtainable from the human mammary gland. The apparatus consists of a special tube which fits the aluminum shield of the medical centrifuge (Fig. 220) and a 5-c.c. pipet. Owing to its narrow stem, the tube is difficult to fill and to clean; and for this reason the Whitman modification, in which the stem is removable, is to be preferred. Exactly 5 c.c. of the milk are introduced into the tube by means of the pipet,

FIG. 220.—Centrifuge tube for milk analysis.

and 1 c.c. of a mixture of equal parts of concentrated hydrochloric acid and amyl-alcohol is added and well mixed. The tube is filled to the o mark with concentrated sulphuric acid, adding a few drops at a time and agitating constantly. This is revolved in the centrifuge at 1000 revolutions a minute for three minutes, or until the fat has separated. The percentage is then read off upon the stem, each small division representing 0.2 per cent. of fat.

Proteins.—*T. R. Boggs' Modification of the Esbach Method.*—This is applied as for urinary albumin (see p. 157), substituting Boggs' reagent for Esbach's. The reagent is prepared as follows:

> (1) Phosphotungstic acid.................. 25 Gm.;
> Distilled water........................ 125 c.c.;

> (2) Concentrated hydrochloric acid......... 25 c.c.;
> Distilled water........................ 100 c.c.

When the phosphotungstic acid is completely dissolved, mix the two solutions. This reagent is quite stable if kept in a dark glass bottle.

Before examination, the milk should be diluted according to the probable amount of protein, and allowance made in the subsequent reading. For human milk the optimum dilution is 1 : 10; for cows' milk, 1 : 20. Dilution must be accurate.

Lactose.—The protein should first be removed by acidifying with acetic acid, boiling, and filtering. The copper methods may then be used as for glucose in the urine (see pp. 166, 167); but it must be borne in mind that lactose reduces copper more slowly than glucose, and longer heating is, therefore, required; and that 10 c.c. of Fehling's solution (or 25 c.c. of Benedict's) are equivalent to 0.0676 Gm. lactose (as compared with 0.05 Gm. glucose).

Detection of Preservatives.—Formalin is the most common preservative added to cows' milk, but boric acid is also used.

To detect formalin, add a few drops of dilute ferric chlorid solution to a few cubic centimeters of the milk, and run the mixture gently upon the surface of some strong sulphuric acid in a test-tube. If formaldehyd be present, a bright red ring will appear at the line of contact of the fluids. This is not a specific test for formaldehyd, but nothing else likely to be added to the milk will give it.

To detect boric acid, Goske's method as used by the Chicago Department of Health, is simple and satisfactory: Mix 2 c.c. of concentrated hydrochloric acid with 20 c.c. of the milk and place in a 50-c.c. beaker. In this suspend a long strip of turmeric paper (2 cm. wide), so that its end reaches to the bottom of the beaker. Allow to remain about half an hour. The liquid will rise by capillarity, and if boric acid be present a red-brown color will appear at the junction of the moist and dry portions of the paper. If this is touched with ammonia, a bluish-green slate color develops. A rough idea of the amount of boric acid may be had by comparing the depth of color with that produced by boric acid solutions of known strength.

SYPHILITIC MATERIAL

In 1905 Schaudinn and Hoffmann described the occurrence of a very slender, spiral micro-organism in the lesions of syphilis. This they named *Spirochæta pallida*, because of its low refractive power and the difficulty with which it takes up staining reagents. The name was later changed to *Treponema pallidum*. Its etiologic relation to syphilis is now universally admitted. It is found in primary, secondary, and tertiary lesions, but is not present in the latter in sufficient numbers to be of value in diagnosis

Treponema pallidum is an extremely slender, spiral,

motile thread, with pointed ends. There is a flagellum at each end, but it is not clearly seen in ordinary preparations. The organism varies considerably in length, the average being about 7 μ, or the diameter of a red blood-corpuscle; and it exhibits three to twelve, sometimes more, spiral curves, which are sharp and regular and resemble the curves of a corkscrew (Figs. 158, 221, 222). It is so delicate that it is difficult to see even

FIG. 221.—*Treponema pallidum* (× 1000) (Leitz 1/12 oil-immersion objective and Leitz dark-ground condenser). The parasite has the same appearance as in India ink preparations.

in well-stained preparations; a high magnification and careful focusing are, therefore, required. Upon ulcerated surfaces it is often mingled with other spiral micro-organisms, which adds to the difficulty of its detection. The most notable of these is *Spirochæta refringens*, described on page 462.

Treponema pallidum is most easily demonstrated in chancres and mucous patches, although the skin lesions —papules, pustules, roseolous areas—often contain large numbers. Tissue-juice from the deeper portions of the lesions is the most favorable material for examination, because the organisms are commonly more abundant than upon ulcerated surfaces and are rarely accompanied by other micro-organisms. After cleansing, the surface is gently scraped with a curet or rubbed briskly with a swab of cotton or gauze. In a few moments serum will exude and very thin smears are then made from it. In transferring the serum from the lesion to the slide or cover-glass it is convenient to use a capillary pipet. The rubbing should not be so vigorous as to bring much blood, because the corpuscles may hide the treponema; but a few red corpuscles are an advantage as an aid in locating favorable fields and as a check upon the quality of the staining. Best fields are those with the clearest background and with a few red corpuscles, which must be well-stained, well-preserved, and not shrunken.

Staining Methods.—Giemsa's stain (see p. 313) is the most widely used and is perhaps the best (see Fig. 222). It is best purchased ready prepared. Smears are fixed in absolute alcohol for fifteen minutes. Ten drops of the stain are added to 10 c.c. of faintly alkaline distilled water (1 drop of a 1 per cent. solution of potassium carbonate to 10 c.c. of the water), and the fixed smear is immersed on edge in this diluted stain for one to three hours or longer. It is then rinsed in distilled water, dried, and mounted. More intense staining may be obtained and the time shortened by conducting the process in the incubator. In well-stained specimens *Treponema pallidum* is reddish; most other micro-organisms, bluish. If desired, Giemsa's stain may be used as described for blood (see p. 313), but the organisms do not then stand out quite so clearly.

It is a waste of time to search for treponemata in films in which the leukocytes and the red corpuscles are not well-stained. The nuclei of the former should be dark purple; the latter should be deep copper-red or salmon-colored when the stain is used as for blood, and deep slate-blue when alkali has been added.

FIG. 222.—*Treponema pallidum, spirochæta refringens,* and two red blood-corpuscles in a smear from a chancre (× 1200). From a preparation stained with Giemsa's stain, with alkali, for three hours. The treponemata were purplish red; refringens, bluish-purple; red corpuscles, deep slate-blue.

Wright's blood-stain, used in the manner already described (see p. 310) except that the diluted stain is allowed to act upon the film for fifteen minutes, gives fair results. Wright now recommends the following: In a test-tube mix 10 c.c. distilled water, 1 c.c. Wright's stain and 1 c.c. of a 0.1 per cent. solution of potassium carbonate. Heat to boiling and cover the preparation with the hot solution.

After three or four minutes pour off the fluid. Repeat this twice. Rinse, dry and mount.

Silver Method.—The silver impregnation method has long been used for tissues. Stein has applied it to smears as follows:

1. Dry the films in the incubator at 37°C. for three or four hours.

2. Immerse in 10 per cent. silver nitrate solution, in diffuse daylight, for some hours, until the preparation takes on a metallic luster.

3. Wash in water, dry, and mount.

The organisms are black against a brownish background.

India-ink Method.—A small drop of India-ink of good grade (Günther and Wagner's "Chin-Chin liquid pearl" or Grübler's "nach Burri" recommended) is mixed on a slide with 1 or 2 small drops of serum from the suspected lesion. The mixture is then spread over the slide and allowed to dry. After drying, it is examined with an oil-immersion lens. Micro-organisms, including *Treponema pallidum*, appear clear white on a brown or black background, much as they do with the dark ground condenser (see Fig. 221). If desired, the mixture of ink and serum may be covered with a cover-glass and examined in the fresh state, the living organisms being thus demonstrated. Because of its extreme simplicity this method has been favorably received. It cannot, however, be absolutely relied upon, since, as has been pointed out, many India-inks contain wavy vegetable fibrils which might easily mislead a beginner, and sometimes, indeed, even an experienced worker. Instead of India-ink, collargol, diluted 1:20 with water and thoroughly shaken, has been recommended.

Dark ground illumination (see p. 24) may be used to study the living organisms in fresh tissue juices. This offers a satisfactory means of diagnosis, but since the instrument is expensive the practitioner will rely upon one or more of the staining methods just enumerated.

Method of Oppenheim and Sachs.—Very thin air-dried films are stained for from thirty seconds to three minutes with phenol-gentian-violet (saturated alcoholic solution of gentian-violet, 10 c.c.; 5 per cent. phenol, 90 c.c.). Previous fixation is not necessary.

SEMEN

Absence of spermatozoa is a more common cause of sterility than is generally recognized. In some cases they are present, but lose their motility immediately after ejaculation.

Semen should be kept warm until examined. When it must be transported any considerable distance, the method suggested by Boston is convenient: The fresh semen is placed in a small bottle, to the neck of which a string is attached. This is then suspended from a button on the trousers, so that the bottle rests against the skin of the inguinal region. It may be carried in this way for hours. When ready to examine, place a small quantity upon a warmed slide and apply a cover. The spermatozoa are readily seen with a 4 mm. objective (see Fig. 74). Normally, they are abundant and in active motion.

Detection of semen in stains upon clothing, etc., is often important. The finding of spermatozoa, after soaking the stain for an hour in normal salt solution or dilute alcohol and teasing in the same fluid, is absolute proof that the stain in question is semen although it is not possible to distinguish human semen from that of the lower animals in this way. A little eosin added to the fluid will bring the spermatozoa out more clearly.

Florence's Reaction.—The suspected material is softened with water, placed upon a slide with a few drops of the reagent, and examined at once with a medium power of the microscope. If the material be semen, there will be found dark-brown crystals (Fig. 223) in the form of rhombic platelets resembling hemin crystals, or of needles, often grouped in clusters. These crystals can

FIG. 223.—Seminal crystals (medium size) (× 750) from a stain on clothing. A single thread ⅛ inch long was used in the test, the stain being three years and four months old (Peterson and Haines).

also be obtained from crushed insects, watery extracts of various internal organs, and certain other substances, so that they are not absolute proof of the presence of semen. Negative results, upon the other hand, are conclusive, even when the semen is many years old.

The reagent consists of iodin, 2.54 Gm.; potassium iodid, 1.65 Gm.; and distilled water, 30 c.c.

DIAGNOSIS OF RABIES

In view of the brilliant results attending prophylactic treatment by the Pasteur method, early diagnosis of rabies (hydrophobia) in animals which have bitten persons is extremely important.

The most reliable means of diagnosis is the production of the disease in a rabbit by subdural or intracerebral injection of a little of the filtrate from an emulsion of the brain and medulla of the suspected animal. The diagnosis can, however, usually be quickly and easily made by microscopic demonstration of Negri bodies. Whether these bodies be protozoan in nature and the cause of the disease, as is held by many, or whether they be products of the disease, it is certain that their presence is pathognomonic.

Negri bodies are sharply outlined, round, oval, or somewhat irregular structures which vary in size, the extremes being 0.5 and 18 μ. They consist of a hyalin-like cytoplasm, in which when properly stained one or more chromatin bodies can usually be seen. They are situated chiefly within the cytoplasm of the large cells of the central nervous system, the favorite location being the multipolar cells of the hippocampus major (Ammon's horn). In many cases they suggest red blood-corpuscles lying within nerve-cells.

Probably the best clinical method of demonstrating Negri bodies is the impression method of Langdon Frothingham, which is carried out as described below.

1. Place the dog's brain[1] upon a board about 10 inches

---

[1] For Dr. Frothingham's method of removing a dog's brain see American Journal of Public Hygiene for February, 1908.

square, and divide into two halves by cutting along the median line with scissors.

2. From one of the halves cut away the cerebellum and open the lateral ventricle, exposing the Ammon's horn.

3. Dissect out the Ammon's horn as cleanly as possible.

4. Cut out a small disk at right angles to the long axis of the Ammon's horn, so that it represents a cross-section of the organ.

5. Place this disk upon the board near the edge, with one of the cut surfaces upward.

6. Press the surface of a thoroughly clean slide upon the disk and lift it suddenly. The disk (if its exposed surface has not been allowed to become too dry) will cling to the board, leaving only an impression upon the slide. Make several similar impressions upon different portions of the slide, using somewhat greater pressure each time. Impressions are also to be made from the cut surface of the cerebellum, since Negri bodies are sometimes present in the Purkinje cells when not found in the Ammon's horn.

7. Before the impressions dry, immerse in methyl-alcohol for one-half to two minutes.

8. Cover with Van Gieson's methylene-blue-fuchsin stain, warming gently for one-half to two minutes. This stain, as modified by Frothingham, is as follows. It remains effective for three or four days:

> Tap-water................................. 20 c.c.;
> Saturated alcoholic solution basic fuchsin....... 1 drop;
>   ("Fuchsin f. Bac.," Grübler).
> Saturated aqueous solution methylene-blue..... 1 drop.
>   ("Methylenblau f. Bac., Koch." Grübler).

9. Wash in water and dry with filter-paper. Examine with a low power to locate the large cells in which the bodies are apt to be found, and study these with an oil-immersion lens.

PLATE XII

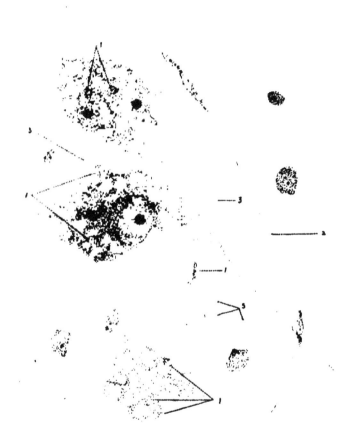

Nerve-cells containing Negri bodies.

Hippocampus impression preparation, dog. Van Gieson stain; × 1000. 1, Negri bodies; 2, capillary; 3, free red blood-corpuscles (courtesy of Langdon Frothingham).

The Negri bodies are stained a pale pink to purplish red, and frequently contain small blue dots (Plate XII). The nerve-cells are blue, and red blood-corpuscles are colorless or yellowish-copper colored.

When the work is finished, the board with the dissected brain is sterilized in the steam sterilizer.

Demonstration of Negri bodies by this method is quicker and, possibly, more certain than by the study of sections. It has the decided advantage over the smear method that the histologic structure is retained. One or more of the impressions generally shows the entire cell arrangement almost as well as in sections, and it is very easy to locate favorable fields with a 16-mm. objective.

CHAPTER VIII

BACTERIOLOGIC METHODS

BACTERIOLOGY has become so important a part of medicine that some knowledge of bacteriologic methods is imperative for the present-day practitioner. It has been the plan of this book to describe the various bacteria and bacteriologic methods with the subjects to which they seemed to be particularly related. The tubercle bacillus and its detection, for example, are described in the chapters upon Sputum and Urine; blood-cultures are discussed in the chapter upon Blood. There are, however, certain methods, notably the preparation of media and the study of bacteria by cultures, which do not come within the scope of any previous section, and an outline of these is given in the present chapter.

I. APPARATUS

Much of the apparatus of the clinical laboratory is called into use. Only the following need special mention:

1. **Sterilizers.**—Two are required.

The *dry*, or *hot-air sterilizer*, is a double-walled oven similar to the detached ovens used with gas and gasolene stoves. It has a hole in the top for a perforated cork with thermometer. The oven of any stove, even without a thermometer, will answer for many purposes.

Ordinarily the heat should be sufficient to slightly brown but not char paper or cotton and should be continued for one-half to one hour.

The *steam sterilizer* is preferably of the Arnold type, opening either at the top or the side. An *autoclave*, which sterilizes with steam under pressure, is very desirable, but not necessary. An aluminum pressure cooker (Fig. 224) is a very satisfactory substitute for the autoclave. It costs about fifteen dollars.

Fig. 224.—Aluminum pressure cooker, an efficient and comparatively inexpensive substitute for an autoclave.

2. **Incubator.**—This is the most expensive piece of apparatus which will be needed. It is made of copper, and has usually both a water- and an air-jacket surrounding the incubating chamber. It is provided with thermometer, thermo-regulator, and some source of heat, usually a Koch safety Bunsen burner. With a little ingenuity one can rig up a drawer or a small box, in which a fairly constant temperature can be maintained by means of an electric light. The degree of

heat can be regulated by moving the drawer in or out, or holes can be made in which corks may be inserted and removed as needed. A Thermos bottle has been suggested as a temporary makeshift. Upon occasion cultures may be kept warm by carrying them in an inside pocket.

The gas-heated copper incubators are now fast being displaced by the cheaper and more satisfactory wooden incubators in which electricity is the source of heat.

3. **Culture-tubes and Flasks.**—For most work ordinary test-tubes, 125 × 19 mm. without flange, are satisfactory. For special purposes a few 100 × 13 mm. and 150 × 19 mm. tubes may be needed. Heavy tubes, which do not easily break, can be obtained, and are especially desirable when tubes are cleaned by an untrained assistant. The tubes are usually stored in wire baskets.

Flasks of various sizes are needed. The Erlenmeyer type is best. Quart and pint milk bottles and 2-ounce wide-mouthed bottles will answer for most purposes.

4. **Platinum Wires.**—At least two of these are needed. Each consists of a piece of platinum wire about 8 cm. long, fixed in the end of a glass or metal rod. One is made of about 22 gage wire and its end is curled into a loop 2 to 3 mm. in diameter. The other wire is somewhat heavier and its tip is hammered flat.

Lyon recommends the use of No. 20 nichrome wire as nearly equal to platinum and very much cheaper. He makes a handle of No. 8, or thicker, aluminum wire, sawing an oblique notch in the end, inserting the nichrome wire, and hammering the aluminum over it.

5. **Pipets, etc.**—In addition to the graduated pipets with which every laboratory is supplied, there are a number of forms which are generally made from glass tubing as needed. One of the simplest of these is made as follows: A section of glass tubing, about 12 cm. long and 5 mm. in diameter, is grasped at the ends, and its center is heated in a concentrated flame. A blast-lamp is best, but a Bunsen burner will usually answer, particularly if fitted with a "wing" or "fish-tail"

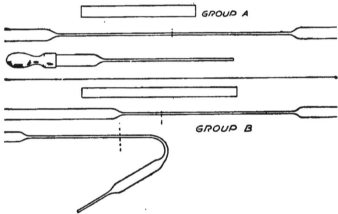

FIG. 225.—Process of making pipets (Group A) and Wright's capsule (Group B). The dotted lines indicate where the glass is to be broken.

attachment. When the glass is thoroughly softened it is removed from the flame, and, with a steady, but not rapid pull, is drawn out as shown in Fig. 225. The slender portion is scratched near the middle with a file and is broken to make two pipets, which are then fitted with rubber nipples. Two conditions are essential to success: the glass must be thoroughly softened and it must be removed from the flame before beginning to pull.

36

A nipple can be made of a short piece of rubber tubing, one end of which is plugged with a glass bead.

This pipet has many uses about the laboratory. When first made it is sterile and may be used to transfer cultures. With a grease-pencil mark about 2 cm. from its tip (see Fig. 228), it is useful for measuring very small quantities of fluid, as in making dilutions for the Widal test and in counting bacteria in vaccines. Mett's tubes for pepsin estimation may be made from the capillary portion. The capillary portion also makes a very satisfactory blood-lancet if the center is heated in a low flame and the two ends pulled quickly apart.

. Another useful device is the Wright capsule, which is made as shown in Fig. 225. Its use is illustrated in Fig. 230. After the straight end is sealed, the curved portion may be hooked over the aluminum tube of the centrifuge, and the contained blood or other fluid sedimented; but the speed should not be so great as to break the capsule.

ıl. STERILIZATION

All apparatus and materials used in bacteriologic work must be sterilized before use.

Glassware, metal, etc., are heated in the hot-air sterilizer at 150°C. for one hour, at 180°C. for half an hour, or at 200°C for five minutes. Flasks, bottles, and tubes are plugged with cotton before heating. Petri dishes may be wrapped in paper in sets of three. Pipets and glass or metal hypodermic syringes are placed in cotton-stoppered test-tubes.

Culture-media and other fluids must be sterilized by steam. Exposure in an autoclave to a temperature of

110°C. (6 pounds' pressure) for one-half hour or of 121°C (about 15 pounds' pressure) for fifteen minutes is sufficient. With the Arnold sterilizer, which is more commonly used, the intermittent plan must be adopted, since steam at ordinary pressure will not kill spores. This consists in steaming for thirty to forty-five minutes on three or four successive days. Spores which are not destroyed upon the first day develop into the vegetative form and are destroyed at the next heating. Gelatin media must not be exposed to steam for more than twenty minutes at a time, and must then be removed from the sterilizer and cooled in cold water, otherwise the gelatin may lose its power to solidify.

Cotton and **gauze** are sterilized by either hot air or steam, preferably the latter.

III. PREPARATION OF CULTURE-TUBES

New tubes should be washed in a very dilute solution of nitric acid, rinsed in clear water, and allowed to drain dry.

Tubes which contain dried culture-media are cleaned with a test-tube brush after boiling in a strong solution of washing-soda. They are then rinsed successively in clear water, acidulated water, and clear water, and allowed to drain.

The well-known **bichromate cleaning fluid** is very valuable for cleaning glassware of all kinds. It consists of:

Potassium bichromate................ 100 Gm.;
Concentrated sulphuric acid.......... 120 c.c.;
Water .... 1000 c.c.

Glass-ware may be placed in this solution for one day or longer and then rinsed thoroughly and dried.

The tubes are now ready to be plugged with raw cotton—the "cotton batting" of the dry goods stores. This is done by pushing a wad of cotton into each tube to a depth of about 3 cm. with a glass rod. The plugs should fit snugly, but not too tightly, and should project from the tube sufficiently to be readily grasped by the fingers. The tubes are next placed in wire baskets and heated in an oven for about one-half hour at 150°C. in order to mold the stoppers to the shape of the tubes. The heating should not char the cotton, although a slight browning does no harm. The tubes are now ready to be filled with culture-media.

IV. CULTURE-MEDIA

For a careful study of bacteria a great variety of culture-media is required, but only a few—bouillon, agar or solidified blood-serum, and gelatin—are much used in routine work. A great deal of work can be done with a single medium, for which purpose solidified blood-serum is probably best. The ordinary culture-media, put up in tubes ready for use, can be purchased through any pharmacy, and some can be obtained in powder or tablet form ready to be dissolved in the appropriate amount of water.

Preparation of Culture-media.—

BEEF INFUSION

Hamburger steak, lean.................. 500 Gm.;
Tap-water............................. 1000 c.c.

Mix well; let soak about twenty-four hours in an ice-chest, and squeeze through cheese-cloth. This infusion is not used by itself, but forms the basis for various media. "Double strength" infusion, used in making agar-agar, requires equal parts of the meat and water.

INFUSION BOUILLON

Beef infusion............................ 1000 c.c.;
Peptone (Witte)........................ 10 Gm.;
Salt.................................... 5 Gm.

Boil until dissolved; bring to original volume with water; adjust reaction, and filter.

BEEF EXTRACT BOUILLON

Liebig's extract of beef.................. 3 Gm.;
Peptone................................ 10 Gm.;
Salt.................................... 5 Gm.;
Tap-water.............................. 1000 c.c.

Heat until all ingredients are dissolved, cool, and beat in the whites of two eggs; bring slowly to the boiling-point again; boil briskly for five minutes and filter. It is not usually necessary to adjust the reaction.

AGAR-AGAR

Preparation of this medium usually gives the student much trouble. There should be no difficulty if the directions are carefully carried out.

Agar-agar, powdered or in shreds.......... 15 Gm.;
Tap-water.............................. 500 c.c.

Boil until thoroughly dissolved and add—

Peptone................................ 10 Gm.;
Salt.................................... 5 Gm.

When these have dissolved, replace the water lost in

boiling, cool to about 60°C., and add 500 c.c. double-strength beef infusion. Bring slowly to the boil, adjusting the reaction meanwhile, and boil for at least five minutes. Filter while *hot* through a moderately thick layer of absorbent cotton wet with *hot* water in a *hot* funnel. A piece of coarse wire gauze should be placed in the funnel underneath the cotton to give a larger filtering surface. This medium will be clear enough for ordinary work. If an especially clear agar is desired, it can be filtered through paper in an Arnold sterilizer.

Agar can also be made by boiling 15 Gm. of powdered agar in 1000 c.c. of bouillon until dissolved, replacing the water lost in boiling, and filtering through paper in a sterilizer. It can be cleared with egg if desired.

GLYCERIN AGAR-AGAR

To 1000 c.c. melted agar add 60 to 70 c.c. glycerin.

BLOOD AGAR-AGAR

The simplest way to prepare this is to smear a drop of blood, obtained by puncture of the finger, over the surface of an agar-slant, and to incubate over night to make sure of sterility. It is used chiefly for growing the influenza bacillus. It may be noted that the bacillus will not grow well on blood from a person who has recently recovered from influenza.

A blood-agar prepared as follows is more satisfactory: Melt 5 c.c. sterile agar in a culture-tube, cool to 45°C. in a water-bath, add 1 c.c. human blood (about 15 drops), and mix well. Cool in an inclined position or pour into Petri plates. Incubate twelve to twenty-four hours to make sure of sterility.

GELATIN

Dissolve 100 to 120 Gm. "golden seal" gelatin in 1000 c.c. nutrient bouillon with as little heat as possible, adjust the reaction, cool, beat in the whites of two eggs, bring slowly to the boiling point, boil for a few minutes, and filter hot through filter-paper wet with hot water. Sterilize in an Arnold sterilizer for twenty minutes upon three successive days and cool in cold water after each heating. Keep at room temperature between heatings.

SUGAR MEDIA

Any desired sugar may be added to bouillon, agar, or gelatin in proportion of 10 Gm. to the liter. Dextrose is most frequently required. When other sugars are added, media made from beef-extract should be used, since those made from beef-infusion contain enough dextrose to cause confusion.

The various sugars may also be added to Dunham's peptone medium and Hiss' serum-water-litmus.

LÖFFLER'S BLOOD-SERUM

Dextrose-bouillon (1 per cent.)............... 1 part;
Blood-serum.............................. 3 parts.

Mix and tube. Place in an inspissator at the proper slant for three to six hours at 80° to 90°C. When firmly coagulated, sterilize in the usual way. A large "double-cooker" makes a satisfactory inspissator. The tubes are placed in the inner compartment upon a layer of cotton at the proper slant, a lid with perforation for a thermometer is applied, and the whole is weighted down in the water of the outer compartment.

Blood-serum is obtained as follows: Beef or pig-blood is collected in a bucket at the slaughter-house and placed in an ice-chest until coagulated. The clot is then gently loosened from the wall of the vessel. After about twenty-four hours the serum will have separated nicely and can be siphoned off. It is then stored in bottles with a little chloroform until needed. Red cells, if abundant, darken the medium, but do no harm.

Solidified blood-serum is probably the most satisfactory medium for general purposes. Nearly all pathogenic organisms grow well upon it.

EGG MEDIUM

This has been recommended as a substitute for solidified blood-serum. In a mortar grind thoroughly the white and yolk of one egg with 10 to 15 c.c. of 1 per cent. dextrose bouillon. Place in tubes, inspissate, and sterilize as described for solidified blood-serum.

LITMUS MILK

Fresh milk is steamed in an Arnold sterilizer for half an hour, and placed in the ice-chest over night. The milk is siphoned off from beneath the cream, and sufficient aqueous solution of litmus or, preferably, azolitmin is added to give a blue-violet color. It is then tubed and sterilized.

POTATO

Cylinders about ½ in. diameter are cut from potato and split obliquely. These wedge-shaped pieces are soaked over night in running water and placed, broad ends down, in large tubes, in the bottom of which is placed a little cotton saturated with water. They are sterilized for somewhat longer periods than ordinary media.

DUNHAM'S PEPTONE SOLUTION

Peptone	10 Gm.;
Salt	5 Gm.;
Water	1000 c.c.

Dissolve by boiling; filter, tube, and sterilize.

This medium is used to determine indol production. To a twenty-four- to forty-eight-hour-old culture is added 5 to 10 drops of concentrated, chemically pure sulphuric acid and 1 c.c. of 1:10,000 solution of sodium nitrite. Appearance of a pink color shows the presence of indol. A pink color before the nitrite is added shows the presence of both indol and nitrites.

HISS' SERUM-WATER MEDIA

Blood-serum	1 part;
Water	3 parts.

Warm and adjust reaction to +0.2 to +0.8. Add litmus or azolitmin solution to give a blue-violet color. Finally, add 1 per cent. of inulin or any desired sugar. The inulin medium is very useful in distinguishing between the pneumococcus and streptococcus.

BILE MEDIUM

Ox- or pig-bile is obtained at the slaughter-house, tubed, and sterilized. This is used especially for growing typhoid bacilli from the blood during life. The following is probably as satisfactory as fresh bile and is more convenient:

Inspissated ox-bile (Merck)	30.0 Gm.;
Peptone	2.5 Gm.;
Water	250.0 c.c.

Dissolve, place in tubes, and sterilize.

Reaction of Media.—The chemical reaction of the medium exerts a marked influence upon the growth of bacteria. It is adjusted after all ingredients are dissolved by adding sufficient caustic soda solution to overcome the acidity of the meat and other substances used. In general, the most favorable reaction lies between the neutral points of litmus and phenolphthalein, representing a very faint alkalinity to litmus. In routine work it is usually sufficient to test with litmus-paper. When greater accuracy is demanded, the following method should be used: After all ingredients are dissolved and the loss during boiling has been replaced with water, 10 c.c. of the medium are transferred to an evaporating dish, diluted with 40 c.c. of water, and boiled for three minutes to drive off carbon dioxid. One c.c. of 0.5 per cent. alcoholic solution of phenolphthalein is then added, and decinormal sodium hydroxid solution is run in from a buret until the neutral point is reached, indicated by the appearance of a permanent pink color. The number of cubic centimeters of decinormal solution required to bring this color indicates the number of cubic centimeters of *normal* sodium hydroxid solution which will be required to neutralize 100 c.c. of the medium. The standard reaction is $+1.5$, which means that the medium must be of such degree of acidity that 1.5 c.c. of normal solution would be required to neutralize 100 c.c. This corresponds to faint alkalinity to litmus. Most pathogenic bacteria grow better with a reaction of $+1.0$ or $+0.8$. Example: If the 10 c.c. which were titrated required 2 c.c. of decinormal solution to bring the pink color, the reaction is $+2$; and 0.5 c.c. of normal sodium hydroxid must be added to each 100 c.c. of the medium to reduce it to the standard, $+1.5$.

Tubing Culture-media.—The finished product is stored in flasks or distributed into test-tubes. This is done by means of a funnel fitted with a section of rubber tubing with a glass tip and a pinch-cock. Great care must be exercised, particularly with media which solidify, not to smear any of them upon the inside of the mouth of the tube, otherwise the cotton stopper will stick. Tubes are generally filled to a depth of 3 or 4 cm. For stab-cultures a greater depth is required.

After tubing, all culture-media must be sterilized as already described. Agar-tubes are cooled in a slanting position to secure the proper surface for inoculation.

Storage of Culture-media.—All media should be stored in a cool place, preferably an ice-chest. Evaporation may be prevented by covering the tops of the tubes with tin-foil or with the rubber caps which are sold for the purpose; or the cotton stopper may be pushed in a short distance and a cork inserted.

V. STAINING METHODS

In general, bacteria are stained to determine their morphology, their reaction with special methods (*e.g.*, Gram's method), and the presence or absence of certain structures, as spores, flagella, and capsules. Staining methods for various purposes have been given in previous chapters and can be found by consulting the Index. The formulæ of the staining fluids are given in the Appendix.

Method of Staining for Morphology.—The following method is used when one wishes to detect the presence of bacteria or to study their morphology. It is applicable both to films from cultures and to smears from pus or other pathologic material. Any simple bacterial stain may be used but Loeffler's methylene blue or Pappenheim's pyronin-methyl-green will generally be found more satisfactory.

1. Make a thin smear upon a slide or cover-glass. Heavy wax-pencil marks across the slide will limit the stain to any portion desired.

2. Dry in the air, or by warming high above the flame, where one can comfortably hold the hand.

3. "Fix" by passing the preparation, film side up, rather slowly through the flame of a Bunsen burner: a cover-glass three times, a slide about twelve times. One can learn to judge the proper temperature by touching the glass to the back of the hand at intervals. If the film takes on a brownish discoloration, most marked about the edges, it has been scorched and is worthless. Smears can also be fixed by flaming with alcohol, as described for blood-films (see p. 307), or by soaking for one to three minutes in a saturated solution of mercuric chlorid and rinsing well. The last avoids all possibility of spoiling the preparation by scorching.

4. Apply the stain for the necessary length of time, generally one-quarter to one minute.

5. Wash in water.

6. Dry by waving high above a flame or by blotting with filter-paper.

7. Mount by pressing the cover, film side down, upon a drop of Canada balsam or immersion-oil on a slide. Slides may be examined with the oil-immersion lens without a cover-glass.

Gram's Method.—This is a very useful aid in differentiating certain bacteria and should be frequently resorted to. It is very easy and should not be the bug-bear which it apparently is to many students. It depends upon the fact that when treated successively with gentian-violet and iodin, certain bacteria (owing to formation of insoluble compounds) retain the stain when subsequently treated with alcohol, whereas others quickly lose it. The former are called *Gram-positive;* the latter, *Gram-negative.* In order to render Gram-negative organisms visible, some contrasting counterstain is commonly applied, but this is not a part of Gram's method proper.

1. Make smears, dry, and fix by heat or mercuric chlorid.

2. Apply carbol-gentian-violet, anilin-gentian-violet or formalin-gentian-violet two to five minutes. The last is probably least satisfactory.

3. Wash with water.

4. Apply Gram's iodin solution one-half to two minutes.

5. Wash in alcohol until the purple color ceases to come off. This is conveniently done in a watch-glass. The preparation is placed in the alcohol, face downward, and one edge is raised and lowered with a needle. As long as any color is coming off, purple streaks will be seen diffusing into the alcohol where the surface of the fluid meets the smear. If forceps be used, beware of stain which may have dried upon them. The thinner portions of smears from pus should be practically colorless at this stage. Microscopically, the nuclei of pus-corpuscles should retain little or no color. *If the smears resist decolorization, the gentian-violet and iodin solution should be applied for a shorter time, say, one-half minute each.*

6. Apply a contrast stain for one-half to one minute. The stains commonly used for this purpose are an aqueous or alcoholic solution of Bismarck brown and a weak solution

of fuchsin. A 1 per cent. aqueous solution of safranin is better. In the writer's experience, Pappenheim's pyro-nin-methyl-green mixture is still more satisfactory; it brings out Gram-negative bacteria sharply, and is especially desirable for intracellular Gram-negative organisms like the gonococcus and influenza bacillus, since the bacteria are bright red and nuclei of cells blue.

7. Wash in water, dry, and mount in balsam.

The more important bacteria react to this staining method as follows:

GRAM STAINING (Deep purple)	GRAM DECOLORIZING (Colorless unless a counterstain is used)
Staphylococcus.	Gonococcus.
Streptococcus.	Meningococcus.
Pneumococcus.	Micrococcus catarrhalis.
Bacillus diphtheriæ.	Bacillus of influenza.
Bacillus tuberculosis.	Typhoid bacillus.
Bacillus of anthrax.	Bacillus coli communis.
Bacillus of tetanus.	Spirillum of Asiatic cholera.
Bacillus aërogenes capsulatus.	Bacillus pyocyaneus.
	Bacillus of Friedländer.
	Koch-Weeks bacillus.
	Bacillus of Morax-Axenfeld.

Möller's Method for Spores.—Bodies of bacteria are blue, spores are red.

1. Make thin smears, dry, and fix.
2. Wash in chloroform for two minutes.
3. Wash in water.
4. Apply 5 per cent. solution of chromic acid one-half to two minutes.
5. Wash in water.
6. Apply carbolfuchsin and heat to boiling.
7. Decolorize in 5 per cent. solution of sulphuric acid.
8. Wash in water.

9. Apply 1 per cent. aqueous solution of methylene-blue one-half minute.

10. Wash in water, dry, and mount.

Huntoon's Method for Spores.—This method is simple and appears to be very reliable. Spores are deep red, bodies of bacteria are blue.

1. Make a rather thick smear, dry, and fix in the usual way.

2. Apply as much of the stain as will remain on the cover-glass, and steam over a flame for one minute, replacing the stain lost by evaporation.

3. Wash in water. The film is bright red.

4. Dip the preparation a few times into a weak solution of sodium carbonate (7 or 8 drops of saturated solution in a glass of water). Too long application of the carbonate will cause the spores to be blue.

5. The instant the film turns blue, rinse well in water.

6. Dry, mount, and examine.

Preparation of Stain.—

    (1) Acid fuchsin (Grübler).................. 4 Gm.;
        Aqueous solution acetic acid (2 per cent.). 50 c.c.

    (2) Methylene-blue (Grübler)......:........ 2 Gm.;
        Aqueous solution acetic acid (2 per cent.). 50 c.c.

Mix the two solutions, let stand for fifteen minutes, and filter off the voluminous precipitate through moistened filter-paper. The filtrate is the staining fluid. It keeps several weeks, but requires filtration when a precipitate forms.

Löffler's Method for Flagella.—The methods for flagella are applicable only to cultures. Enough of the growth from an agar-culture (which should not be more than eighteen to twenty-four hours old) to produce faint cloudiness is added to distilled water. A small drop of this is placed on a cover-glass, spread by tilting, and dried quickly. The covers must be absolutely free from grease. To insure this, they may be warmed in concentrated sulphuric acid, washed in water, and kept in a mixture of alcohol and strong ammonia. When used they are dried upon a fat-free cloth. Covers may be dried without touching them with the fingers by rubbing between two blocks of wood covered with several layers of lint-free cloth.

1. Fix by heating the cover over a name while holding in the fingers.

2. Cover with freshly filtered mordant and gently warm for about a minute.

The mordant consists of:

> Aqueous solution of tannic acid (20 per cent.)..... 10 c.c.;
> Saturated solution ferrous sulphate, cold......... 5 c.c.;
> Saturated aqueous or alcoholic solution gentian-
>   violet................................... 1 c.c.

3. Wash in water.

4. Apply freshly filtered anilin-gentian-violet, warming gently for one-half to one minute.

5. Wash in water, dry, and mount in balsam.

VI. METHODS OF STUDYING BACTERIA

The purpose of bacteriologic examinations is to determine whether bacteria are present or not, and, if present, their species and comparative numbers. In general, this is accomplished by: (1) direct microscopic examination; (2) cultural methods; (3) animal inoculation.

1. Direct Microscopic Examination.—Every bacteriologic examination should begin with a microscopic study of smears from the pathologic material, stained with a general stain, by Gram's method, and often by the method for the tubercle bacillus. This yields a great deal of information to the experienced worker, and in many cases is all that is necessary for the purpose in view. It will at least give a general idea of what is to be expected, and may determine future procedure.

2. Cultural Methods.—1. **Collection of Material.**—Material for examinations must be collected under absolutely aseptic conditions. It may be obtained with a platinum wire—which has been heated to redness just previously and allowed to cool—or with a swab of sterile cotton on a stiff wire or wooden applicator. Such swabs may be placed in cotton-stoppered test-tubes, sterilized, and kept on hand ready for use. Fluids which contain very few bacteria, and hence require that a considerable quantity be used, may be collected in a sterile hypodermic syringe or one of the pipets described on page 561. The method of obtaining blood for cultures is given on pp. 346, 347.

2. **Inoculating Media.**—The material is thoroughly distributed over the surface of some solid medium, solidified blood-serum being probably the best for routine work. When previous examination of smears has shown that many bacteria are to be expected, a second tube should be inoculated from the first, and a third from the second, so as to obtain isolated colonies in at least one of the tubes. The platinum wire must be heated to redness *before* and *after* each inoculation. When only a few organisms of a single species are ex-

37

pected, as is the case in blood-cultures (see p. 347), a considerable quantity of the material is mixed with a fluid medium.

3. **Incubation.**—Cultures are placed in an incubator which maintains a uniform temperature, usually of 37.5°C., for eighteen to twenty-four hours, and the growth, if any, is studied as described later. Gelatin will melt with this degree of heat, and must be incubated at about room-temperature.

4. **Study of Cultures.**—When the original culture contains more than one species, they must be separated, or obtained in "pure culture," before they can be studied satisfactorily. To accomplish this it is necessary to so distribute them on solid media that they form separate colonies, and to inoculate fresh tubes from the individual colonies. In routine work the organisms can be sufficiently distributed by drawing the contaminated wire over the surface of the medium in a series of streaks. If a sufficient number of streaks be made, some of them are sure to show isolated colonies. Another method of obtaining isolated colonies is to inoculate the water of condensation of a series of tubes, the first from the second, the second from the third, etc., and, by tilting, to flow the water once over the surface of the medium. One or more of these tubes will be almost sure to show nicely separated colonies.

In order to determine the species to which an organism belongs it is necessary to consider some or all of the following points:

1. Naked-eye and microscopic appearance of the colonies on various media.

2. Comparative luxuriance of growth upon various

media. The influenza bacillus, for example, can be grown upon media containing hemoglobin, but not on the ordinary media.

3. Morphology, special staining reactions, and the presence or absence of spores, flagella, and capsules. Staining methods for these purposes have been given.

4. Motility. This is determined by observing the living organism with an oil-immersion lens in a hanging-drop preparation, made as follows: A small drop of a bouillon culture or of water of condensation from an agar or blood-serum tube is placed upon the center of a cover-glass; and over this is pressed the concavity of a "hollow-ground slide" previously ringed with vaselin. The slide is then turned over so as to bring the cover-glass on top. In focusing, the edge of the drop should be brought into the field. Great care must be exercised not to break the cover by pushing the objective against it.

It is not always easy to determine whether an organism is or is not motile, since the motion of currents and the Brownian motion which affects all particles in suspension are sometimes very deceptive.

5. Production of chemical changes in the media. Among these are coagulation of milk; production of acid in milk and various sugar media to which litmus has been added to detect the change; production of gas in sugar media, the bacteria being grown in fermentation tubes similar to those used for sugar tests in urine; and production of indol.

6. Ability to grow without free oxygen.

5. **Anaërobic Methods.**—Some bacteria, the "obligate anaërobes," will not grow unless free oxygen is ex-

cluded. This may be accomplished in various ways. Perhaps the most convenient is the following method of J. H. Wright. After the culture medium in the test-tube has been inoculated, push the cotton stopper in until its top is about 1.5 cm. below the mouth of the tube. Fill in the space above the stopper with dry pyrogallic acid and pour on it just enough strong solution of sodium hydroxid to dissolve it. Finally, insert a rubber cork and seal with paraffin.

7. Effects produced when inoculated into animals.

3. Animal Inoculation.—In clinical work this is resorted to chiefly to detect the tubercle bacillus. The method is described on page 534.

For the study of bacteria in cultures, a small amount of a pure culture is injected subcutaneously or into the peritoneal cavity. The animals most used are guinea-pigs, rabbits, and mice. For intravenous injection, the rabbit is used because of the easily accessible marginal vein of the ear.

VII. CHARACTERISTICS OF SPECIAL BACTERIA

Owing to the great number of bacterial species, most of which have not been adequately studied, positive identification of an unknown organism is often a very difficult problem. Fortunately, however, only a few are commonly encountered in routine work, and identification of these with comparative certainty presents no great difficulty. Their more distinctive characteristics are outlined in this section.

1. Staphylococcus pyogenes aureus.—The morphology and staining reactions (described on p. 516) and the appearance of the colonies are sufficient for diagnosis. Colonies on solidified blood-serum and agar are rounded, slightly elevated, smooth and shining, and vary in color from light yellow to deep orange. Young colonies are sometimes white.

2. Staphylococcus pyogenes albus.—This is similar to the above, but colonies are white. It is generally less virulent.

3. Staphylococcus pyogenes citreus.—The colonies are lemon yellow; otherwise it resembles the white staphylococcus.

4. Streptococcus pyogenes.—The morphology and staining reactions have been described (see p. 516). The chains are best seen in the water of condensation and in bouillon cultures. The cocci are not motile. Colonies on blood-serum are minute, round, grayish, and translucent. Litmus milk is usually acidified and coagulated, although slowly. The streptococcus rarely produces acid in Hiss' serum-water-litmus-inulin medium (see p. 569).

Some strains of streptococcus are capable of hemolyzing red blood corpuscles, and this property is utilized in classification. Hemolysis is manifested by the appearance of a wide clear zone around the colonies when grown upon blood-agar. Many non-hemolyzing streptococci produce a narrow green zone upon this medium, and these are grouped under the name *Streptococcus viridans*. Such streptococci are less actively virulent than the hemolyzing type, being most frequently associated with chronic inflammations.

5. Pneumococcus.—The only organism with which this is likely to be confused is the streptococcus. The distinction is often extremely difficult.

Detection of the pneumococcus in fresh material has been described (see p. 85). In cultures it frequently forms long chains. Capsules are not present in cultures except upon special media. They show best upon a serum-medium like that described for the gonococcus, but can frequently be seen in milk. Colonies on blood-serum resemble those of the streptococcus. Colonies on blood-agar show a green zone like those of *Streptococcus viridans*. The pneumococcus usually promptly acidifies and coagulates milk and acidifies and coagulates Hiss' serum-water with inulin. The latter property is very helpful in diagnosis.

6. Micrococcus catarrhalis grows readily at room temperature and on ordinary media where it forms large, white, dry colonies with irregular edges and elevated centers. This readily distinguishes it from the gonococcus and meningococcus, which it closely resembles in morphology and staining reactions.

7. Gonococcus.—Its morphology and staining peculiarities are given on page 518. These usually suffice for its identification, cultural methods being rarely undertaken. In cultures the chief diagnostic point is its failure to grow on ordinary media. To grow it, the most convenient medium is made by adding ascitic or hydrocele fluid (which has been obtained under aseptic conditions) to melted agar in proportion of 1 part of serum to 3 parts of agar. The agar in tubes is melted and cooled to about 45°C.; the serum is added with a pipet and mixed by shaking; and the tubes are again cooled in a slanting position. Colonies upon this medium are minute, grayish, and translucent.

8. Diplococcus intracellularis meningitidis.—It grows poorly or not at all on plain agar. On Löffler's blood-serum, upon which it grows fairly well, colonies are round, colorless or hazy, flat, shining, and viscid-looking. It quickly dies out.

9. Diphtheria Bacillus.—The diagnosis is usually made from a study of stained smears from cultures upon blood-serum, grown for twelve to eighteen hours. Its morphology and staining peculiarities are characteristic when grown on this medium (see p. 538). The bacilli are non-motile and Gram-positive. The colonies are round, elevated, smooth, and grayish.

10. Typhoid and Colon Bacilli.—These are medium-sized, motile, Gram-negative, non-spore-bearing bacilli. Upon blood-serum they form rounded, grayish, slightly elevated, viscid looking colonies, those of the colon bacillus being somewhat the larger. They do not liquefy gelatin. They represent the extremes of a large group with many intermediate types. They are distinguished as follows:

Typhoid Bacillus	*Colon Bacillus*
Actively motile.	Much less active.
Growth on potato usually invisible.	Growth distinctly visible as dirty gray or brownish slimy layer.
No gas produced in glucose media.	Produces gas.
Growth in litmus milk produces no change.	Litmus milk pink and coagulated.
Produces no indol in Dunham's peptone medium.	Produces indol. (For test, see p. 569.)
Agglutinates with serum from typhoid-fever patient. (Recently isolated bacilli do not agglutinate well.)	Does not agglutinate with typhoid serum.

11. Bacillus of Influenza. Diagnosis will usually rest upon the morphology and staining peculiarities, described on page 89, and upon the fact that the bacillus will not grow on ordinary media, but does grow upon hemoglobin-containing media. It can be grown upon agar-slants which have been smeared with a drop of blood from a puncture in the finger. Before inoculation these slants should be incubated to make sure of sterility. The colonies are difficult to see without a hand lens. They are very minute, discrete, and transparent, resembling small drops of dew.

12. Bacillus of Tuberculosis.—The methods of identifying this important organism have been given (see pp. 76, 235). Cultivation is not resorted to in routine clinical work. It grows very slowly and only on certain media. It is Gram-positive and non-motile.

CHAPTER IX

PREPARATION AND USE OF VACCINES

BACTERIAL vaccines, sometimes called "bacterins," which within recent years have come to play an important rôle in therapeutics, are suspensions of definite numbers of dead bacteria in normal salt solution. While in many cases, notably in gonorrhea and tuberculosis, ready prepared or "stock" vaccines are satisfactory, it is usually desirable and often imperative for best results to use vaccines which are especially prepared for each patient from bacteria which have been freshly isolated from his own lesion. These latter are called "autogenous vaccines." Only through them can one be certain of getting the exact strain of bacterium which is producing the disease.

I. PREPARATION OF VACCINE

1. Preparation of Materials.—A number of 2-ounce wide-mouthed bottles are cleaned and sterilized. Each is charged with 50 c.c. freshly filtered normal salt solution (0.85 per cent. sodium chlorid in distilled water), and is capped with a new rubber nursing-nipple, without holes, inverted as shown in Fig. 226. A small section of hollow wire or a hypodermic needle is thrust through the cap near the edge to serve as an air vent, and the bottle and contents are sterilized in an autoclave. If an autoclave is not at hand, successive steamings in an

Arnold sterilizer will answer, provided it is not opened between steamings. After sterilization, the pieces of wire are pulled out and the holes sealed with collodion.

Most workers use a smaller bottle with less salt solution and with a cotton stopper; and, after the solution has been sterilized, apply a specially made rubber "vaccine-bottle cap." Instead of the cotton stopper, a

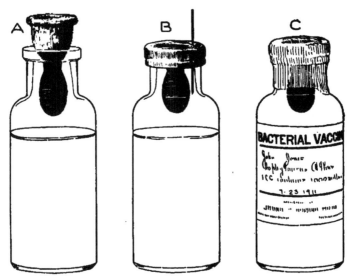

FIG. 226.—Vaccine bottles: A, Cap ready to be applied; B, ready for sterilization; C, finished vaccine.

sheet of paper which is placed over the top, folded closely about the neck of the bottle, and held in place with a rubber band may advantageously be used as a temporary cap. The first method calls for an unnecessarily large quantity of fluid (which is no real objection), but has certain slight advantages: the nursing nipples are easily obtained at any pharmacy; the rubber is not put upon the stretch as is the case with some caps, and is, therefore, self-sealing; no cotton-lint falls into the salt solution before the cap is applied; and the cap offers a concavity which may be filled with 80 per cent. alcohol for sterilizing before the needle is plunged through.

A number of test-tubes, each charged with 10 c.c. of normal salt solution and plugged with cotton, are also prepared and sterilized.

2. Obtaining the Bacteria.—A culture on some solid medium is made from the patient's lesion, and a pure culture is obtained in the usual way. This preliminary work should be carried through as quickly as possible in order that the bacteria may not lose virulence by long growth upon artificial media. If for any reason there is much delay, it is best to begin over again, the experience gained in the first trial enabling one to carry the second through more rapidly. When a pure culture is obtained, a number of tubes of blood-serum or agar—10 or 12 in the case of streptococcus or pneumococcus, 4 or 5 in the case of most other organisms—are planted and incubated over night or until a good growth is obtained.

3. Making the Suspension.—A few cubic centimeters of the salt solution from one of the 10-c.c. salt-tubes is transferred by means of a sterile pipet to each of the culture-tubes, and the growth thoroughly rubbed up with a stiff platinum wire or a glass rod whose tip is bent at right angles. The suspension from the different tubes, usually amounting to about 10 c.c., is then collected in one large tube (size about 150 × 19 mm.); and the upper part of the tube is drawn out in the flame of a blast lamp or Bunsen burner, as indicated in Fig. 227, *B*, a short section of glass tubing being fused to the rim of the tube to serve as a handle. It is then stood aside, and when cool the narrow portion is sealed off.

The resulting hermetically sealed capsule is ·next thoroughly shaken for ten to twenty minutes to break up all clumps of bacteria. Some small pieces of glass

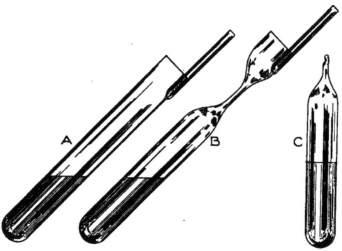

FIG. 227.—Process of making hermetically sealed capsules containing liquid.

or a little clean sterile sand may be introduced to assist in this, but with many organisms it is not necessary.

4. Sterilization.—The capsule is placed in a water-bath at 60°C. for forty-five minutes. This can be done in an ordinary rice-cooker, with double lid, through which a thermometer is inserted. When both compartments are filled with water it is an easy matter to maintain a uniform temperature by occasional application of a small flame. The time and temperature are important:

too little heat will fail to kill the bacteria, and too much will destroy the efficiency of the vaccine.

When sterilization is complete the capsule is opened, a few drops are planted on agar or blood-serum, and the capsule is again sealed.

5. Counting.—When incubation of the planted tube has shown the suspension to be sterile it is ready for counting. Of the two methods given here the latter is the more accurate.

Wright's Method.—There must be ready a number of clean slides; a few drops of normal salt solution on a slide or in a watch-glass; a blood-lancet, which can be improvised from a spicule of glass or a pen; and two capillary pipets with squarely broken off tips and wax-pencil marks about 2 cm. from the tip (Fig. 228). These are easily made by drawing out a piece of glass tubing, as described on page 561.

It is necessary to work quickly. After thorough shaking, the capsule is opened and a few drops forced out upon a slide. Any remaining clumps of bacteria are broken up with one of the pipets by holding it against and at right angles to the slide, and alternately sucking the fluid in and forcing it out. The pipet is most easily controlled if held in the whole hand with the rubber bulb between

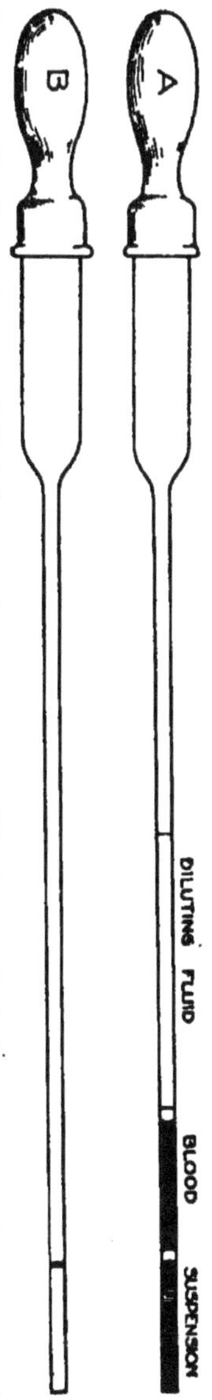

Fig. 228.—Capillary pipets: A, Filled for counting a vaccine by Wright's method; B, empty, showing wax-pencil mark. The slender portion should be narrower than here represented.

DILUTING FLUID

BLOOD

SUSPENSION

the thumb and the side of the index-finger. A finger is then
pricked until a drop of blood appears; and into the second
pipet are quickly drawn successively: 1 or 2 volumes
normal salt solution (or, better, a 1 per cent. solution of
sodium citrate, which prevents coagulation); a small bubble
of air; 1 volume of blood; a small bubble of air; and, finally,
1 volume of bacterial suspension. (A "volume" is measured
by the distance from the tip of the pipet to the wax-
pencil mark.) The contents of the pipet are then forced
out upon a slide and thoroughly mixed by sucking in and
out, care being taken to avoid bubbles; after which the
fluid is distributed to a number of sides and spread as in
making blood-smears.

The films are stained with Wright's blood-stain or,
better, by a few minutes' application of carbol-thionin,
after fixing for a minute in saturated mercuric chlorid
solution. With an oil-immersion lens both the red cells
and the bacteria in a number of microscopic fields are
counted. The exact number is not important; for convenience 500 red cells may be counted. From the ratio
between the number of bacteria and of red cells, it is
easy to calculate the number of bacteria in 1 c.c. of the
suspension, it being known that there are 5000 million
red corpuscles in a cubic centimeter of normal human blood.
If there were twice as many bacteria as red corpuscles in
the fields counted, the suspension would contain 10,000
million bacteria per cubic centimeter.

Hemacytometer Method.—This is carried out in the same
manner as a blood count, using any convenient dilution,
usually 1:200. A weak carbol-fuchsin or gentian-violet
solution, freshly filtered, may be used as diluting fluid, but
the following solution, recommended by Callison is better:

Hydrochloric acid...................... 2 c.c.;
Mercuric chlorid (0.2 per cent. solution).... 100 c.c.;
Acid fuchsin (1 per cent. aqueous solution), to color.

The color should be just deep enough not to obscure the ruled lines.

A very thin cover-glass must be used; and, after filling, the counting-chamber must be set aside for an hour or more to allow the bacteria to settle. Mallory and Wright advise the use of the shallow Helber chamber manufactured by Zeiss for counting blood-plates, but many 2-mm. oil-immersion objectives have sufficient working distance to allow the use of the regular counting-chamber, provided a very thin cover is used. The heavy cover with central excavation is recommended.

6. Diluting.—The amount of the suspension which, when diluted to 50 c.c., will give the strength desired for the finished vaccine having been determined, this amount of salt solution is withdrawn with a hypodermic syringe from one of the bottles already prepared, and is replaced with an equal amount of suspension. One-tenth cubic centimeter of trikresol or lysol is finally added and the vaccine is ready for use. To prevent possible leakage about the cap, the neck of the bottle is dipped in melted paraffin. The usual strengths are: staphylococcus, 1000 million in 1 c.c.; most other bacteria, 100 million in 1 c.c.

II. METHOD OF USE

Vaccines are administered subcutaneously, usually in the arm or abdominal wall or between the shoulder-blades. The technic is the same as for an ordinary hypodermic injection. The syringe is usually sterilized by boiling. The site of the injection may be mopped with alcohol or may be touched with a pledget of cotton saturated with tincture of iodin or liquor cresolis compositus. The rubber cap of the container is sterilized by filling the concavity with alcohol for some minutes, usually while the syringe is being sterilized, or simply placing a drop of liquor cresolis compositus upon it. The bottle is then inverted and well shaken, when the needle is plunged through the rubber and the desired quantity withdrawn. The hole seals itself. A satisfactory syringe is the comparatively inexpensive Luer 1 c.c. "Tuberculin" syringe graduated in hundredths of a cubic centimeter.

III. DOSAGE

Owing to variations, both in virulence of organisms and susceptibility of patients, no definite dosage can be assigned. Each case is a separate problem. Wright's original proposal was to regulate the size and frequency of dose by its effect upon the opsonic index, but this is beyond the reach of the practitioner. The more widely used "clinical method" consists in beginning with a very small dose and cautiously increasing until the patient shows either improvement or some sign of a "reaction," indicated by headache, malaise, fever, exacerbation of local disease, or inflammatory reaction at the site of injection. The reaction indicates that the dose has been too large. The beginning dose of staphylococcus is about 50 million; the maximum, 1000 million or more. Of most other organisms the beginning dose is 5 million to 10 million; maximum, about 100 million. Ordinarily, injections are given once or twice a week; very small doses may be given every other day.

IV. THERAPEUTIC INDICATIONS

The therapeutic effect of vaccines depends upon their power to produce active immunity: they stimulate the production of opsonins and other antibacterial substances which enable the body to combat the infecting bacteria. Their especial field is the treatment of subacute and chronic localized infections, in some of which they offer the most effective means of treatment at our command. In most chronic infections the circulation of blood and lymph through the diseased area is very sluggish, so that the antibodies, when formed, cannot readily reach the seat of disease. Ordinary measures which favor circulation in the diseased part should, therefore, accompany the vaccine treatment. Among these may be mentioned incisions and drainage of abscesses, dry cupping, application of heat, Bier's hyperemia, etc., but such measures should not be applied during the twenty-four hours succeeding an injection, since the first effect of the vaccine may be a temporary lowering of resistance. Vaccines are of little value, and, in general, are contraindicated in very acute infections, particularly in those which are accompanied by much systemic intoxication, for in such cases the power of the tissues to produce antibodies is already taxed to the limit. It is true, nevertheless, that remarkably beneficial results have occasionally followed their use in such acute conditions as malignant endocarditis, but here they should be tried with extreme caution.

Probably best results are obtained in staphylococcus infections, although pneumococcus, streptococcus, and colon bacillus infections usually respond nicely. Among clinical conditions which have been treated successfully with vaccines are furunculosis, acne vulgaris, infected operation-wounds, pyelitis, cystitis, subacute otitis media, osteomyelitis, infections of nasal accessory si-

muses, etc. Vaccine treatment of the mixed infection is
doubtless an important aid in tuberculosis therapy,
and in some cases the result is brilliant. When, as is
common, several organisms are present in the sputum,
a vaccine is made from each, excepting the tubercle
bacillus, of which autogeneous vaccines are not used in
practice. To avoid the delay and consequent loss of
virulence entailed by study and isolation of the several
varieties, many workers make the bacterial suspension
directly from the primary cultures. The resulting vaccines contain all strains which are present in the sputum
in approximately the same relative numbers. Although
open to criticism from a scientific standpoint, this
method offers decided practical advantages in many
cases.

V. PROPHYLACTIC USE OF VACCINES

It has been shown that vaccines are useful in preventing as well as curing infections. Their value has been
especially demonstrated in typhoid fever. Three
doses of about 500 million, 1000 million, and 1000
million typhoid bacilli, respectively, are given about
seven to ten days apart. Results in the army, where
the plan has been tried on a large scale, show that
such vaccination is effective, protecting the individual
for six months to a year, or longer.

VI. TUBERCULINS

Tuberculins contain the products of tubercle bacilli
or their ground-up bodies, the latter class being practically vaccines. They are undoubtedly of great value
in the treatment of localized tuberculosis, particularly

of bones, joints, and glands; and are of rather indefinite though certainly real value in chronic pulmonary tuberculosis, especially when the disease is quiescent. The best known are Koch's old tuberculin (T. O.), bouillon filtrate (B. F.), triturate residue (T. R.), bacillary emulsion (B. E.) and purified tuberculin (Endotin). There seems to be little difference in the actions of these, although theoretically T. R. should immunize against the bacillus and B. F. against its toxic products. The choice of tuberculin is much less important than the method of administration. The making of autogenous tuberculins is impracticable, hence stock preparations are used in practice.

Since the dose is exceedingly minute, the tuberculin as purchased must be greatly diluted before it is available for use. A convenient plan is to use the rubber-capped bottles of sterile salt solution described for vaccines (see p. 585), adding sufficient tuberculin to give the desired strength, together with 0.1 c.c. trikresol to insure sterility. The practitioner should bear in mind that while tuberculin is capable of good, it is also capable of great harm. Everything depends upon the dosage and plan of treatment. Probably a safe beginning dose for a pulmonary case is 0.00001 mg.; for gland and bone cases, about 0.0001 mg. The intervals are about one week or, rarely, three days, when very small doses are given. The dose is increased steadily, but with *extreme caution;* and should be diminished or temporarily omitted at the first indication of a "reaction," of which, in general, there are three forms:

(*a*) *General.*—Elevation of temperature (often slight), headache, malaise.

(*b*) *Local.* Increase of local symptoms, amount of sputum, etc.

(*c*) *Stick.*—Inflammatory reaction at site of injection.

Treatment is usually continued until a maximum dose of 1 mg. is reached, the course extending over a year or more.

VII. TUBERCULIN IN DIAGNOSIS

The tissues of a tuberculous person are sensitized toward tuberculin, and a reaction (see preceding section) occurs when any but the most minute quantity of tuberculin is introduced into the body. Non-tuberculous persons exhibit no such reaction. This is utilized in the diagnosis of obscure forms of tuberculosis, the test being applied in a number of ways:

1. **Hypodermic Injection.**—After first determining the patient's normal temperature variations, Koch's old tuberculin is used in successive doses, several days apart, of 0.001, 0.01, and 0.1 mg. A negative result with the largest amount is considered final. The reaction is manifested by fever within eight to twenty hours after the injection. A rise in temperature of 1°F. is generally accepted as positive. The method involves some danger of lighting up a latent process, and has been largely displaced by safer, although perhaps less reliable, methods.

2. **Calmette's Ophthalmo-reaction.**—One or 2 drops of 0.5 per cent. purified old tuberculin are instilled into one eye. Tuberculin ready prepared for this purpose is on the market. If tuberculosis exists anywhere in the body, a conjunctivitis is induced within twelve to twenty-four hours. This generally subsides within a

few days. The method is now rarely used since it is not without some, though slight, risk of injury to the eye. It is absolutely contraindicated in the presence of any form of ocular disease. A second instillation should not be tried in the same eye.

3. **Moro Reaction.**—A 50 per cent. ointment of old tuberculin in lanolin is rubbed into the skin of the abdomen, a piece about the size of a pea being required. Dermatitis, which appears in twenty-four to forty-eight hours, indicates a positive reaction. The ointment can be purchased ready for use.

4. **Von Pirquet's Method.**—This is the most widely used of the tuberculin tests. Two small drops of old tuberculin are placed on the skin of the front of the forearm, about 2 inches apart, and the skin is slightly scarified, first at a point midway between them, and then through each of the drops. A convenient scarifier is a piece of heavy platinum wire, the end of which is hammered to a chisel edge. A wooden toothpick with a chisel-shaped end is also convenient. This is held at right angles to the skin, and rotated six to twelve times with just sufficient pressure to remove the epidermis without drawing blood. In about ten minutes the excess of tuberculin is gently wiped away with cotton. No bandage is necessary. A positive reaction is shown by the appearance in twenty-four to forty-eight hours of a papule with red areola, which contrasts markedly with the small red spot left by the control scarification.

These tests have very great diagnostic value in children, especially those under three years of age, but are often misleading in adults, positive reactions occurring in many apparently healthy individuals. Negative tests are very helpful in deciding against the existence of tuberculosis.

VIII. CUTANEOUS TEST FOR SYPHILIS

Noguchi has prepared a substance called *luetin*, which produces a cutaneous reaction in syphilis similar to the tuberculin skin reaction in tuberculosis. Luetin consists of ground cultures of *Treponema pallidum* sterilized and preserved with trikresol. It can be purchased through any pharmacy.

A small drop (0.05 c.c.) of luetin is injected into the skin (not under it) of one arm. A similar preparation without the treponema is injected into the skin of the other arm as a control. A positive reaction usually begins within forty-eight hours and consists of an inflammatory induration, papule, or pustule. It is sometimes delayed three or even four weeks.

The test is positive in late secondary, tertiary, latent, and hereditary syphilis, but is usually absent in primary and early secondary cases. In general paralysis and tabes dorsalis it is inconstant. Compared with the Wassermann reaction it is more constant in tertiary and latent syphilis, while the Wassermann is more constant in primary and secondary cases. Unlike the Wassermann, the reaction does not disappear with treatment, but persists probably until a complete cure is effected.

Recent work has shown that luetin will produce the typical reaction in non-syphilitic persons who are taking potassium iodid or other drugs containing iodin; also that syphilitic persons will react to intradermal injections of agar and certain other substances in practically the same way as they react to luetin.

IX. SCHICK TEST FOR IMMUNITY TO DIPHTHERIA

By means of this test, which was introduced by Schick in 1913, it is possible to select from a group of individuals those who are immune to diphtheria by virtue of natural or artificial immunity. In an epidemic of diphtheria, therefore, it is of very great value as a means of determining who shall and who shall not receive prophylactic injections of antitoxin. The reaction is not applicable to the diagnosis of diphtheria. As performed by Koplik and Unger the test is as follows:

An ordinary hypodermic needle is bent at a slight angle (about 170 degrees) ¼ inch from the tip and is mounted in a handle which leaves only ¼ inch projecting. To make the test, cleanse an area upon the skin of the forearm, and encircle the forearm with the thumb and index-finger, holding the skin tense between them. Dip the needle into pure undiluted diphtheria toxin and immediately insert its unshielded quarter-inch *into*, not *through* the skin, the bend making it easy to insert intradermally. No injection is necessary, as enough of the toxin adheres to the needle.

No reaction occurs in those who are immune to diphtheria. In those who are not immune a distinct red spot about 1 to 2 cm. in diameter appears at the site of injection within eight hours. This is followed by marked induration, reaching its height in about forty-eight to seventy-two hours.

CHAPTER X

SERODIAGNOSTIC METHODS [1]

I. IMMUNITY

WITH exception of the last, the diagnostic methods here described depend on one or another law of immunity. These laws are customarily described in terms of Ehrlich's side-chain theory. It is not practicable to undertake a detailed discussion of the theory here, and I shall, accordingly, confine myself to such discussion and definition of the bodies concerned as will enable the reader to undertake the reactions himself with a reasonably intelligent conception of their mechanism.

Acquired immunity, that form of immunity resulting from an attack of a given disease, depends upon the formation within the body, under the influence of the disease-producing agent, or "antigen," of bodies possessing the power to neutralize the poisons produced by the antigen, or to destroy or otherwise affect the antigen itself. Since the action of these bodies is specific (*i.e.*, they act only on the particular antigen whose presence has led to their production), the search for them may be resorted to for diagnostic purposes whenever they can be found more easily

[1] By Ross C. Whitman, B.A., M.D., Professor of Pathology in the University of Colorado.

than can the antigen itself. With certain exceptions, to be noted later, the presence of one or other of these bodies may be regarded as pathognomonic of the corresponding disease.

The several "immune bodies" act by means of different mechanisms, by virtue of which they may be classified in three groups—the three orders of receptors of Ehrlich's side-chain theory. With the first group we are not immediately concerned.

1. *Receptors of the First Order.*—These are receptors which serve simply as connecting links between the disease-producing agent (or, rather, of its toxin) and the tissues. Under the influence of the antigen they are produced in excess, and are finally set free in the circulation. Here they seize upon and, so to speak, saturate the free valence of the antigen, while it also is still free in the blood and lymph, in such a way as to leave the latter no chemical affinities by means of which it may combine with similar bodies still in relation with the cells. The antigen is thus rendered inert. This order of receptors includes only the antitoxins; for example, those of diphtheria and tetanus.

2. *Receptors of the Second Order.*—These have a combining group similar to that of the first order, and, in addition, a group possessing a ferment-like action, by means of which the characteristic action of the body is effected. The ferment, or zymophore, group is readily destroyed by heat, so that serum to be used for any of the purposes included in the group must not be heated. The group includes the agglutinins, responsible for the several applications of the Widal reaction; the precipitins, responsible for one of the biologic methods to be

described later for the forensic identification of blood-stains; and the opsonins.

3. *Receptors of the Third Order.*—These bodies consist of two combining affinities only.   One of these combines with suitable analogous groups of the antigen, the other combines with a substance which is called complement because it "complements" or supplements or completes the specific action of the immune substance. Complement is normally present in the blood, but is unable to act upon the antigen without the mediation and aid of the immune body.   The latter is, therefore, called the amboceptor or 'tween body.   It is (relatively) thermostabile and keeps practically indefinitely under suitable conditions.   It is to be remembered that this is the specific immune substance whose presence or absence is indicative of the presence or absence of the corresponding disease.   The native, normally present complement is (relatively) thermolabile, being destroyed in a few minutes by a temperature of 54° to 56°C., and keeps only a few hours under the best conditions. It is non-specific, and within certain limits the complement of one species may be substituted for that of another.   Thus, in the Wassermann reaction, complement containing fresh serum from guinea-pigs is usually substituted for the normally present complement of the patient's serum, after the latter has been destroyed.

This group contains the lysins and the bodies responsible for the various applications of the complement-deviation method to the diagnosis of syphilis (Wassermann reaction), gonorrhea, typhoid fever, forensic identification of blood, etc.

II. APPARATUS

Before the description of the several tests is taken up, I shall give, to save space, a list of general equipment needed for such work. Special apparatus needed for some of the tests will be mentioned in connection with these.

1. *Centrifuge.*—While the usual small electric or water-driven instrument can be employed, a larger machine, capable of holding 4 or 8 tubes of about 50-c.c. capacity, is desirable.

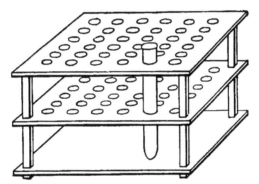

FIG. 229.—Convenient test-tube rack for serum work.

2. *Scales*, about 0.1 to 100 Gm. capacity.

3. *Microscope*, magnifying 50 to 750 diameters.

4. *Incubator* at 37°C.

5. *Water-bath*, to be regulated as required.

6. A large number of *test-tubes*, about 120 by 16 mm.

7. *Test-tube racks* to accommodate the above. A double row of holes is very convenient. Still better is a special rack (Fig. 229), made of copper or zinc, with six rows of holes, six to each row. A sheet of metal

midway between top and bottom contains holes to correspond, so that tubes are held without danger of tipping. The rack holds tubes enough for 18 Wassermann reactions, if but one antigen is used. A similar rack, made round, and with the two lower sheets of metal small enough to go through the circular opening of the water-bath, while the top sheet is larger, so as to rest on the edge of the opening, is also very convenient.

8. *Volumetric pipets*, 0.1 c.c. in one-hundredths, and 10 c.c. in one-tenths. The graduation should start near the point where the emptying of the pipet is stopped by capillarity.

9. *Capillary pipets* (see Fig. 228), made from soft-glass tubing, as described on page 561. The tube should be of such a size that the ordinary medicine-dropper nipple will fit it snugly. Such pipets are useful for a variety of purposes. After being used once they should be thrown away.

10. *Glass Capsules.*—These may be purchased or, with a little practice, can be readily prepared from the same sort of tubing by drawing out a piece at both ends, and sealing in the flame. If desired, one end may be bent over to form a hook at the point where the narrowing begins (see Figs. 230 and 231).

11. An *all-glass syringe*, such as the Luer, about 5-c.c. capacity, with a fairly large needle, say 19 or 20 gage, preferably of platinum.

III. REACTIONS BASED UPON IMMUNE BODIES OF THE SECOND ORDER
A. THE WIDAL REACTION

The test may be employed for the diagnosis of a variety of infections, *e.g.*, typhoid, paratyphoid, bacillary dysentery, the plague, Asiatic cholera, epidemic meningitis, etc. In clinical work it is used only for the diagnosis of typhoid and paratyphoid infections.

1. **Materials Required.**—The following especial equipment is needed:

1. A *homogeneous suspension* of the *organism* or organisms suspected of causing the disease. Such suspensions may be purchased from the manufacturers of

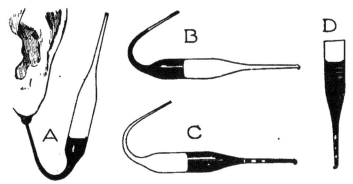

Fig. 230.—Method of obtaining blood in a Wright capsule: A, Filling the capsule; the long arm should be held more nearly horizontal than is here represented; B, the bulb has been warmed and the capillary end sealed in the flame; C, cooling of the capsule has drawn the blood to the sealed end; D, the serum has separated, and the top of the capsule has been broken off.

biologic preparations, or may be prepared by the worker himself. In the latter case twenty-four-hour-old agar-slant cultures (preferably attenuated by long-continued growth on culture-media) should be washed off with normal salt solution (0.85–0.9 per cent. sodium chlorid), containing either 0.5 per cent. phenol or 0.1 per cent. formalin, and shaken until the suspension is as uniform as possible. Dilute by adding more of the salt solution until the suspension is but slightly milky, and preserve

in the Ice-box.   Such a suspension will keep for months.
Shake thoroughly before using.   The suspension will
settle less rapidly if 10 per cent. lactose is added to it.
Suspensions which show any tendency to spontaneous
agglutination cannot, of course, be used.

2. Instead of the suspension of killed bacteria, *living
young cultures* (not over twenty-four hours old) of attenuated organisms can be employed.

3. About 0.1 c.c. of the *patient's serum*.   This may be
obtained by pricking the cleansed finger or ear rather

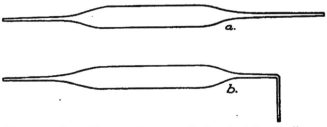

FIG. 231.—A satisfactory glass capsule for obtaining small quantities of blood, as for the Widal test.   The straight tube (*a*) is more
convenient to carry in the hand-bag than is Wright's capsule.   It
may be bent as shown in the lower figure by brief application of a
match flame at the bedside.   After the tube is filled the ends may be
sealed with the match flame.

deeply and collecting the blood in one of the capsules
above mentioned, as is indicated in Figs. 230 and 231.
More than one capsule should be at hand, so that a fresh
one may be substituted if the first is plugged by fibrin
before enough blood is obtained.

When the capsules are not at hand, blood may be obtained in little vials such as can be made by breaking off the
lower ½ inch of the tubes which have contained peptoniz-
ing powder.   Vials in which hypodermic tablets are sold can
be used, but are somewhat too narrow.   They must, of

course, be well cleaned. One of these is filled to a depth of about ¼ inch from a puncture in the ear, and is then set aside for a few hours. When the clot has separated it is picked out with a needle, leaving the serum.

Sufficient blood may also be collected by allowing drops to dry on glass or unglazed paper (without heating), to be afterward macerated in water. In this case, however, dilutions can only be made approximately.

4. *Slides*, preferably hollow ground, *cover-glas· vaselin*.

2. **Methods.**—Two methods of performing the test will be described:

(1) **Macroscopic Method.**—Separate the clot and serum in the capsule by centrifugation, nick the wall of the capsule a short distance above the serum with a file, and break the capsule at this point. Pipet off the serum, place it in a clean test-tube, and add 9 volumes of salt solution. Counting the drops of serum as they fall from the capillary pipet, and adding nine times the number of drops of salt solution will give sufficiently accurate dilution. Now place a number of very small test-tubes in a rack, and add to each one *except the first* 0.5 c.c. of salt solution by means of a volumetric pipet. Then place in the first and second tubes *only* 0.5 c.c. of the diluted blood-serum. Shake the second tube, and with the pipet transfer 0.5 c.c. to the third tube. Shake this tube and transfer 0.5 c.c. to the fourth tube, and so on, to the end. Discard 0.5 c.c. from the last tube. One tube, to serve as control, should contain only 0.5 c.c. of salt solution, without any serum. The volumetric pipet should be thoroughly rinsed out with salt solution after each transfer. One

thus arrives at a series of dilutions of the serum, as follows: 1-10, 1-20, 1-40, 1-80, 1-160, 1-320, 1-640, etc. Now add to each tube a like amount (0.3 to 0.5 c.c.) of the suspension of killed bacteria. This doubles the dilution of the serum in each of the tubes. Mix all the tubes thoroughly by shaking, and place the rack in a moderately warm place or in the incubator for eight to twelve hours. In those tubes in which the reaction is positive there will be found a sediment consisting of agglutinated bacteria at the bottom of the test-tube, with a clear supernatant fluid. The control tube and the negative tubes will be cloudy and without sediment.

Dead cultures of typhoid bacilli, together with all apparatus necessary for performing the macroscopic test, are put up at moderate cost by various firms under the names of typhoid diagnosticum, typhoid agglutometer, etc. Full directions accompany these outfits.

Bass and Watkins have described a modification of the macroscopic method (using very concentrated suspensions of the bacilli) by which the test can be applied at the bedside. Agglutination occurs within a few minutes. The apparatus has been put upon the market by Parke, Davis & Co.

(2) **Microscopic Method.**—Arrange a series of dilutions of the blood-serum as above, or, if dried blood is used, macerate the dried clot with salt solution or tap-water. In the latter case, unless the size of the original drop of blood is known, the color is the only guide as to the degree of dilution. A light amber color will roughly correspond to a dilution of 1-50. From such a dilution others can be prepared. On the center of each of

several clean cover-glasses place a loopful of each of the several dilutions, employing a platinum loop of about 2 mm. diameter. With the same loop add to each droplet of diluted serum a loop from a twelve- to twenty-four-hour-old bouillon culture of the organism in question, or of a suspension in salt solution prepared from a young agar-slant culture. This doubles the dilution of serum in each case. One cover-glass containing no serum should be prepared to serve as a control. Press over each cover-glass a hollow-ground slide previously ringed with vaselin. Turn the slide over so as to bring the cover-glass on top. Drying is prevented and the cover-glass held in place by the vaselin.

When hollow-ground slides are not at hand, a drop each of the diluted serum and the bacterial suspension may be placed in the center of a heavy ring of vaselin on an ordinary slide and a cover-glass applied to this. Vaselin containing an antiseptic must not, of course, be used for this purpose.

Place the slides in a moderately warm place or in the incubator at 37°C. for two hours. Examine under the oil-immersion lens or, better, the high-power dry lens of the microscope, *using very subdued light.* Yellow (artificial) light gives a clearer view than does white light. In the negative slides and in the control the organism will be found moving freely (if a motile species) and not clumped; while in the positive slides the organisms are found motionless and gathered in tangled masses and balls, *i.e.,* they are agglutinated (Fig. 232). Pseudoreactions, in which there are a few small clumps of organisms whose motion is not entirely

39

lost, together with many freely moving organisms
scattered throughout the field, should not mislead.

Jagic suggests that, after agglutination has taken
place, a drop of the suspension be mixed with a drop
of India-ink and spread upon a slide. In this way a
permanent record may be kept. To insure sterility the
preparation may be fixed by heat.

3. **Interpretation of Results.**—Experience has shown
that not much significance attaches to reactions occur-

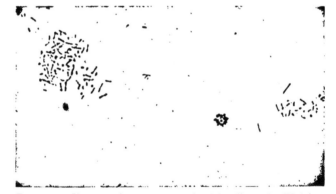

FIG. 232.—Showing clumping of typhoid bacilli in the Widal reaction. At one point a crenated red blood-corpuscle is seen (Wright and
Brown).

ring in two hours with dilutions of serum less than
1–75 or 1–100. With killed organisms the dilution may
be somewhat lower than when living organisms are employed. On the other hand, recently isolated virulent
cultures are, in general, more resistant to agglutination
than old attenuated ones. A number of other disease
conditions may give rise to a positive reaction with the
typhoid bacillus, notably infections with closely related

organisms, such as the colon bacillus. (In such cases, if tests are made with several species, the species agglutinated in the highest serum dilution may *generally*, but not always, be regarded as the cause of the infection.) Agglutination of typhoid bacilli may also, though rarely, occur in diseases of the liver, particularly those accompanied by jaundice, and in pneumococcus infections. The Widal test is of no value if the patient has received anti-typhoid vaccination within several years previously.

In typhoid the average time of first appearance of the reaction in the dilutions above recommended is the fourteenth to the fifteenth day of actual disease, which corresponds roughly to the eighth or tenth day of apparent disease. In doubtful cases the test should be repeated at frequent intervals, and no disappointment should be felt if, as sometimes (though rarely) happens, the reaction does not appear until the twentieth to the twenty-fifth day of the disease. It is evident, therefore, that its value for early diagnosis is much less than that of the blood-culture (see p. 346). After the Widal reaction first appears it remains throughout the whole course of the disease and often persists for years.

B. BIOLOGIC IDENTIFICATION OF UNKNOWN PROTEINS

This includes the differentiation of human and animal blood, detection of meat adulteration, etc., by means of the **precipitin test** (method of Uhlenhuth).

1. **Materials Required.**—The following equipment is needed:

1. *Blood-serum* of an animal (rabbit) highly immunized against the protein to be determined. Im-

munize several rabbits by several intravenous or intraperitoneal injections of, for example, human blood, or
better, blood-serum.   Placental blood may be used, or
the blood may be obtained as for the Wassermann
reaction.   The doses should be 2 or 3 c.c. and should be
given at four- or five-day intervals.   After the fourth or
fifth dose draw 2 or 3 c.c. of blood from an ear vein, separate the serum, and determine its strength as follows:

Prepare dilutions of (in this case) human blood-serum
in the proportions 1–1000, 1–5000, 1–10,000, 1–20,000,
etc., using physiologic salt solution as a diluent.   Place
2 c.c. each of the several dilutions in a series of test-
tubes.   To each tube add 0.1 c.c. of the rabbit's serum,
without shaking.   A distinct cloud should appear in the
lowest dilution (1–1000) within a minute or two, rapidly
deepening to a heavy flocculent precipitate; the reaction
develops somewhat more slowly in the higher dilutions,
but no reaction is significant which occurs after more
than twenty minutes.

If the titration results as above described, anesthetize
the rabbit *while it is in a fasting condition*, as otherwise
the serum is apt to be opalescent; remove the anterior
breast wall under aseptic conditions; take out the left
lung and open the heart, so as to allow the animal to
bleed to death into its pleural cavity.   Cover the body
with sterile towels wet with antiseptic solution   After
clotting has occurred, pipet the serum into sterile bottles, and add $\frac{1}{10}$ volume of 5 per cent. carbolic acid as
a preservative.   If the serum is opalescent, it cannot
be used; if cloudy, it must be filtered clear through a
sterile Berkefeld filter.   Sometimes the cloudiness can
be removed by simple sedimentation.   The titration

above described should be repeated and verified before the serum is used for making the test proper.

Other sera immune to horse, dog, sheep, beef, fowl, etc., may, of course, be prepared in the same way.

2. A *solution* of the *unknown substance* in physiologic salt solution. The stock dilution should be about 1–1000. If made from a dried clot this can only be approximate. The following criteria may be used:

(*a*) It should be almost completely colorless by transmitted light.

(*b*) It should give only a slight cloudiness when heated with a little nitric acid.

(*c*) It should, nevertheless, foam freely on shaking. The solution must be made perfectly clear—by filtration if necessary.

2. **Method.**—Arrange a series of test-tubes and charge them as follows:

Tube No. 1—2 c.c. of the unknown solution (diluted 1–1000) plus 0.1 c.c. of immune serum.

Tube No. 2—2 c.c. of normal salt solution plus 0.1 c.c. of immune serum.

Tube No. 3—2 c.c. of a 1–1000 dilution of known serum of the species corresponding to that suspected to be present in the unknown material plus 0.1 c.c. of immune serum.

Tube No. 4—2 c.c. of a 1–1000 dilution of a serum from a species different from that suspected to be present in the unknown material plus 0.1 c.c. of immune serum.

Tube No. 5—2 c.c. of the unknown solution alone.

When the first and third tubes give a positive reaction, as above defined, and all the others a negative reaction, the presence of the protein of the species tested for is established. It must be remembered that *shaking must not be employed.* When only limited amounts of material are available, the test can be made by contact in capillary tubes.

Meat adulteration may be recognized by the same method. Usually it is a question of horse flesh sold as beef or as sausage. Remove about 30 Gm. of the meat from the deeper portions of the specimen with a clean sterile knife, free as much as possible from fat, chop up on a clean board, and, if salted, extract several times in the course of ten minutes with sterile distilled water. Cover the 30 Gm. of freshened chopped meat with about 50 c.c. of 0.85 per cent. salt solution, and allow it to stand three hours at room-temperature or over night in the ice-box. Pipet off the supernatant fluid, and clarify and dilute for use according to the criteria given above for preparing extracts of the unknown substance. If acid to litmus, it is to be neutralized before use by adding an excess of an insoluble alkali, such as magnesium oxid, and filtering.

The immune serum is prepared as above by injecting rabbits with, in this case, horse-serum. It must have a titer of at least 1–20,000. That is, it must give a reaction with the homologous serum in a dilution of the latter of that degree.

3. **Interpretation of Results.**—These reactions are very closely specific, and are fully established for forensic purposes. Doubt can only arise as between the proteins of very closely related species, and this can be practically always removed by the use of adequate controls.

The "Typhoid Diagnosticum" of Ficker is based on the same principle. A filtrate of an autolyzed culture is used instead of the suspension of killed typhoid bacilli described on page 605. The dilutions are prepared as for the Widal test, but a positive reaction is indicated by a precipitate instead of by agglutination.

C. OPSONINS

That phagocytosis plays an important part in the body's resistance to bacterial invasion has long been recognized. According to Metchnikoff, this property of leukocytes resides entirely within themselves, depending upon their own vital activity. The studies of Wright and Douglas, upon the contrary, indicate that the leukocytes are impotent in themselves, and can ingest bacteria only in the presence of certain substances which exist in the blood-plasma. These substances have been named *opsonins*. Their nature is undetermined. They probably act by uniting with the bacteria, thus preparing them for ingestion by the leukocytes; but they do not cause death of the bacteria, nor produce any appreciable morphologic change. They appear to be more or less specific, a separate opsonin being necessary for phagocytosis of each species of bacteria. There are, moreover, opsonins for other formed elements—red blood-corpuscles, for example. It has been shown that the quantity of opsonins in the blood can be greatly increased by inoculation with dead bacteria.

To measure the amount of any particular opsonin in the blood Wright has devised a method which involves many ingenious and delicate technical procedures. Much skill, such as is attained only after considerable training in laboratory technic, is requisite, and there are many sources of error. It is, therefore, beyond the province of this work to recount the method in detail. In a general way it consists in: (*a*) Preparing a mixture of equal parts of the patient's blood-serum, a suspension of the specific micro-organism, and a suspension of washed leukocytes; (*b*) preparing a similar mixture, using serum of a normal person; (c) incubating both mixtures for a definite length of time; and (d) making smears from each, staining, and examining with an oil-immersion objective. The number of bacteria which have been taken up by a definite number of leukocytes is counted, and the average number of bacteria per leukocyte is calculated; this gives the "phagocytic index." The phagocytic index of the blood under investigation, divided by that of the normal blood, gives the *opsonic index* of the former, the opsonic index of the normal blood being taken as 1. Simon regards the percentage of leukocytes which have ingested bacteria as a more accurate measurement of the amount of opsonins than the number of bacteria ingested, because the bacteria are apt to adhere and be taken in in clumps.

Because of its simplicity the clinical laboratory worker will prefer some modification of the Leishman method, which uses the patient's own leukocytes. It is, perhaps, as accurate as the original method of Wright, although variations in the leukocyte count have been shown to affect the result. Two pipets like those shown in Fig. 228 are used.

1. Make a suspension of the specific organism by mixing a loopful of a young agar culture with 1 c.c. of a solution containing 1 per cent. sodium citrate and 0.85 per cent. sodium chlorid. Thoroughly break up all clumps by sucking the fluid in and forcing it out of one of the capillary pipets held vertically against the bottom of the watch-glass.

2. Puncture the patient's ear, wipe off the first drop of blood, and from the second draw blood into the other pipet to the wax-pencil mark, let in a bubble of air, and draw in the same amount of bacterial suspension.

3. Mix upon a slide by drawing in and forcing out of the pipet.

4. Draw the mixture high up in the pipet, seal the tip in the flame, and place in the incubator for fifteen minutes.

5. Repeat steps (2), (3) and (4) with the blood of a normal person.

6. After incubation, break off the tip of the pipet, mix the blood-bacteria mixture, and spread films on slides.

7. Stain with Wright's blood-stain or carbol-thionin.

8. With an oil-immersion lens count the bacteria which have been taken in by 100 leukocytes, and calculate the average number per leukocyte. Divide the average for the patient by the average for the normal person. This gives the opsonic index. If in the patient's blood there was an average of 4 bacteria per leukocyte, and in the normal blood 5 bacteria per leukocyte, the opsonic index would be $\frac{4}{5}$ or 0.8.

Wright and his followers regarded the opsonic index as an index of the power of the body to combat bacterial invasion. They claimed very great practical importance for it as an aid to diagnosis and as a guide to treatment by the vaccine method. This method of treatment consists in increasing the amount of protective substances in the blood by injections of normal salt suspensions of dead bacteria of the same species as that which has caused and is maintaining the morbid process, these bacterial suspensions being called "vaccines." Vaccine Therapy (Chapter IX) has probably taken a permanent place among our methods of treatment of bacterial infections, particularly of those which are strictly local, but the opsonic index is now little used either as a measure of resisting power or as an aid to diagnosis and guide to treatment.

IV. REACTIONS BASED UPON IMMUNE BODIES OF
THE THIRD ORDER

The reactions of this group comprise the various applications of the Wassermann reaction or, more properly, of the Bordet-Gengou phenomenon of complement-fixation or deviation. Since the reaction involves three active substances—viz., antigen (the substance inducing the immune reaction); the specific amboceptor, or immune substance; and the non-specific complement —it is possible to so adjust matters that, any two factors being known, the third may be determined either qualitatively or (roughly) quantitatively. Practically, the method is employed chiefly for determining the presence of the middle term, or amboceptor. It may be applied to the diagnosis of any disease the antigen of which is known and which can be obtained in suitable form. This includes syphilis, gonorrhea, tuberculosis, echinococcus and cysticercus diseases, trichiniasis, typhoid fever, and pneumococcus, meningococcus and staphylococcus infections, etc. In several of these, other and simpler methods are, however, available. The method as applied to the first three diseases above mentioned is given below.

The method employed is based upon the fact that if suitable quantities of antigen, amboceptor (*i.e.*, patient's serum containing the same), and complement are mixed together and warmed gently in the incubator, a supposedly chemical, firm union of the three takes place. The mere fact of combining in this way produces, however, no visible change in the fluid. It is necessary, therefore, to test for free complement by adding the two other units of another immune system which also requires the presence of complement, and which will produce a visible reaction if free complement is present. A "hemolytic system" is used for this purpose. The mixture is then incubated a second time. If the three units of the first system have combined (in other words, if the patient's serum contains syphilis antibody), and not otherwise, the complement is "fixed" or "deviated" during the first incubation period, so that it is no longer available to assist in completing the second and visible reaction represented by the hemolytic system. As will be seen later, an elaborate system of controls is needed.

A. COMPLEMENT DEVIATION TEST FOR SYPHILIS
THE WASSERMANN REACTION

Of the many modifications of the Wassermann reaction, but one, the standard form of the reaction, will be given.

1. **Materials Required.**—The following reagents are needed:

1. *Syphilitic Antigen.*—The reaction, originally supposed to depend upon the presence in the patient's serum of true syphilis antibodies, is now known to depend instead on a disorder of lipoid metabolism characterized by the presence of serum-foreign lipoids in the serum. Accordingly, solutions of lipoids from various sources can be used for the test. The following may be recommended:

(*a*) Grind or chop the liver and spleen of a syphilitic fetus. Place in a suitable vessel and add 4 to 10 parts of absolute ethyl alcohol. (The amount of alcohol varies in the hands of different workers.) Extract for three or four days in the incubator or for one to two weeks at room-temperature, with frequent vigorous shakings. Filter through paper. The filtrate constitutes the stock solution, which is diluted with salt solution for use, as described later.

(*b*) Grind in a mortar with quartz sand one or more guinea-pig hearts, previously weighed, place in a suitable receiver, and add 10 c.c. of absolute alcohol for each gram of heart tissue. Complete the preparation as above. This solution can be purchased from the various biologic houses

(*c*) The "fortified" or cholesterinized antigen of Swift is prepared from (*b*) by dividing a given lot in half, saturating one of the halves with Merck's cholesterin in the incubator, placing in water at 15°C. for two hours, and finally filtering and mixing with the other half.

2. *Antisheep Amboceptor.*—This can now be obtained so readily in the market that the somewhat elaborate method of preparation may be omitted here.

3. *Sheep's Red Blood-cells.*—Where a slaughter-house is available, it constitutes the most convenient source of supply. A sterile bottle (about 100-c.c. capacity), containing some glass beads, bits of glass rod, or steel shavings, is carried to the slaughter-house. After the first gush of blood from the slaughtered animal has cleansed the wound, the bottle is filled not quite full with blood. It is then stoppered and the bottle kept in motion for ten or fifteen minutes or until defibrination is complete. For use, "wash" the cells thoroughly free from serum by filling centrifuge tubes about one-quarter full of defibrinated blood, and adding 0.9 per cent. sodium chlorid solution to the top. Centrifuge

thoroughly and pipet off the supernatant fluid. Again fill with salt solution, mix, centrifugate, and remove the supernatant fluid. Repeat *at least* three times. Finally, prepare a 5 per cent. emulsion by adding 1 volume of the cells, thoroughly packed by centrifugation, to 19 volumes of salt solution. This is the standard against which the strength of all other solutions is measured or titrated, as described below.

The following method, a modification by W. W. Williams of that of Rous and Turner, furnishes cells which remain serviceable for about two weeks—a much longer period than do those preserved by the customary method.

Place in a quart Mason jar the following:

Granulated sugar	11.0 Gm.;
Sodium citrate (Merck)	8.0 Gm.;
Gelatin, bacteriological	2.0 Gm.;
Distilled water	350.0 c.c.;
Liquefied phenol	1.5 c.c.

Sterilize jar and contents in the autoclave at 15 pounds for ten minutes. At the abattoir fill to within about an inch of the top with fresh sheep's blood. Preserve in the ice-box. Small portions may be removed and washed for use as required.

4. *Complement.*—Stun a fasting guinea-pig by a blow at the base of the skull, cut the throat, and collect the blood in a clean, dry dish. The complement will be destroyed if, by cutting the esophagus, stomach contents become mixed with the blood. The serum may be allowed to separate spontaneously over night in the ice-box, or be separated just before use by centrifugation. *Serum more than twenty-four hours old is worthless as complement.*

5. *Patient's Serum* —About 5 c.c. of blood will suffice. A convenient method consists in applying an Esmarch bandage to the upper arm, after cleansing the flexor surface of the elbow with alcohol or tincture of iodin. If the patient opens and closes the fist vigorously a few times the veins become more prominent. Insert the needle of the syringe above described above or alongside the vein and at an acute angle to the skin surface. Once through the skin, a little practice will enable one to quickly find the way into the vein. Slow withdrawal of the plunger will quickly fill the syringe. If the vein is a large one the blood will flow into the syringe, driving the plunger ahead of it. Remove the bandage before withdrawing the syringe to avoid a hematoma. Withdraw the needle quickly, and have the patient or an assistant apply fairly firm pressure over the punctured vein for a minute or two. In the meantime empty the syringe into a scrupulously clean test-tube, and immediately wash out the syringe and needle thoroughly with water, followed, especially if the needle is of steel, by alcohol. If blood is given time to clot in the needle or syringe the instrument is practically ruined. The needle should, of course, be sterilized by boiling before use. The syringe should be clean and *dry* (as otherwise hemolysis will take place), but need not be sterilized.

After an hour or two, separate the clot, if necessary, from the test-tube wall with a clean wire, and either complete the separation of the serum at once by centrifugation or place in the ice-box over night. Transfer the serum with a capillary pipet to a second clean test-tube.

Before the test is made the serum is "inactivated" (*i.e.*, the native complement present is destroyed) by immersing the tube for half an hour in the water-bath at 55° to 56°C.

Unless a considerable number of sera are to be examined simultaneously, known positive and known negative control sera must be prepared in the same way.

2. **The Titrations.**—The strength of the complement and antisheep amboceptor must be determined on each occasion of its use. The antigen must be titrated every few weeks.

(1) **Titration of the Complement.**—The complement may be used undiluted or in varying dilutions of from 40 to 10 per cent. The greater the dilution, of course, the greater the accuracy with which it can be titrated. Assuming that it is to be used in a 40 per cent. dilution (1 part of complement serum to 1½ parts of salt solution), arrange a series of test-tubes somewhat as follows:

Tube No. 1—0.02 c.c. complement serum plus 1 c.c. 5 per cent. sheep blood-cells and 1½ units amboceptor.[1]

[1] One unit of amboceptor is the amount required to bring about solution of 1 c.c. of the 5 per cent. red cell emulsion, in the presence of 1 unit of complement, in one hour at incubator temperature. In the same way 1 unit of complement is the amount required to bring about solution in the presence of one unit of amboceptor under the same conditions. In the above experiment the 1½ units of amboceptor is only approximate. It is assumed that the worker has purchased amboceptor in 1-c.c. vials, guaranteed to contain 1000 units, and actually containing a slight excess over that amount. For use this is diluted with 100 parts of salt solution; 0.1 c.c. will then contain something over 1 unit. On the first occasion of its use, 0.15 c.c. may be accepted for titration purposes, the aim being to use a moderate excess to allow for the chance of deterioration and slight variations in the strength of the blood emulsion. On each later occasion the approximate value is known from the last previous titration. Amboceptor dilutions keep well in the ice-box, but may undergo a very abrupt deterioration at the end of about six months.

Tube No. 2—0.04 c.c. complement serum plus 1 c.c. 5 per cent. sheep blood-cells and 1½ units amboceptor.

Tube No. 3—0.06 c.c. complement serum plus 1 c.c. 5 per cent. sheep blood-cells and 1½ units amboceptor.

Tube No. 4—0.08 c.c. complement serum plus 1 c.c. 5 per cent. sheep blood-cells and 1½ units amboceptor.

Tube No. 5—0.10 c.c. complement serum plus 1 c.c. 5 per cent. sheep blood-cells and 1½ units amboceptor.

Tube No. 6—0.12 c.c. complement serum plus 1 c.c. 5 per cent. sheep blood-cells and 1½ units amboceptor.

Make up all tubes to a like volume (1.5 or 2 c.c.). Mix thoroughly by gentle shaking, and place in the incubator (preferably standing in a dish of water, since this insures rapid and uniform heating to incubator temperature) at 37°C. for one hour. The tube containing the smallest amount of complement which shows *complete* solution of the red cells (the solution bright red, perfectly clear, and free from sediment) contains one unit of complement. Twice this amount is used in making the test proper, to allow for the rapid deterioration which takes place and for the small amount of complement directly absorbed by the antigen.

(2) **Titration of the Amboceptor.**—Arrange tubes as follows:

Tube	Complement	Red cells (5 per cent.)	Amboceptor (1–100 dilution)
No. 1	1½ units[1]	1.0 c.c.	0.06 c.c.
No. 2	1½ units	1.0 c.c.	0.08 c.c.
No. 3	1½ units	1.0 c.c.	0.10 c.c.
No. 4	1½ units	1.0 c.c.	0.12 c.c.
No. 5	1½ units	1 0 c.c.	0.14 c.c.
No. 6	1½ units	1.0 c.c.	0.16 c.c.

Bring all tubes to a like volume, mix, and incubate for one hour. The tube containing the smallest amount

[1] As determined in the previous titration.

of amboceptor which causes complete hemolysis contains one unit. *Two* units are used for the test proper.

(3) **Titration of the Antigen.**—The stock solution is to be diluted freshly for use with salt solution. This makes a milky fluid. The amount of dilution will vary with the strength of the stock solution as determined by the following tests. For the latter a 10 per cent. dilution may be employed.

Arrange test-tubes as follows:

Tube	Antigen (10 per cent.)	Red cells (5 per cent.)
No. 1	0.1 c.c.	1.0 c.c.
No. 2	0.2 c.c.	1.0 c.c.
No. 3	0.3 c.c.	1.0 c.c.
No. 4	0.4 c.c.	1.0 c.c.
No. 5	0.5 c.c.	1.0 c.c.
No. 6	0.6 c.c.	1.0 c.c.

Bring all tubes to a like volume. Mix and incubate. The amount used in making the test proper must not be more than one-half the smallest amount which causes hemolysis in the above. A modified form of this titration is repeated each time the antigen is used.

At the same time with the above arrange six test-tubes as follows:

Tube	Antigen (10 per cent.)	Complement
No. 1	0.1 c.c.	2 units
No. 2	0.2 c.c.	2 units
No. 3	0.3 c.c.	2 units
No. 4	0.4 c.c.	2 units
No. 5	0.5 c.c.	2 units
No. 6	0.6 c.c.	2 units

Bring all tubes to a like volume, mix, and incubate. Then add to all tubes 1 c.c. of 5 per cent. red cell emulsion and 2 units amboceptor solution. Mix and reincu-

40

bate. It the antigen is "anticomplementary" it will prevent hemolysis in one or more of the tubes. The amount used for the test proper must not exceed one-half the smallest amount showing such action.

The antigen must also be shown to react with known positive sera, and the amount required to produce a reaction determined. For this purpose an abundant supply of serum from a patient with active secondary syphilis (still better, from several such patients) is obtained, and the complete reaction carried out as described below, employing varying amounts of the antigen dilution, *e.g.*, 0.04, 0.06, 0.08, 0.1, 0.12, 0.14 c.c., etc. For the test proper an amount is used in the greatest possible excess of that amount which gives a positive reaction, but which complies, nevertheless, with the requirements mentioned above as to hemolytic and anticomplementary action.

3. **Errors and Their Causes.**—1. Dirty glassware unquestionably is responsible for most of the errors. No control can eliminate an error caused by a single dirty test-tube among a hundred clean ones. The hemolytic system will be completely destroyed in such a tube, with the likelihood of interpreting the result as a positive reaction, or as due to an anticomplementary patient's serum, etc., depending on what it happens to be used for. Glassware need not be sterile, but *must be absolutely clean*. Never allow used tubes to stand and dry out. Immediately the work is finished wash with soap and water, rinse thoroughly with clean water followed by dilute (10 per cent.) nitric acid, then caustic soda solution, then several changes of distilled water. Place in a basket and dry in a dry-air sterilizer

if one is available.   New glassware should be prepared
for use in the same way.   Tubes may be kept stored
in the oven or set aside in a clean dust-proof cupboard.
Tubes used for this purpose should never be used for
any other.   Other glassware should be cleaned with
equal care, and preserved from any other use.

2. The patient's serum may be "anticomplementary,"
i.e., it may have the power to combine with or absorb
complement in the absence of antigen.   Serum which
has been kept too long, or which has been inactivated
at a temperature above 56°C., is apt to exhibit this
property.   The anticomplementary property may sometimes be made to disappear by renewed inactivation.
If this fails, the serum must be discarded.   A control of
this property is included in setting up the test.

3. The antigen may become anticomplementary or
hemolytic, or both.   When this happens it must be discarded.   A control for this is set up each time the
antigen is used.

4. The hemolytic system may fail to function for a
variety of reasons.   A control of this is furnished by
the titration of complement and amboceptor above
described.

4. **The Test Proper.**—After all preparations have
been completed and the titrations satisfactorily performed, one proceeds to set up the test proper.   We
may suppose that we are dealing with at least three
sera—viz., the patient's serum and known positive and
known negative controls.   In a rack having two rows
of holes arrange test-tubes as shown on page 629.   Mix
by gentle shaking and place in the incubator for
one hour.   The rack should stand with the tubes

Immersed in water to about the level of the contents. Then add to all the tubes except the last one, already containing blood-cells, 2 units of antisheep amboceptor dilution and 1 c.c. of 5 per cent. sheep red cell emulsion. Mix as before and incubate for two hours. The tubes in the back row show for each serum tested whether any of them is anticomplementary. They should all show complete solution of the red cells. The last two tubes show whether the antigen is hemolytic or anticomplementary respectively. The first, containing complement, should show complete hemolysis. The second, containing only antigen and red cells, should show no solution. Assuming that these controls are all satisfactory, one turns to the tubes in the front row. The known positive control shows no solution of the red cells, the complement having been deviated or bound during the first incubation, and hence being not available for the reaction with the red cells. For the opposite reason the known negative serum will show complete solution. The unknown serum will behave like the first or the second, according as it is positive or negative.

It is apparent that an excess of complement may convert a positive reaction into a negative, while a deficiency may cause a negative serum to behave like a (more or less) positive one. It is to avoid this contingency that the *unknown* quantity of complement present in the serum of the patient is removed by inactivation, to be replaced by an accurately measured amount of guinea-pig complement.

With **cerebrospinal fluid** the reaction is carried out in the same way, except that this fluid *must not be inactivated*, and is used in larger amounts. When enough

	I Unknown serum, inactivated	II Known positive serum, inactivated	III Known negative serum, inactivated
Back row	Tube No. 1 Serum.............. 0.2 c.c. Complement......... 2.0 units Salt solution to make.. 2 c.c.	Tube No. 1 Serum.............. 0.2 c.c. Complement......... 2.0 units Salt solution to make.. 2 c.c.	Tube No. 1 Serum.............. 0.2 c.c. Complement......... 2.0 units Salt solution to make.. 2 c.c.
Front row	Tube No. 2 Serum.............. 0.2 c.c. Complement......... 2.0 units Antigen,[1] quantity indicated by titration. Salt solution to make.. 2 c.c.	Tube No. 2 Serum.............. 0.2 c.c. Complement......... 2.0 units Antigen, quantity indicated by titration. Salt solution to make.. 2 c.c.	Tube No. 2 Serum.............. 0.2 c.c. Complement......... 2.0 units Antigen, quantity indicated by titration. Salt solution to make.. 2 c.c.
Placed as convenient	Antigen control tube No. 1 Twice the above amount of antigen plus 2.0 units complement plus salt solution to make........ 2 c.c. Antigen control tube No. 2 Twice the above amount of antigen plus 1 c.c. 5 per cent. red cell suspension plus salt solution to make.......... 2 c.c.		

[1] It will be found convenient to arrange the dilution of the stock antigen in such a way that the amount required for the test is contained in an easily measured volume unit, e.g., 0.1 c.c. The same device may, of course, be adopted with reference to the complement.

fluid is available the test should be set up so as to give a reading for 0.4, 0.6, 0.8, and 1 c.c. When economy of material is necessary a reading should be obtained for 0.5 and 1 c.c. Further modifications may, of course, be imposed by the exigencies of the case.

Several degrees of the reaction are recognized and are customarily indicated as follows:

Complete inhibition (cells intact with colorless supernatant fluid),	++++	or 4 +.
Almost complete inhibition,	+++	or 3 +.
About one-half complete inhibition,	++	or 2 +.
Slight inhibition,	+	or 1 +.
No inhibition.	o.	

5. **Interpretation of Results.**—1. Jaundice and marked alcoholism may convert a positive reaction into a negative one.

2. Scarlet fever, leprosy, active malaria, and malignant tumors may cause a positive reaction.

3. The reaction is negative in primary syphilis, but becomes rapidly and strongly positive as the general manifestations of the disease develop. During this stage only a strongly positive reaction should be regarded as significant. In late and especially in latent syphilis the reaction again grows weaker. More significance may, therefore, attach to weak reactions in such cases.

4. A positive reaction quickly becomes negative under specific treatment, to recur if treatment is inadequate. Apparently cured cases may show a positive reaction six months to a year after a "provocative" dose of salvarsan.

5. The behavior of the blood is no guide as to the

condition of the central nervous system. Recent investigations have shown that the central nervous system becomes involved very early in practically all cases, and the organisms so located are peculiarly inaccessible to attack by present methods. No case may be regarded as cured until both blood and cerebrospinal fluid show a persistent normal condition.

Routine Methods.—The labor involved in carrying out the somewhat elaborate details of the method as above outlined may be materially lightened by systematizing the work somewhat as follows: On the day before the tests are to be made prepare an abundance of clean glassware and the red cell emulsion; see that the water-bath is properly regulated; and, if desired, bleed one or more guinea-pigs for complement, and place the blood in the ice-box. In the morning proceed as follows:

1. Set up the complement titration, place in the incubator, and mark the time.

2. Pipet off the sera to be tested into clean test-tubes, and place in the water-bath to inactivate. Mark the time.

3. Arrange in the rack the tubes needed for the test, the antigen control, and the amboceptor titration.

4. By the time inactivation is complete the complement titration will be nearly or quite finished. Forty-five minutes will suffice for the latter. Now set up the tests proper, with the controls and the amboceptor titration.

5. At the end of the hour the titration may be read, and the indicated amount, and the red cells, added to all tubes. Two hours later the final result is read and recorded. Glassware should immediately be washed and put away for the next occasion. A little experience will enable one to make from twenty-five to fifty tests between 9 a.m. and 3 p.m.

B. Complement Deviation Test for Gonorrhea
Method of Schwartz and McNeil

The method as given below represents minor modifications of the original method suggested by experience in the writer's laboratory. The antigen is an autolysate of a large number of strains of the gonococci. It may be obtained from Parke, Davis & Co. For use, dilute with 9 parts of salt solution. The amount used for the test is one which gives a strong positive reaction with a known positive serum or with the antigonococcic serum of Torrey (also marketed by Parke, Davis & Co.), provided this amount is not anticomplementary. In our experience 0.15 c.c. of a 10 per cent. dilution has met these conditions

The complement is used in a 10 per cent. dilution. Complement and amboceptor are titrated against 0.1 c.c. of 5 per cent. sheep cell emulsion, instead of 1 c.c. The same quantity of red cell emulsion is, of course, also used in making the test.

The patient's serum is used inactivated. In the original method the test is carried out with 0.05, 0.10, and 0.15 c.c. In our experience 0.05 c.c. is almost invariably negative, while 0.15 c.c. is almost invariably anticomplementary. We have, therefore, used only 0.1 c.c.[1]

In other respects the test is carried out exactly like the test for syphilis. The reaction is negative during the acute stages of the disease, but is useful in determining the presence of a focus of chronic infection. Its chief importance lies in the fact that it becomes negative

[1] My assistant, Dr. T. F. Walker, has kindly furnished me with these data, based on an unusually extensive experience in my laboratory with the method.

in a short time (probably about two weeks) after a cure is completed.

C. COMPLEMENT DEVIATION TEST FOR TUBERCULOSIS
METHOD OF HAMMER

The antigen is a mixture of Koch's old tuberculin and an extract of tuberculous granulation tissue freed as much as possible from other tissue. Tissues from a surgical lesion, such as the knee, are most suitable. Cover the tissue with 4 parts of alcohol and extract for three to five days. Filter, and dilute the filtrate with 3 parts of salt solution for use. Test 0.4, 0.2, and 0.1 c.c. of this against 0.1 c.c. of known positive serum. Or, cover the tissue with 9 parts of acetone and extract for ten days. Filter, and evaporate to dryness at 37°C. Take up the residue in an equal volume of alcohol and dilute for use with 10 volumes of salt solution. Titrate as above. In either case the dose used is the largest, twice which is not anticomplementary.

Now add to 9 volumes of the diluted extract 1 volume of old tuberculin, and repeat the titration as above. The dose is determined according to the same rule. A certain proportion of cases will react with one or other of the antigens alone, but the larger percentage of positive results will be obtained with the mixed antigen.

Arrange the tubes as for the Wassermann reaction. In all the tubes place 1 c.c. of 5 per cent. complement serum. To the front tubes add the titrated dose of antigen. In each pair of tubes, front and rear, place 0.1 c.c. of the several sera respectively, inactivated at 56°C. for thirty minutes. Bring all the tubes to a like volume, mix, and let stand for three hours, at room-temperature. Add 2 units of amboceptor and 1 c.c. of 5 per cent. red cell emulsion. Mix, and place in the incubator for one hour. The tests are then ready for the final reading.

METHOD OF CRAIG

Craig has recently proposed a modification which gives in his hands a smaller proportion of positive results in non-tuberculous cases. The antigen is made by growing several strains of the human bacillus on an alkaline bouillon containing a teaspoonful of aseptically removed egg-white and egg-yolk for each 250 c.c. of bouillon. When growth is well advanced add an equal amount of 95 per cent. alcohol, and shake in a shaking machine for twelve hours; then place in the incubator for twenty-four hours, shake again for six hours, and finally filter through a very fine filter-paper or a Berkefeld filter. The mixed filtrates constitute the antigen, which is usually used without diluting. The antigen must be kept constantly in the ice-box and amounts needed for making tests removed aseptically as required. It is titrated, without diluting, for antigenic, anticomplementary, and hemolytic properties as given under the Wassermann reaction. *One* antigenic unit is used for the test. The serum to be tested is collected as for the Wassermann reaction, and inactivated at 56° for thirty minutes. Four capillary drops are used for the test. The complement used is fresh guinea-pig serum diluted with 1½ parts salt solution, and titrated against 0.1 c.c. of 10 per cent. washed *human* red cells, of which 2 units (usually about 0.1 c.c.) are used. The antihuman amboceptor may be purchased, or may be prepared by injecting each of several well-grown, healthy rabbits with three doses of thoroughly washed human red cells, made up after the final centrifugation with salt solution to about one-half the original volume of the blood. The first dose is 5 c.c. given under the skin of the abdomen; the second and third doses are of 3 c.c. given in the marginal vein of the ear. These doses are given at intervals of five or six days, and nine days after the last dose a small amount of blood is drawn from the ear of each rabbit and titrated. If one or two capillary drops of a 1–40 dilution completely hemolyzes 0.1 c.c. of 10 per cent. suspension of washed human red cells, in one hour in the incubator, in the presence of 1 unit of complement, the amboceptor is strong enough, and the rabbit should be killed while in a fasting condition, by cutting the carotid artery. Collect the blood in a clean dish, and after the serum has separated, preserve, preferably by impregnating a suitable filter-paper with it. Craig uses Schleicher and Schüll, No. 597.

The following summarizes the differences in the technic of the test and that of the Wassermann reaction already described:

(*a*) Four capillary drops of the (inactivated) serum to be tested, instead of 0.2 c.c.

(*b*) 0.1 c.c. of 10 per cent., or 1.0 c.c. of 1 per cent. washed *human* red cells, instead of 1.0 c.c. of 5 per cent. washed sheep cells.

(*c*) Antihuman, instead of antisheep amboceptor.

(*d*) Two units of 40 per cent. guinea-pig serum, as titrated against (*b*) and (*c*) (usually 0.1 c.c.).

(*o*) *One* antigenic unit of the above described antigen. The stock antigen is usually used undiluted.

V. COBRA-VENOM TEST FOR SYPHILIS

Of the several cobra-venom reactions, the method of Weil, for the diagnosis of syphilis, possesses the greatest practical value, and is here given. It appears to depend upon the same disturbance of lipoid metabolism which is responsible for the Wassermann reaction. It is known that syphilis is characterized by a withdrawal of lipoids from their chief depots, viz.: the central nervous system and the red blood-cells, with a marked increase of the same in the fluid part of the blood. Since it is also known that the hemolytic action of the cobra venom depends upon its activation by lecithin, in other words, upon a lecithin-venom complex in which the lecithin serves as complement, it may fairly be assumed that the loss of lipoids by the red cells is responsible for the *increased resistance to hemolysis by cobra venom* upon which Weil's reaction is based.

1. **Materials Required.**—1. The *cobra venom* may be obtained from Poulenc Frères, Paris. Weil's stock solution is a 0.5 per cent. solution in 0.9 per cent. salt solution, made, of course, very accurately. It deteriorates very rapidly unless kept frozen. For this reason I have used very successfully the solvent usually employed for the purpose of other reactions in this group, viz., a 1 per cent. solution of venom in equal parts of distilled water and chemically pure glycerin. Before it is used, this should be allowed to stand several days in the ice-box, where it keeps extraordinarily well.

2. The *blood-cells to be tested*. Have ready normal

salt solution to which 2 per cent. sodium citrate is *freshly* added, and which has been cooled in the ice-box. Into about 10 c.c. of this, contained in a graduated centrifuge tube, discharge about 2 c.c. of the patient's blood. *Do not shake.* The red cells must stand in contact with the citrate solution over night in the ice box before proceeding to the test. Wash at least four times with 0.9 per cent. salt solution. The last washing of all bloods in a series is done at the same speed and for the same length of time. Accurate and uniform dilution of the cells is, of course, an absolute essential to obtain comparable readings. Pipet off the last wash-water and make up to a 4 per cent. emulsion by adding 24 volumes of solution to the 1 volume of cells as read in the graded tube before they are disturbed The salt solution used for washing and diluting should be ice cold and the final emulsion should be placed on ice several hours before the test is made.

2. **Method.**—From the stock solution of venom prepare the following solutions for the test: 1–10,000, 1–20,000, 1–30,000, 1–40,000. Arrange a suitable rack with 4 tubes for each test. In the respective tubes of each row place 1 c.c. of the several venom solutions and 1 c.c. of the cell emulsion. Incubate for one hour at 37°C. Mix thoroughly by gentle shaking and place in the ice-box over night. In the morning again mix thoroughly and make the final reading an hour later. The result will depend on comparison with known normal cells. Something like the following may be anticipated:

No hemolysis at 1–10,000 = strongly positive.
Moderate hemolysis at 1–20,000 = positive.
Partial hemolysis at 1–30,000 = negative.
Complete hemolysis at 1–40,000 = hypersensitive.

The test appears later in the disease than the Wassermann reaction, and yields a higher percentage of positive results in late latent syphilis. Furthermore, it yields less quickly to treatment. It is unquestionably an important aid to diagnosis and treatment in the class of cases indicated.

APPENDIX

I. STAINING SOLUTIONS

In this section are given the formulæ for staining fluids which have general use, particularly for identification of bacteria. Blood-stains and others which are used only for special purposes are discussed in the body of the book and may be found by consulting the Index.

1. Carbol Thionin.—Saturated solution thionin in 50 per cent. alcohol, 20 c.c.; 2 per cent. aqueous solution phenol, 100 c.c.

This stain is especially useful in counting bacteria for standardization of vaccines (see p. 590). It can be used as a general stain. In blood work it is used for the malarial parasite and for demonstration of basophilic degeneration of the red cells. The fluid is applied for one-half to three minutes, after fixation by heat, or about a minute in saturated aqueous solution of mercuric chlorid or 1 per cent. formalin in alcohol.

2. Fuchsin.—This dye should not be confused with acid fuchsin. Its solutions are generally made with phenol as a mordant and they are then very powerful bacterial stains, with a strong tendency to over-staining. They are used chiefly for the tubercle bacillus.

Czaplewski's carbol-fuchsin is superior to the widely used Ziehl solution in that it acts more quickly and is permanent. To 1 Gm. fuchsin and 5 c.c. liquefied phenol, add 50 c.c. glycerol with constant stirring; and finally add 50 c.c. water, mix well, and filter.

3. Gentian Violet.—The combinations given below are powerful bacterial stains which have their chief use in Gram's method. They may be used interchangeably, but the solution with phenol is probably most serviceable. Formalin-gentian-violet remains good for years but is less satisfactory for Gram's method than the others. *Methyl violet may be substituted for gentian violet in these formulæ, and is preferable.*

Anilin-gentian violet.—Ehrlich's formula is the one generally used, but this keeps only a few weeks. Stirling's solution, which keeps much better and seems to give equal results, is as follows: gentian violet, 5 Gm.; alcohol, 10 c.c.; anilin oil, 2 c.c.; water, 88 c.c.

Czaplewski's Carbol-gentian-violet.—To 1 Gm. gentian violet and 5 c.c. liquefied phenol, add 50 c.c. glycerol with constant stirring; finally add 50 c.c. water, mix well, and filter.

Formalin-gentian-violet consists of 5 per cent. solution formalin, 75 parts; saturated alcoholic solution gentian violet, 25 parts.

4. Hematoxylin is one of the best nuclear stains available. There are many combinations, most of which require weeks or months for "ripening." The following is a good solution which is ready for use as soon as made.

Harris' Hematoxylin.—Dissolve 1 Gm. hematoxylin crystals in 10 c.c. alcohol. Dissolve 20 Gm. ammonia alum in 200 c.c. distilled water with the aid of heat and add the alcoholic hematoxylin solution. Bring the mixture to a boil and add half a gram of mercuric oxide.

As soon as the solution assumes a dark purple color, remove the vessel from the flame and cool quickly in a basin of cold water.

5. Iodin is used as a part of Gram's method and as a special stain for various purposes. For starch, a very weak solution is desirable; for *Leptothrix buccalis*, a strong solution such as Lugol's. The solutions deteriorate upon long standing.

Gram's Iodin Solution.—Iodin, 1 Gm.; potassium iodid, 2 Gm.; water, 300 c.c.

Lugol's solution (*Liquor Iodi Compositus*, U. S. P.) consists of iodin, 5 Gm.; potassium iodid, 10 Gm.; water, 100 c.c. Gram's iodin solution may be made from this by adding fourteen times its volume of water.

6. Methylene-blue is a widely used basic dye which does not readily over-stain. The following solutions are useful:

Gabbet's Stain.—This is used in Gabbet's method for the tubercle bacillus. It consists of methylene-blue, 2 Gm.; water, 75 c.c.; concentrated sulphuric acid, 25 c.c.

Loeffler's alkaline methylene-blue is one of the most useful bacterial stains for general purposes. The solution is applied at room temperature for 30 seconds to three minutes, and is followed by rinsing in water. Fixation may be by heat or chemicals. The stain is composed of 30 parts of a saturated alcoholic solution of methylene-blue and 100 parts of a 1:10,000 aqueous solution of caustic potash. It keeps indefinitely.

Pappenheim's methylene-blue solution is used as decolorizer and contrast stain in Pappenheim's method for the tubercle bacillus. Dissolve 1 Gm. corallin

41

(roselic acid) in 100 c.c. absolute alcohol; saturate with methylene-blue; and add 20 c.c. glycerol.

7. Pyronin.—Used in strong aqueous solution, this is useful as a contrast stain in Gram's method, but results are more satisfactory when the dye is combined with methyl green.

Pappenheim's Pyronin-methyl-green Stain.—This solution colors bacteria red and nuclei of cells blue. It is, therefore, especially useful for intracellular bacteria like the gonococcus and the influenza bacillus. It is a good stain for routine purposes, is a most excellent contrast stain for Gram's method, and is also used to demonstrate Döhle's inclusion bodies in the blood. It colors the cytoplasm of lymphocytes bright red, and has been used as a differential stain for these cells. The solution is applied cold for one-half to five minutes. It consists of saturated aqueous solution methyl-green, 3 to 4 parts, and saturated aqueous solution pyronin, 1 to 1½ parts. It is a good plan to keep these solutions in stock and to mix a new lot of the staining fluid about once a month. If it stains too deeply with either dye, the proper balance is attained by adding a little of the other.

8. Simple Bacterial Stains.—A simple solution of any basic anilin dye (methylene-blue, basic fuchsin, gentian-violet, etc.) will stain nearly all bacteria. These simple solutions are not much used in the clinical laboratory, because other stains, such as Löffler's methylene-blue and Pappenheim's pyronin-methyl-green stain, which serves the purpose even better, are at hand.

9. Sudan III is a valuable stain for fat, to which it gives an orange color. Scharlach R is a similar but stronger dye, and may be substituted to advantage. They may be used as a saturated solution in 70 per cent. alcohol or in the following combination.

Herxheimer's Sudan III consists of equal parts of 70 per cent. alcohol and acetone saturated with Sudan III (or Scharlach R).

II. OFFICE LABORATORY EQUIPMENT

It is not to be expected that a physician in active practice will make routine use of all the methods described in this book. Although he will need nearly all of them for the study of his more difficult cases, his daily laboratory work will probably be limited to a few simple procedures. With this in mind the following list of laboratory procedures is suggested as the minimum with which a physician should be thoroughly familiar and upon which he may build as his practice requires. The methods are selected because of their simplicity and practical usefulness.

METHODS FOR OFFICE ROUTINE

SPUTUM

Careful inspection (see p. 59).
Simple microscopic examination unstained (see p. 63).
Examination for tubercle bacilli (see p. 76).

URINE

Reaction (see p. 106).
Specific gravity (see p. 107).
Calculation of total solids (see p. 110).
Phenolsulphonephthalein test of kidney function (see p. 112).

Albumin, qualitative:
  Roberts's ring test (see p. 154).
  Purdy's heat test (see p. 156).
Albumin, quantitative:
  Esbach's test (see p. 157).
Sugar, qualitative:
  Benedict's test (see p. 163).
Sugar, quantitative:
  Benedict's method (see p. 167), or Roberts's yeast method
    (see p. 170).
Acetone, Lange's or Rothera's test (see p. 176).
Diacetic acid, Gerhardt's test (see p. 177).
Bile, Gmelin's test (see p. 180).
Hemoglobin, benzidin test (see p. 182).
Indican, Obermayer's test (see p. 134).
Microscopic examination (see p. 198).

BLOOD

Coagulation time, simple method (see p. 258).
Hemoglobin, Dare or Sahli method (see pp. 266 to 268).
Red corpuscle count (see p. 272).
Color index calculation (see p. 284).
Leukocyte count (see p. 294).
Differential leukocyte count (see p. 324).
Microscopic examination of stained films for pathological red cells
    and malarial parasites (see pp. 315, 357).

STOMACH CONTENTS

Careful inspection (see p. 398).
Total acidity, Töpfer's method (see p. 408).
Free hydrochloric acid, Töpfer's method (see p. 410).
Lactic acid, Kelling's test (see p. 403).
Microscopic examination (see p. 414).

FECES

Careful inspection (see p. 424).
Occult blood, benzidin test after extraction with ether (see p. 429).
Microscopic examination:
  (a) for parasites or their ova (see p. 443).
  (b) to ascertain state of digestion (see pp. 437, 444).

SERUM METHODS

Widal test by macroscopic method, using one of the commercial outfits (see p. 608).

MISCELLANEOUS

Microscopic examination of pus:

(a) simple stain (see p. 571).

(b) Gram's method (see p. 572).

Puncture fluids:

(a) Careful inspection (see pp. 520 and 524).

(b) Microscopic examination for bacteria and differential cell count (see p. 521).

Syphilitic material for spirochetes, Giemsa stain or India-ink method (see pp. 550–553).

Milk:

Fat, Leffmann-Beam method (see p. 546).

Protein, by calculation (see p. 545).

A list of equipment which is sufficient for all the above-mentioned methods (and for many others in addition) is given below. The total cost, exclusive of the furniture, but including a first class microscope, simple mechanical stage, and Sahli hemoglobinometer will be about $130.00 in normal times.[1] There is no real economy in purchasing instruments of inferior quality.

A. FURNITURE

A table, with drawer, and a few shelves for bottles and glassware constitute the only really essential laboratory furniture even for fairly extensive work. When a special room is not available these may stand behind a screen in the physician's consulting room. Gas and running water are very desirable, but not absolutely necessary.

[1] The entire outfit, with ready-prepared reagents and staining solutions, can be purchased of Paul Weiss, 1620 Arapahoe St., Denver; A. H. Thomas Co., West Washington Square, Philadelphia, and probably many other supply houses.

The shelves may conveniently take the form of a shallow case without doors, which stands upon the back of the table and which, in addition to the shelves, has two tall compartments, one for the combined buret and filter stand, the other for the microscope in its case. If space allows, however, it will be found more satisfactory to keep the microscope under a glass bell-jar (or pasteboard cover, p. 41) on a stand or small table before a window. It is thus always ready for use and is away from the neighborhood of corroding chemicals. A stool or chair of the proper height (see p. 37) should be at hand.

A convenient reservoir for wash water is a large bottle, which stands upon the top shelf and from which water is siphoned by means of a rubber tube with a medicine-dropper tip and a Mohr pinch cock. The glass tip should hang directly over a miniature sink consisting of a large glass funnel whose stem passes through the table top and drains by means of a rubber tube into an earthen jar below. All staining should be done over this funnel-sink, the slides being supported upon a rack consisting essentially of two small rods about 2 inches apart placed across the top of the funnel.

The following wood finish is extensively used for table tops in the laboratories of this country. It gives an ebony-black surface which resists practically all reagents. The wood must be new, or at least not painted, varnished, or waxed.

Solution No. 1.

Copper sulphate	125 Gm.;
Potassium chlorate (or permanganate)	125 Gm.;
Water	1000 c.c.

Solution No. 2.

Anilin oil	120 c.c.;
Hydrochloric acid, concentrated	180 c.c.;
Water	1000 c.c.

Apply two coats of Solution No. 1, hot, and then two coats of Solution No. 2, without heating, allowing each coat to dry thoroughly before the next is applied. When the last coat is dry remove the excess of the chemicals by rubbing with a coarse cloth. Finally rub thoroughly with a mixture of equal parts of turpentine and linseed oil.

B. APPARATUS

1 Basin of white enameled ware.

2 Beakers with lip, about 50 c.c. capacity. Small coffee cups may be substituted.

1 Blood-lancet or some substitute, as a Hagedorn needle (see p. 253).

1 Bunsen burner with rubber tubing, the small "micro" burner being especially satisfactory. An alcohol lamp will answer.

1 Buret, 25 c.c. capacity, preferably with Schellbach stripe. An accurate 10 c.c. graduated pipet may be used for much work but is not so satisfactory as the buret.

1 Centrifuge, hand, electric or water power (see Figs. 33, 34). The last is cheap and satisfactory. Metal shields with flat bottoms and rubber cushions are preferable to the ordinary conical aluminum shields because they allow the use of ordinary test-tubes as well as conical centrifuge tubes.

1 Esbach tube (see Fig. 42).

1 Pack filter papers, round, about 12 cm. in diameter, good quality.

1 Funnel, glass, about 7 cm. in diameter.

4 feet Glass tubing, about 7 or 8 mm. outside diameter, of soft glass for making urine pipets, etc.

1 Graduate, cylinder, 100 c.c., double graduations. This is used chiefly for making solutions.

1 Hemacytometer (see Figs. 98, 102, 103). Probably the most satisfactory outfit consists of a Buerker-type counting chamber with Neubauer ruling, a "red pipet" and a "white pipet."

1 Hemoglobinometer. The Dare will probably be found most convenient if the price is not prohibitive; otherwise the Sahli (or Kuttner) is recommended. A Tallquist book should be carried in the hand-bag.

1 Pack lens-cleaning paper. Two rows of stitching, $\frac{1}{2}$ inch apart, may be run across the middle of the package on the sewing machine and the package then cut into little booklets of convenient size.

1 Box labels for bottles. Denison's No. A-4 is a useful size.

1 Box labels for slides.

1 Mechanical stage, attachable (see p. 49).

4 Medicine droppers: two, labeled "Stain" and "Water" respectively, to be reserved for use with Wright's blood-stain; one, which delivers the proper sized drop, to be reserved for the quantitative sugar estimation.

1 Eye-piece micrometer. The card-board micrometer made as described on p. 44 will answer for most clinical work.

1 Microscope equipped as described on p. 48.

50 Micro cover-glasses, No. 2 thickness. The 22 mm. squares are most convenient for general purposes.

1 Box ($\frac{1}{2}$ gross) micro slides, 75 × 25 mm., clear white glass, medium thickness, ground edges.

1 Pencil, wax, for writing on glass, red or blue.

1 Petri dish with cover, about 15 cm. in diameter.

1 Pipet, 10 c.c., graduated.

1 Rule, celluloid, 6 inches and 15 cm. These are sold by

Bausch and Lomb Optical Co. and Spencer Lens Co. for five cents each.

1 Stand for filter, buret, etc. Convenient, but not absolutely essential.

1 Stomach tube. The Rehfuss type is required if fractional method of examination is employed and is best for all purposes.

1 Test-glass, conical   A wine-glass will serve.

12 Test-tubes, size about 125 × 16 mm., without flange.

1 Test-tube brush, bristle, with tuft at tip.

1 Test-tube rack holding six tubes.

1 Urinometer with cylinder. Must have wide graduations. Test with distilled water.

1 Widal test outfit, macroscopic method. Satisfactory outfits are sold under various trade names, "Typhoid agglutometer," etc.

1 Box wooden toothpicks.

C. REAGENTS AND STAINS

All staining solutions and many reagents are best kept in small dropping bottles, of which the "flat-topped T. K." pattern is most satisfactory. Other reagents may be kept in ordinary round prescription bottles of 4 to 8 ounces' capacity. Bottles containing highly volatile reagents should be sealed with paraffin if not in constant use; while those containing strong caustic soda solutions should have rubber stoppers.

Most staining solutions and chemical reagents can be purchased ready prepared. For the physician who does only a small amount of work the "Soloid" tablets manufactured by Burroughs, Welcome & Co. are convenient and satisfactory. These tablets have only to be dissolved in a specified amount of the appropriate fluid to produce the finished solution. Most of the stains and many of the commoner reagents are supplied in this form.

If, however, his time permits the physician will find it more satisfactory and much more economical to prepare his own solutions, with exception of normal solutions and a very few stains.

REAGENTS

50 c.c. Acid, acetic, glacial, 99½ per cent. Other strengths can be made from this as desired.

50 c.c. Acid, hydrochloric, C.P., Sp. Gr. 1. 16. Contains about 32 per cent. HCl. An approximate decinormal solution for use with the Sahli hemoglobinometer can be made by adding 12 c.c. of this acid to 988 c.c. distilled water.

50 c.c. Acid, nitric, C.P. Yellow nitric acid can be made from this by adding a splinter of pine (match stick) or allowing it to stand in the sunlight for a short time.

50 c.c. Acid, sulphuric, C.P.

50 c.c. Alcohol, amylic, C.P. Used in the estimation of fat in milk.

200 c.c. Alcohol, ethylic (grain alcohol). This is ordinarily about 93 to 95 per cent. and other strengths can be made as desired. Whenever the word "alcohol" is used in the text without qualification, this alcohol is meant.

100 c.c. Alcohol, methylic, Merck's "Reagent," for making Wright's blood stain. May be omitted if the stain is purchased ready prepared.

100 c.c. Ammonium hydroxid (strong ammonia) Sp. Gr. 0.9.

200 c.c. Benedict's solution for qualitative sugar test (see p. 163).

100 c.c. Benedict's solution for quantitative sugar estimation (see p. 167).

10 Gm. benzidin. Specify "for blood test."

1 tube Canada balsam in xylol. Necessary only if permanent mounts are to be made.

100 c.c. Chloroform, U. S. P. .

100 c.c. Diluting fluid for red corpuscle count, Hayem's preferred (see p. 279).

100 c.c. Diluting fluid for leukocyte count (see p. 300).

30 c.c. Dimethyl-amido-azo-benzol, 0.5 per cent. alcoholic solution.

100 c.c. Esbach's solution (see p. 157).

200 c.c. Ether, sulphuric, U. S. P.

30 c.c. Ferric chlorid, 10 per cent. aqueous solution.

100 c.c. Formalin (40 per cent. solution of formaldehyd gas). The expression "10 per cent. formalin" means 1 part of this 40 per cent. solution and 9 parts of water making a 4 per cent. solution of formaldehyd gas.

50 c.c. Hydrogen peroxid, U. S. P.

1 Vial litmus paper, Squibb, red.

1 Vial litmus paper, Squibb, blue.

25 c.c. Lugol's solution (*Liquor Iodi Compositus*, U. S. P.). Gram's iodin solution (see p. 641) can be made from this by adding 14 times its volume of water.

50 Gm. Magnesium sulphate, C.P., for making Roberts' solution for albumin in urine.

100 c.c. Obermayer's reagent for indican (see p. 134).

25 c.c. Oil of cedar for immersion. A sufficient quantity is usually supplied with the microscope when purchased.

25 c.c. Phenolphthalein, 1 or 0.5 per cent. solution in alcohol.

2 Ampoules phenolsulphonephthalein.

50 Gm. Sodium chlorid, C.P., for Purdy's albumin test. Table salt may be used but is not so good.

1000 c.c. Sodium hydroxid, decinormal solution. The practitioner will find it best to purchase this solution ready prepared. Most chemical supply houses carry it in stock.

For rough clinical work 41 grams of Merck's "Sodium hydrate by alcohol" from a freshly opened bottle may be dissolved in 1000 c.c. distilled water. This makes a normal solution and must be diluted with 9 volumes of water to make the decinormal solution.

25 Gm. Sodium nitroprussid, C.P., crystals.

50 Gm. Talc, purified (*Talcum purificatum*, U. S. P.) or diatomaceous earth (*Kieselguhr*) for clearing urine.

2000 c.c. Water, distilled. In some regions ordinary tap-water answers for practically all purposes.

STAINS

It will be found most satisfactory to have on hand a stock of dry stains (which keep well) and to make solutions as needed. Ordinarily the smallest quantity obtainable in an unbroken package should be purchased. The following dry stains make up a fairly complete stock for the clinical laboratory: Fuchsin, basic; gentian violet; methylene blue, B. X.; methyl green; pyronin; and Wright's stain. Wright's stain is obtainable in 1-Gm. vials, the others in 10-Gm. vials. The most frequently used solutions, which can be purchased in 25-c.c. bottles, are:

Carbol-fuchsin (see p. 639).

Carbol-gentian-violet (see p. 640).

Giemsa's stain (see p. 313). This is not necessary if the India-ink method for spirochetes is used.

Löffler's alkaline methylene-blue (see p. 641).

Pappenheim's methylene-blue contrast stain for tubercle bacilli (see p. 641).

Pappenheim's pyronin-methyl-green stain (see p. 642).

Wright's blood stain (see p. 309). Much of the solution on the market is unsatisfactory.

III. WEIGHTS, MEASURES, ETC., WITH EQUIVALENTS

METRIC

Meter (unit of length) : Millimeter (mm.) $= \frac{1}{1000}$ meter.
Centimeter (cm.) $= \frac{1}{100}$ meter.
Kilometer $= 1000$ meters.
Micron (μ) $= \frac{1}{1000}$ millimeter.
Gram (unit of weight) : Milligram (mg.) $= \frac{1}{1000}$ gram.
Kilogram (kilo.) $= 1000$ meters.
Liter (unit of capacity) : Cubic Centimeter $= \frac{1}{1000}$ liter. Same measure as milliliter (ml.).

1 Millimeter $= \begin{cases} 0.03937 \; (\frac{1}{25} \text{ approx.}) \text{ in.} \\ 1000 \text{ microns.} \end{cases}$

1 Centimeter $= \begin{cases} 0.3937 \; (\frac{2}{5} \text{ approx.}) \text{ in.} \\ 0.0328 \text{ feet.} \end{cases}$

1 Meter $= \begin{cases} 39.37 \text{ in.} \\ 3.28 \text{ feet.} \end{cases}$

1 Micron (μ) $= \begin{cases} \frac{1}{25000} \text{ in.} \\ 0.001 \text{ millimeter.} \end{cases}$

Gram $= \begin{cases} 15.43 \text{ grains.} \\ 0.563 \text{ dram} \\ 0.035 \text{ ounce} \\ 0.0022 \text{ pound} \end{cases}$ Avoir.
$\begin{cases} 0.257 \text{ dram} \\ 0.032 \text{ ounce} \\ 0.0027 \text{ pound} \end{cases}$ Apoth.

1 Kilogram $= \begin{cases} 35.27 \text{ ounce (Avoir.).} \\ 2.2 \text{ pound (Avoir.).} \end{cases}$

1 Liter $= \begin{cases} 1.056 \; (1 \text{ approx.}) \text{ quart.} \\ 61.02 \text{ cu. inches.} \\ 1000 \text{ cu. centimeters.} \end{cases}$

1 Sq. Millimeter $= 0.00155$
1 Sq. Centimeter $= 0.1550$
1 Sq. Meter $= 1550$
1 Sq. Meter $= 10.76$ sq. feet.

1 Cu. Millimeter $= 0.00006$
1 Cu. Centimeter $= 0.0610$ cu. in.
1 Cu. Centimeter $= 0.001$ liter.
1 Cu. Meter $= \begin{cases} 35.32 \text{ cu. feet.} \\ 61025.4 \text{ cu. in.} \end{cases}$

1 Inch $= 25.399$ millimeters.
1 Sq. Inch $= 6.451$ sq. centimeters.
1 Cu. Inch $= 16.387$ cu. centimeters.

1 Foot $= 30.48$ centimeters.
1 Sq. Foot $= 0.093$ sq. meter.
1 Cu. Foot $= 0.028$ cu. meter.

AVOIRDUPOIS WEIGHT

1 Ounce $= \begin{cases} 437.5 \text{ grains.} \\ 16 \text{ drams.} \end{cases}$
1 Pound $= 16$ ounces.

1 Grain $= 0.065 \; (\frac{1}{15} \text{ approx.})$
1 Dram $= 1.77 \; (1\frac{3}{4} \text{ approx.})$
1 Ounce $= 28.35 \; (30 \text{ approx.})$
1 Pound $= 453.59 \; (500 \text{ approx.})$ grams.
1 Pound $= 27.7$ cu. inches.
1 Pound $= 1.215$ lb. Troy.

APOTHECARIES' MEASURE

1 Dram $= 60$ minims.
1 Ounce $= 8$ drams.
1 Pint $= 16$ ounces.
1 Gallon $= 8$ pints.

1 Dram $= 3.70$
1 Ounce $= 29.57$
1 Pint $= 473.1$ cu. centimeters.
1 Gallon $= 3785.4$
1 Gallon $= 231$ cu. inches.

APOTHECARIES' WEIGHT

1 Scruple = 20 grains.	1 Grain = 0.065 ⎫
1 Dram = 3 scruples = 60 grains.	1 Dram = 3.887 ⎪ grams.
1 Ounce = 8 drams = 480 grains.	1 Ounce = 31.10 ⎬
1 Pound = 12 ounces.	1 Pound = 373.2 ⎭

To convert			into		multiply by	
"	"	minims	into	cubic centimeters	multiply by	0.061
"	"	fluidounces	"	cubic centimeters	"	29.57
"	"	grains	"	grams	"	0.0648
"	"	drams	"	grams	"	3.887
"	"	cubic centimeters	"	minims	"	16.23
"	"	cubic centimeters	"	fluidounces	"	0.0338
"	"	grams	"	grains	"	15.432
"	"	grams	"	drams	"	0.257

TEMPERATURE

CENTIGRADE.	FAHRENHEIT.	CENTIGRADE.	FAHRENHEIT.
110°	230°	37°	98.6°
100	212	36.5	97.7
95	203	36	96.8
90	194	35.5	95.9
85	185	35	95
80	176	34	93.2
75	167	33	91.4
70	158	32	89.6
65	149	31	87.8
60	140	30	86
55	131	25	77
50	122	20	68
45	113	15	59
44	111.2	10	50
43	109.4	+5	41
42	107.6	0	32
41	105.8	−5	23
40.5	104.9	−10	14
40	104	−15	+5
39.5	103.1	−20	−4
39	102.2		
38.5	101.3	0.54°	= 1°
38	100.4	1	= 1.8
37.5	99.5	2	= 3.6
		2.5	= 4.5

To convert Fahrenheit into Centigrade, subtract 32 and multiply by 0.555.

To convert Centigrade into Fahrenheit, multiply by 1.8 and add 32.

INDEX

Church and Peterson's
Nervous *and* Mental Diseases

Nervous and Mental Diseases. By ARCHIBALD CHURCH,
M.D., Professor of Nervous and Mental Diseases and Head of
Neurologic Department, Northwestern University Medical School,
Chicago; and FREDERICK PETERSON, M. D., formerly Professor
of Psychiatry in Columbia University, New York. Octavo of 940
pages, with 350 illustrations. Cloth, $5.00 net.

Published October, 1914

EIGHTH EDITION

For this new eighth edition the entire work has been most thoroughly
revised. To show with what thoroughness the authors have revised their
work, we point out that in the nervous section alone over one hundred and
fifty interpolations have been made, and, in addition, well over three hundred
minor corrections. The section on Mental Diseases has been wholly re-
arranged to conform to the latest classification, some obsolete matter struck
out, and much new matter added.

American Journal of the Medical Sciences

"This edition has been revised, new illustrations added, and some new matter, and really
as two books. . . . The descriptions of disease are clear, directions as to treatment definite,
and disputed matters and theories are omitted. Altogether it is a most useful text-book."

Herrick's Neurology

Introduction to Neurology. By C. JUDSON HERRICK,
PH. D., Professor of Neurology in the University of Chicago.
12mo of 360 pages, illustrated. Cloth, $1.75 net.

A KEY TO NEUROLOGY

Professor Herrick's new work will aid the student to organize his knowledge and appreciate the significance of the nervous system as a mechanism
right at the beginning of his study. It is sufficiently elementary to be used
by students of elementary psychology in colleges and normal schools, by
students of general zoölogy and comparative anatomy in college classes, and
by medical students as a guide and key to the interpretation of the larger
works on neurology. **Published September, 1915**

Kaplan's Serology of Nervous and Mental Diseases

Serology of Nervous and Mental Diseases. By D. M. KAPLAN, M. D., Director of Clinical and Research Laboratories, Neurological Institute, New York City. Octavo of 346 pages, illustrated. Published June, 1914. Cloth, $3.50 net.

ILLUSTRATED

This is an entirely new work on this subject. Here you get the newest technic—the application of serology in practice. You get the indications, contra-indications, preparation of patients, technic, after-phenomena, after-care, and disposal of the fluids obtained by lumbar puncture. You get the physical, chemical and cytologic properties of normal and pathologic fluids discussed in *detail*, including the interpretation of findings and bacteriology. You get a full discussion of the serology of all nervous and mental diseases of *non-luetic* etiology—meningeal affections (infectious and non-infectious), brain diseases, cord diseases, nerve affections (including disorders of *internal secretion*), the psychoses, and the intoxications. The serology of every type of *luetic* nervous and mental disease is next presented to you, giving the *Wassermann reaction in detail*, the use of *salvarsan* and *neosalvarsan*.

Elsberg's Surgery of Spinal Cord

Surgery of the Spinal Cord. By CHARLES A. ELSBERG, M. D., Professor of Clinical Surgery, New York University and Bellevue Hospital Medical School. Octavo of 330 pages, with 153 illustrations. Cloth, $5.00 net.

ORIGINAL ILLUSTRATIONS
(Published July, 1916)

There is no other book published like this by Dr. Elsberg. It gives you in clear definite language the diagnosis and treatment of all surgical diseases of the spinal cord and its membranes, illustrating each operation with original pictures. Because it goes so thoroughly into symptomatology, diagnosis, and indications for operation this work appears as strongly to the general practitioner and neurologist as to the surgeon. The first part of the work is devoted to anatomy and physiology of the spinal cord, and to the symptomatology of surgical spinal diseases. The second part takes up operations upon the spine, the cord, and nerve-roots. The third part is given over to surgical diseases of the cord and its membranes—their diagnosis and treatment. Included also are chapters on hematomyelia and spinal gliosis, because in these diseases much harm is done to the fiber tracts by compression. There is also a chapter on *x*-rays in spinal diseases.

Abt's Preparation of Infants' Foods

A PRACTICAL GUIDE FOR THE PREPARATION OF INFANTS' AND CHILDREN'S FOODS. By ISAAC A. ABT, M. D., Professor of Diseases of Children, Northwestern University Medical School. 12mo of 143 pages. Cloth, $1.25 net.   **Published July, 1917**

In this book Dr. Abt gives to young mothers, nurses, and those who care for infants and children minute directions on the preparation of every kind of food that has value in infant feeding. Included also is such practical material as diet-lists for constipation, an outline of a plan for feeding babies, care of nipples and bottles, etc. You are given weights and measures; the mineral constituents and caloric values of foods; and a wealth of recipes for foods of all kinds and purposes.

Hill and Gerstley's Infant Feeding

CLINICAL LECTURES ON INFANT FEEDING. By LEWIS WEBB HILL, M. D., Alumni Assistant in Pediatrics, Harvard Medical School; and JESSE R. GERSTLEY, M. D., Instructor in Pediatrics, Northwestern University Medical School. 12mo of 377 pages, illustrated. Cloth, $2.75 net.   **Published October, 1917**

In these clinics you are given the full details of the Boston method of infant feeding as developed by Dr. Rotch, and of the Chicago method. You are given the theory, use in both normal and abnormal cases, exact quantities and percentages, and concrete clinical examples. The book is equivalent to a postgraduate course in infant feeding. It brings these two systems right to your door.

Herrick and Crosby's Laboratory Neurology

LABORATORY OUTLINES OF NEUROLOGY. By C. JUDSON HERRICK, PH. D., and ELIZABETH CROSBY, PH. D., of the Anatomical Laboratory of the University of Chicago. 12mo of 120 pages, illustrated.
Published January, 1918

Hecker, Trumpp, and Abt on Children

ATLAS AND EPITOME OF DISEASES OF CHILDREN. By Dr. R. HECKER and Dr. J. TRUMPP, of Munich. Edited, with additions, by ISAAC A. ABT, M. D. With 48 colored plates, 144 text-cuts, and 453 pages of text. Cloth, $5.00 net.   **Published April, 1907**

Ruhrah's Diseases of Children

A Manual of Diseases of Children. By JOHN RUHRAH, M. D., Professor of Diseases of Children, College of Physicians and Surgeons, Baltimore. 12mo of 552 pages, fully illustrated. Cloth, $2.75 net. *Published September, 1914*

FOURTH EDITION

The fourth edition makes this work more than ever the ideal desk book for the general practitioner. Although there have been added over one hundred pages of new matter and some sixty new illustrations, the book remains of a handy size and is still flexible.

American Journal of the Medical Sciences

"Treatment has been satisfactorily covered, being quite in accord with the best teaching, yet withal broadly general and free from stock prescriptions."

Griffith's Care of the Baby

The Care of the Baby. By J. P. CROZER GRIFFITH, M. D., Professor of Pediatrics at the University of Pennsylvania. 12mo of 455 pages, illustrated. Cloth, $1.50 net. *Published June, 1915*

SIXTH EDITION

New York Medical Journal

"We are confident if this little work could find its way into the hands of every trained nurse and of every mother, infant mortality would be lessened by at least fifty per cent."

Bandler's The Expectant Mother

This is decidedly a book for the woman preparing for childbirth. It has chapters on menstruation, nourishment of mother during pregnancy, nausea, care of breasts, examination of urine, preparations for labor, care of mother and child after delivery, twilight sleep, and dozens of other matters of great interest to the expectant mother. *Published August, 1916*

12mo of 213 pages, illustrated. By S. WYLLIS BANDLER, M. D., Professor of Diseases of Women, New York Post-Graduate Medical School and Hospital. Cloth, $1.25 net.

Kerley's New Pediatrics

Practice of Pediatrics. By CHARLES GILMORE KERLEY, M. D., Professor of Diseases of Children, New York Polyclinic Medical School and Hospital. Octavo of 913 pages, illustrated.

Published January, 1918

SECOND EDITION

This is an entirely new work—not a revision of Dr. Kerley's earlier work. It is not a cut-and-dried treatise—but the *practice* of pediatrics, giving, of course, fullest attention to *diagnosis* and *treatment*. The chapters on the newborn and its diseases, the feeding and the growth of baby, the care of the mother's breasts, artificial feeding, milk modification and sterilization, diet for older children—form a monograph of 125 pages. Then are discussed in detail every disease of childhood, *telling just what measures should be instituted*, what drugs given, *60 valuable prescriptions* being included. The chapter on *vaccine therapy* is right down to the minute, including every new method of proved value—with the exact technic. There is an excellent chapter on *Gymnastic Therapeutics*, giving explicit directions for the correction of certain abnormalities in which gymnastics have proved efficacious. Another feature consists of *the 165 illustrative cases—case teaching* of the most practical sort.

Grulee's Infant Feeding

Infant Feeding. By CLIFFORD G. GRULEE, M. D., Assistant Professor of Pediatrics at Rush Medical College. Octavo of 326 pages, illustrated, including 8 in colors. Cloth, $3.25 net.

Published September, 1917

THIRD EDITION

Dr. Grulee tells you *how* to feed the infant. He tells you—and *shows* you by clear illustrations—the *technic* of giving the child the breast. Then artificial feeding is thoughtfully presented, including a number of simple formulas. The colored illustrations showing the actual shapes and appearances of stools are extremely valuable.

Kerr's Diagnostics *of* Diseases *of* Children

DIAGNOSTICS OF THE DISEASES OF CHILDREN. By LEGRAND KERR, M. D., Professor of Diseases of Children, Brooklyn Postgraduate Medical School, Brooklyn. Octavo of 542 pages, fully illustrated. Cloth, $5.00 net.

Published February, 1907

Hoxie and Laptad's Medicine for Nurses Second Edition

MEDICINE FOR NURSES AND HOUSEMOHHERS. By GEORGE HOWARD HOXIE, M. D., Physician to the German Hospital, Kansas City; and PEARL L. LAPTAD, formerly Principal of Training-School, University of Kansas. 12mo of 351 pages, illustrated. Cloth, $1.50 net.

Published April, 1913

The purpose of this book is to provide enough information about the nature, cause, and cure of disease to enable you to carry out the doctor's orders intelligently, to know when home treatment is sufficient, and how and what to apply, to recognize when professional advice should be sought, and how to render first aid in emergencies. It is just the manual you need in caring for illness and preventing its spread in your home.

Aikens' Home Nurse's Hand-Book New (2d) Edition

HOME NURSE'S HAND-BOOK. By CHARLOTTE A. AIKENS. 12mo of 303 pages, illustrated. Cloth, $1.50 net. **Published March, 1917**

You will find the entire realm of *home* nursing very completely presented here. The author *knows*, from long active experience, just how to meet emergencies, how to improvise, how to *manage*. She shows you what to have ready in times of sickness, how to make the bed, how to care for, feed, and bathe the patient, how to give home treatments, and what they are; how to disinfect, how to make and apply bandages, how to render first aid in emergencies.

"Several books have been written upon home nursing, but none has been so clear, so simply put, so easy to learn, so easy to teach, so free from matter not to the point, so all-comprehensive."—*Trained Nurse and Hospital Review.*

Galbraith's Personal Hygiene and Physical Training for Women Second Edition, January, 1917

PERSONAL HYGIENE AND PHYSICAL TRAINING FOR WOMEN. By ANNA M. GALBRAITH, M.D., Fellow New York Academy of Medicine. 12mo of 393 pages, with original illustrations. Cloth, $2.25 net.

This is a book for every woman. It tells you how to train your physical powers to their fullest development. For home study and as a text-book this manual has no equal.

"Just the information that every woman should possess. It gives more valuable and practical help than we have ever seen in any other one volume."—*Health Culture.*

Galbraith's Four Epochs of Woman's Life Third Edition

THE FOUR EPOCHS OF WOMAN'S LIFE. By ANNA M. GALBRAITH, M. D. With an Introductory Note by JOHN H. MUSSER, M. D. 12mo of 296 pages. Cloth, $1.50 net. **Published March, 1917**

Women have at last awakened to a sense of the penalties they have paid for their ignorance of those laws of nature which govern their physical being. Maidenhood, marriage, maternity, and menopause—these are the periods of a woman's life fraught with the greatest dangers. Well-ordered education in these subjects lessens suffering and is essential to prevention of disease. Dr. Galbraith discusses the subject in a modest and conclusive manner. *Sex Instruction* is handled most admirably. There is a valuable chapter on *eugenics* and a glossary of medical terms.

"These truths should be known by every woman, and I gladly commend the essay to their thoughtful consideration."—*Dr. John H. Musser, University of Pennsylvania.*

Winslow's Prevention of Disease

PREVENTION OF DISEASE. By KENELM WINSLOW, M. D., formerly Assistant Professor of Comparative Therapeutics, Harvard University. 12mo of 348 pages, illustrated. Cloth, $1.75 net.   November, 1916

This book is a practical guide for the layman, giving him briefly the means to avoid the various diseases described. The chapters on diet, exercise, tea, coffee, and alcohol are of special interest, as is that on the prevention of cancer. There are chapters on the prevention of malaria, colds, constipation, obesity, nervous disorders, tuberculosis, etc. The work is a record of 25 years' active practice.

Brady's Personal Health

PERSONAL HEALTH. A Doctor Book for Discriminating People. By WILLIAM BRADY, M. D., Elmira, N. Y. 12mo of 405 pages. Cloth, $1.50 net.   Published September, 1916

Knowledge is the only protection against sickness. This book will make you more efficient in preventing illness and in caring for those who are ill when you cannot get a doctor, and in co-operating with him when you can get one. It is written in a clear, concise, and simple style. It is wholly understandable. All the mystery in medicine is swept aside and you are told what is most worth while in modern medicine—your personal health and how to maintain it. It is written by a doctor unusually well qualified to present medicine to the lay reader. It is an instructive book, indeed.

Keefer's Military Hygiene

MILITARY HYGIENE AND SANITATION. By LIEUT.-COL. FRANK R. KEEFER, Professor of Military Hygiene, United States Military Academy, West Point. 12mo of 305 pages, illustrated. Cloth, $1.50 net.

This is a concise though complete text-book on this subject, containing chapters on the care of troops, recruits and recruiting, personal hygiene, physical training, preventable diseases, clothing, equipment, water supply, foods and their preparation, hygiene and sanitation of posts and barracks, the troopship, hygiene and sanitation of marches, camps and battlefields, disposal of wastes, tropical and arctic service, venereal diseases, alcohol and other narcotics, and a glossary.   July, 1914

Bergey's Hygiene   Fifth Edition

THE PRINCIPLES OF HYGIENE. By D. H. BERGEY, A. M., M. D., Assistant Professor of Bacteriology, University of Pennsylvania. Octavo of 529 pages, illustrated. Cloth, $3.00 net.   Published September, 1914

"It will be found of value to the practitioner of medicine and the practical sanitarian ; and students of architecture, who need to consider problems of heating, lighting, ventilation, water supply, and sewage disposal, may consult it with profit."
Buffalo Medical Journal.

Owen's Treatment of Emergencies

THE TREATMENT OF EMERGENCIES. By HUBLEY R. OWEN, M. D., Surgeon to the Philadelphia General Hospital. 12mo of 350 pages, with 249 illustrations. Cloth, $2.00 net.   Published June, 1917

Dr. Owen's book is a complete treatment of emergencies. It gives you not only the *actual technic* of the procedures, but, what is equally important, the underlying principles of the treatments, and the reason *why* a particular method is advised. This makes for correctness. You get chapters on fractures of all kinds, going fully into symptoms, treatments, and complications. You get treatments of contusions, of wounds, both lacerated and incised. Particularly strong is the chapter on *gun-shot wounds*, which gives the new treatments that the great European War has developed. You get the principles of hemorrhage—arterial, venous, and capillary—together with its constitutional and local treatments. You get chapters on sprains, strains, dislocations, burns and scalds, treatment of sunburn, chilblain, asphyxiation, convulsions, hysteria, apoplexy, exhaustion, opium poisoning, uremia, and electric shock. You get sections on the making and applying of bandages, varieties, and precautions; and a complete discussion of the various kinds of artificial respiration, including mechanical devices. You are told what equipment is necessary for emergency work, and what household remedies should be always on hand.

Morrow's Immediate Care of Injured   Third Edition

IMMEDIATE CARE OF THE INJURED. By ALBERT S. MORROW, M. D., Major, Medical Reserve Corps, U. S. Army. Octavo of 356 pages, with 242 illustrations. Cloth, $2.75 net.   Published November, 1917

Pyle's Personal Hygiene   The New (7th) Edition

A MANUAL OF PERSONAL HYGIENE: Proper Living upon a Physiologic Basis. By Eminent Specialists. Edited by WALTER L. PYLE, A. M., M. D., Assistant Surgeon to Wills Eye Hospital, Philadelphia. 555 pages, illustrated. Cloth, $1.75 net.   Published August, 1917

To this new edition there have been added, and fully illustrated, chapters on Domestic Hygiene and Home Gymnastics, besides an Appendix containing methods of Hydrotherapy, Mechanotherapy, and First Aid Measures. There is also a Glossary of the mechanical terms used.

"The work has been excellently done, there is no undue repetition, and the writers have succeeded unusually well in presenting facts of practical significance based on sound knowledge."—*Boston Medical and Surgical Journal.*

American Pocket Dictionary   New (10th) Edition

AMERICAN POCKET MEDICAL DICTIONARY. Edited by W. A. NEWMAN DORLAND, M. D. Containing the pronunciation and definition of the principal words used in medicine and kindred sciences, with 64 extensive tables. With 707 pages. Flexible leather, with gold edges. $1.25 net; with patent thumb index, $1.50 net.   September, 1917

Brill's Psychanalysis   Second Edition

PSYCHANALYSIS: Its Theories and Practical Application. By A. A. BRILL, PH. B., M. D., Clinical Assistant in Neurology at Columbia University Medical School. Octavo of 392 pages. Cloth, $1.25 net.
Published May, 1914

Shaw on Nervous Diseases and Insanity

Fifth Edition, published October, 1913

ESSENTIALS OF NERVOUS DISEASES AND INSANITY: Their Symptoms and Treatment. A Manual for Students and Practitioners. By the late JOHN C. SHAW, M. D., Clinical Professor of Diseases of the Mind and Nervous System, Long Island College Hospital, New York. 12mo of 204 pages, illustrated. Cloth, $1.25 net. *Saunders' Compends.*

Golebiewski and Bailey's Accident Diseases

ATLAS AND EPITOME OF DISEASES CAUSED BY ACCIDENTS. By DR. ED. GOLEBIEWSKI, of Berlin. Edited, with additions, by PEARCE BAILEY, M. D., Consulting Neurologist to St. Luke's Hospital, New York. With 71 colored illustrations on 40 plates, 143 text-illustrations, and 549 pages Cloth, $4.00 net. *Saunders' Hand-Atlases.*   **Published 1901**

Bass and Johns' Alveolodental Pyorrhea

ALVEOLODENTAL PYORRHEA. By CHARLES C. BASS, M. D., Professor of Experimental Medicine, and FOSTER M. JOHNS, M. D., Instructor in the Laboratories of Clinical Medicine, Tulane Medical College. Octavo, 168 pages, illustrated. Cloth, $2.50 net.   **Published June, 1915**

Spear's Nervous Diseases

A MANUAL OF NERVOUS DISEASES. By IRVING SPEAR, M. D., Professor of Neurology at the University of Maryland, Baltimore. 12mo of 660 pages, with 172 illustrations. Cloth, $2.75 net. **November, 1916**

Jakob and Fisher's Nervous System   Saunders' Hand-Atlases

ATLAS AND EPITOME OF THE NERVOUS SYSTEM AND ITS DISEASES. By PROFESSOR DR. CHR. JAKOB, of Erlangen. Edited, with additions, by EDWARD D. FISHER, M. D., University and Bellevue Hospital Medical College, New York. With 83 plates and copious text. Cloth, $3.50 net.   **Published 1901**

Hunt's Diagnostic Symptoms of Nervous Diseases

DIAGNOSTIC SYMPTOMS OF NERVOUS DISEASES. By EDWARD L. HUNT, M. D., formerly Instructor in Neurology and Assistant Chief of Clinic, College of Physicians and Surgeons, New York. 12mo of 292 pages, illustrated. Cloth, $2.00 net. **Second Edition, June, 1917**